I0893621

FRONTIERS OF SCIENCE

Frontispiece Vol. I

Frontiers of Science

IMPERIALISM AND NATURAL KNOWLEDGE IN THE GULF SOUTH BORDERLANDS, 1500–1850

CAMERON B. STRANG

Published by the

OMOHUNDRO INSTITUTE OF

EARLY AMERICAN HISTORY AND CULTURE,

Williamsburg, Virginia, and the

UNIVERSITY OF NORTH CAROLINA PRESS,

Chapel Hill

The Omohundro Institute of Early American History and Culture is sponsored by the College of William and Mary. On November 15, 1996, the Institute adopted the present name in honor of a bequest from Malvern H. Omohundro, Jr.

© 2018 The Omohundro Institute of Early American History and Culture
All rights reserved
Manufactured in the United States of America

The University of North Carolina Press has been a member of the Green Press Initiative since 2003.

Frontispiece. From Bernard Romans, *A Concise Natural History of East and West Florida* ... (New York, 1775). Library of Congress, Rare Book and Special Collections Division, F314 .R75

Cover illustration: The eagle map of the United States, "Entered according to act of Congress in the year 1832 by J. Churchman in the Clerk's Office of the Eastern District of Pennsylvania," from *Churchman's Rudiments of National Knowledge* (Philadelphia: E. L. Carey & A. Hart, 1833), courtesy of Library of Congress Geography and Map Division, http://hdl.loc.gov/loc.gmd/g3700.np000151; compass © ANGHI/iStockphoto.com

Library of Congress Cataloging-in-Publication Data
Names: Strang, Cameron B., author. | Omohundro Institute of Early American History & Culture, publisher.
Title: Frontiers of science : imperialism and natural knowledge in the Gulf South borderlands, 1500–1850 / Cameron B. Strang.
Description: Williamsburg, Virginia : Omohundro Institute of Early American History and Culture ; Chapel Hill : University of North Carolina Press, [2018] | Includes bibliographical references and index.
Identifiers: LCCN 2018004390| ISBN 9781469640471 (cloth : alk. paper) | ISBN 9781469640488 (ebook)
Subjects: LCSH: Gulf States—Intellectual life—History. | Nature study—Gulf States—History. | Nature study—Political aspects—Gulf States. | Borderlands—History. | United States—Territorial expansion—History. | Europe—Colonies—History.
Classification: LCC F296 .S77 2018 | DDC 976—dc23
LC record available at https://lccn.loc.gov/2018004390

Parts of Chapters 2 and 7 draw on previously published articles, "Indian Storytelling, Scientific Knowledge, and Power in the Florida Borderlands," *William and Mary Quarterly*, 3d Ser., LXX (2013), 671–700; and "Violence, Ethnicity, and Human Remains during the Second Seminole War," *Journal of American History*, C (2014), 973–994.

For my family

ACKNOWLEDGMENTS

• • •

I worked hard to write this book. But mostly I've just been really lucky.

I was lucky to begin this project under the guidance of Jorge Cañizares-Esguerra and James Sidbury. They have supported me without fail for more than a decade, and I hope this book counts as some small payback for all the hours that they have devoted to my career. Thank you.

I was also lucky to meet an incredibly generous set of friends and colleagues in the places this book took shape. Denise Bossy, Ben Breen, Erika Bsumek, Lina del Castillo, Matt Childs, Paul Conrad, Jesse Cromwell, Surekha Davies, Ann Fabian, Robert Goulding, Sara Gronim, Julie Hardwick, Becca Havens, Rachel Hermann, Bruce Hunt, Jason Jackson, James Jenkins, Brian Jones, Jackie Jones, Neil Kamil, Adam Lewis, Bruce Moran, Christopher Morris, Michele Navakas, Josh Noll, Robert Olwell, Cynthia Radding, Andrés Reséndez, Dan Richter, Adam Rothman, Samantha Seeley, Kyle Shelton, Jim Sweet, and Alan Tully all helped sustain this project by reading parts of the manuscript, giving me a leg up in the profession, or (just as importantly) being friends. In different ways, Christopher Heaney's and Christopher Parsons's fingerprints are all over this book, and they continue to inspire me from the other side of the continent. Marcy Norton gave me a cool home during a sweltering D.C. summer. All of these people deserve thanks for making me better at the scholarly things I do or happier while doing them, but I would not be doing these things at all without the encouragement of historians at the University of New Hampshire: my deep thanks to Jeffrey Bolster, Jan Golinski, Eliga Gould, Nicoletta Gullace, and Julia Rodriguez.

Luck in the archives—and the privilege of working in several archives—made this book what it is. I send my gratitude to the good people at the P. K. Yonge Library (especially James Cusick), the Academy of Natural Sciences of Drexel University, the Historical Society of Pennsylvania, the American Philosophical Society, the Historic New Orleans Collection, the Special Collections Department at Tulane University Library, the New Orleans Public Library, the Louisiana State Museum Collections Historical Center, Louisiana State University's Special Collections Libraries, the Library of Congress, the National Archives, the Jefferson Library at the Robert H. Smith International Center for Jefferson Studies (especially Andrew O'Shaughnessy), the Museo Nacional de Ciencias Naturales, the Archivo del Real Jardín Botánico de Madrid, the Archivo del Museo Naval, the Biblioteca Nacional de España, the Mississippi Department of Archives and History, the Dolph Briscoe Center for American History, the Benson Latin American Collection, the Perry-Castañeda Library, the South Caroliniana Library, the Huntington Library, and the Smithsonian Institution.

I've been really lucky in securing funding for this project, and several organizations deserve special thanks. The Donald D. Harrington Fellows Program, the University of Texas at Austin, the Dianne Woest Fellowship in the Arts and Humanities (Historic New Orleans Collection), the Institute for Southern Studies (University of South Carolina), Louisiana State University's Special Collections Libraries, the McNeil Center for Early American Studies, the W. M. Keck Foundation (Huntington Library), the Robert H. Smith International Center for Jefferson Studies, the New Orleans Center for the Global South's Global South Fellowship (Tulane University), the National Science Foundation, a Cecilia L. Johnson Grant (University of Florida Special Collections Libraries), the Kislak Fellows Program (John W. Kluge Center, Library of Congress), and the Consortium for History of Science, Technology, and Medicine all supported this study during the dissertation stage. The Margaret Henry Dabney Penick Resident Scholar Program at the Smithsonian Institution provided me with a year to begin turning the dissertation into a book, an endeavor that I mostly completed three years later thanks to a Martin L. and Sarah F. Leibowitz membership at the Institute for Advanced Study. I am also grateful to the College of Liberal Arts at the University of Nevada, Reno, for a timely Scholarly and Creative Activities Grant that covered the cost of the images and maps.

I am so lucky to be in the history department at UNR. They have bent over backwards to make me *and* my family welcome, for which I will be forever in their debt. Although all of my colleagues deserve thanks, I am especially grateful to the three chairs who have overseen the department since 2014—Linda Curcio-Nagy, Bruce Moran, and Dennis Dworkin—and to Christopher Church and his family for their friendship.

It has also become abundantly clear to me just how lucky I am to be publishing this book with the Omohundro Institute. Fredrika Teute saw promise in this project when it was still a dissertation and, more importantly, saw many of the (many, many) things I needed to do to make it better. Paul Mapp and Nadine Zimmerli picked up where Fredrika left off. As far as I can tell, they both spent a ridiculous amount of time reading drafts of the manuscript, pondering ways to improve it, and talking through these suggestions with me over the phone. Kaylan Stevenson is copyeditor extraordinaire. If you appreciate the correctness of the prose and the accuracy of the citations, thank her and her team of apprentices. If you find the prose lacking or any facts inaccurate, blame me. Rebecca Wrenn drew the three beautiful maps. To all of you in Williamsburg, thank you. Chuck Grench welcomed this manuscript at University of North Carolina Press, and the suggestions of three outside readers proved invaluable. The insights and enthusiasm of one of these readers, James Delbourgo, were particularly priceless.

Lastly, I have been lucky in love. Bill, Karol, Tyson, Tanya, and all the Kellers have been a bottomless well of warmth and support. Bill and Karol's seemingly unshakable faith in me made it easy to believe that success was always in reach, while their quiet confidence in education, omnivorous wolfing of books, and dedication to each other and their children have marked me to the bone. Baby James lights up my world; it is simply a boundless love. No one has done more for this book or my joy than Renata Keller. She has lived with (and edited, critiqued, and clarified) this project for as long as we have been together. I am constantly floored by her serenity and love, and I am so excited for all the extra time we will have together now that this book is out of my hands. Thank you.

CONTENTS

LIST OF ILLUSTRATIONS

FIGURES

MAPS

ABBREVIATIONS AND SHORT TITLES

AEP	Andrew Ellicott Papers, Library of Congress, Washington, D.C.
AGI, SD	Archivo General de Indias, Documentos … Audiencia de Santo Domingo, MSS 3554, Louisiana and Lower Mississippi Valley Collection, Louisiana State University, Baton Rouge, La.
AHN	Papeles de estado in the Archivo Histórico Nacional (Spain), 1748–1846, Library of Congress, Washington, D.C.
AHR	*American Historical Review*
AMNCN	Archivo de Museo Nacional de Ciencias Naturales, Madrid, Spain
AMNM	Archivo del Museo Naval, Madrid, Spain
ANC	Louisiana Documents from the National Archive of Cuba microfilm, Historic New Orleans Collection, New Orleans
ANSP	Academy of Natural Sciences of Drexel University, Philadelphia
APS	American Philosophical Society, Philadelphia
Baird Papers	Spencer Fullerton Baird Papers, RU 7002, Smithsonian Institution Archives, Washington, D.C.
EFP	East Florida Papers, Library of Congress, Washington, D.C.
"Extracts"	"Extracts from the Letter Book of William Dunbar, from 18 June 1775 to 24 March 1802," ed. B. L. C. Wailes, Misc. Manuscript Collections, MSS3167, Library of Congress, Washington, D.C.
FHSQ	*Florida Historical Society Quarterly*
FHQ	*Florida Historical Quarterly*
HAHR	*Hispanic American Historical Review*
HNOC	Historic New Orleans Collection, New Orleans
HSP	Historical Society of Pennsylvania, Philadelphia
JAH	*Journal of American History*
JER	*Journal of the Early Republic*
LHQ	*Louisiana Historical Quarterly*

LLMVC	Louisiana and Lower Mississippi Valley Collections, Louisiana State University, Baton Rouge, La.
LOC	Library of Congress, Washington, D.C.
LRC	Louisiana Research Collection, Tulane University Special Collections Library, New Orleans
LSM	Louisiana State Museum, New Orleans
LSU	Louisiana State University Special Collections Libraries, Baton Rouge, La.
MDAH	Mississippi Department of Archives and History, Jackson, Miss.
Morton Papers	Samuel George Morton Papers, American Philosophical Society, Philadelphia
MPA, SD	Mississippi Provincial Archives, Spanish Dominion
NARA	National Archives and Records Administration, Washington, D.C.
Nutt Papers	Rush and Haller Nutt Papers, Huntington Library, San Marino, Calif.
NTC	Northwest Territory Collection, Indiana Historical Society, Indianapolis, Ind.
PC-HNOC	Louisiana Documents from the Archivo General de Indias, Papeles Procedentes de Cuba, Historic New Orleans Collection, New Orleans
PC-LLMVC	Archivo General de Indias, Papeles Procedentes de Cuba, MSS 3278, Louisiana and Lower Mississippi Valley Collections, Louisiana State University, Baton Rouge, La.
PCL	Perry-Castañeda Library, University of Texas at Austin
PKY	P. K. Yonge Library of Florida History, University of Florida, George A. Smathers Libraries, Gainesville, Fla.
SCDAH	South Carolina Department of Archives and History, Columbia, S.C.
SI	Smithsonian Institution Archives, Washington, D.C.
Tait Papers	Honorable Charles Tait Papers, Academy of Natural Sciences of Drexel University, Philadelphia
TJP	Thomas Jefferson Papers, Ser. 1, General Correspondence, 1651–1827, Library of Congress, Washington, D.C.
WDP	William Dunbar Papers, Mississippi Department of Archives and History, Jackson, Miss.
WMQ	*William and Mary Quarterly*

FRONTIERS OF SCIENCE

MAP 1.
The Gulf South and New Spain, circa 1600. Drawn by Rebecca Wren

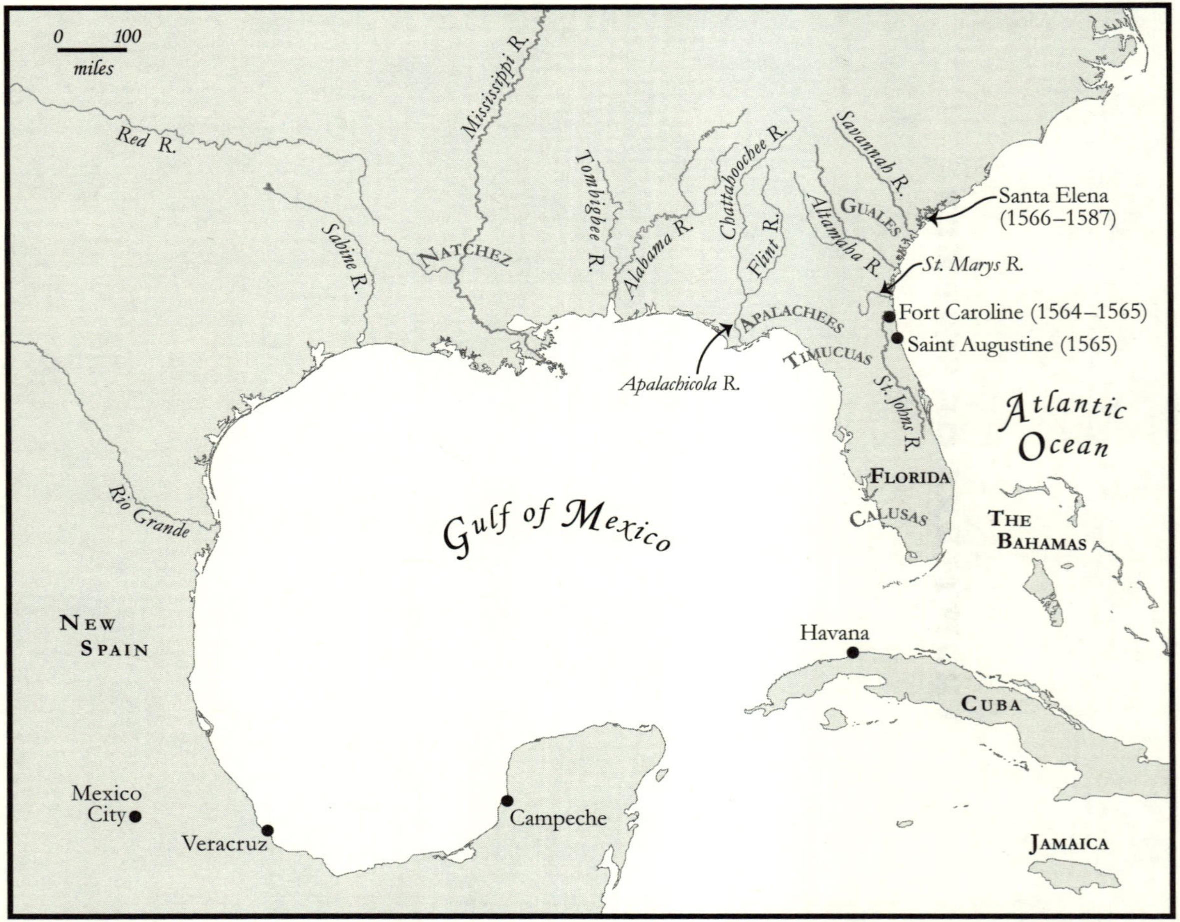

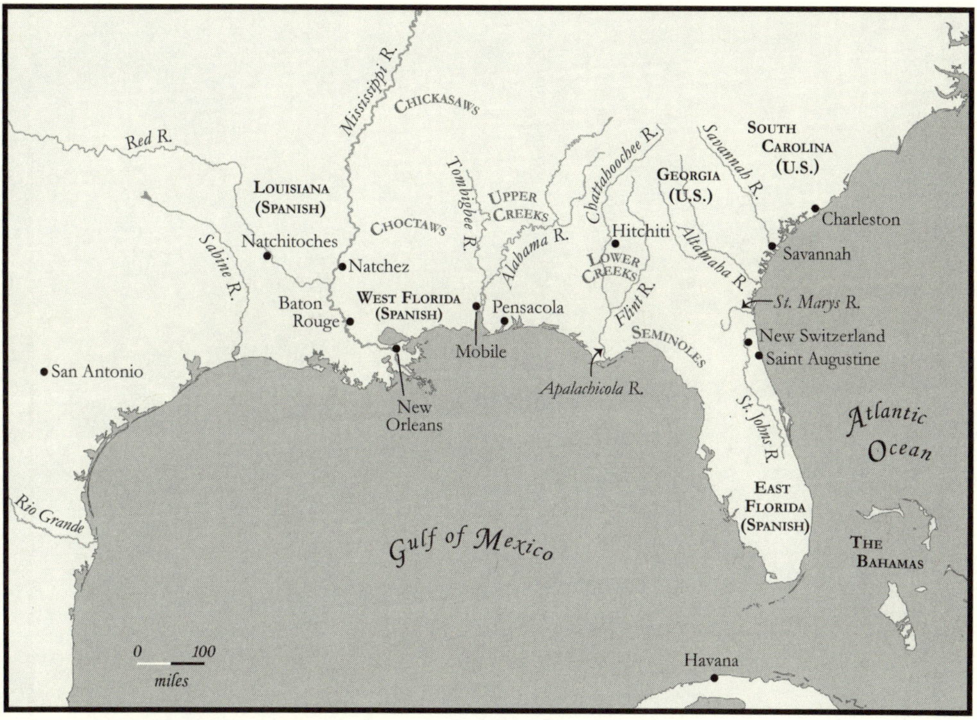

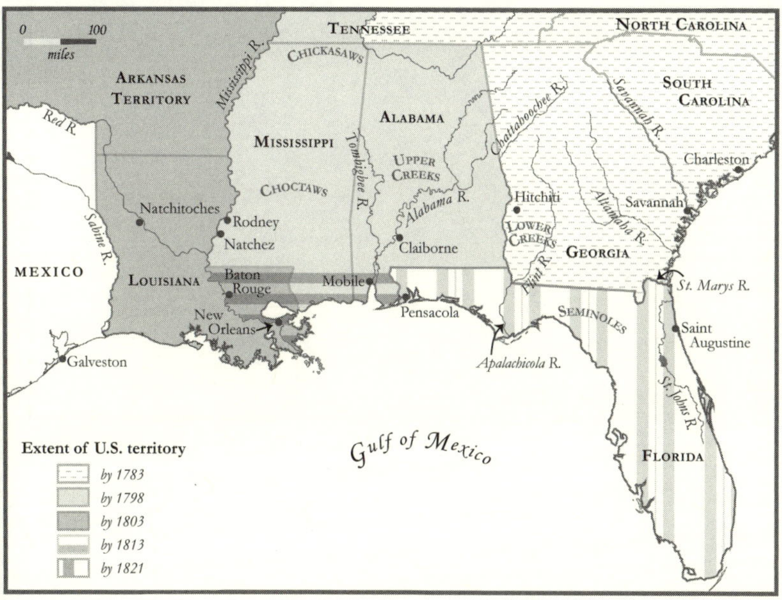

MAP 2.
The Gulf South, New Spain, and the United States, circa 1783. Drawn by Rebecca Wren

MAP 3.
The Gulf South and the United States, circa 1821. Drawn by Rebecca Wren

{ INTRODUCTION }

THE SIGNIFICANCE OF THE FRONTIER IN AMERICAN KNOWLEDGE

• • •

"Convinced as I am that information relative to the situation of any empire now under your particular charge will be always welcome to you (especially if such place be remote) let such information come from whatever person or through whatsoever channel it may." So began the first letter that John Devereux DeLacy, an Irish-born adventurer traveling in the Gulf South, wrote to President Thomas Jefferson in November 1801. DeLacy's letter was a plea for patronage—he hoped Jefferson would fund his work as an explorer—and his decision to emphasize information and empire in the first sentence was no doubt calculated to attract the attention of a busy president devoted to both scientific pursuits and territorial expansion. But it was not particularly clear which empire or empires DeLacy meant. It is possible he was referring to the United States and, specifically, its new lands in the Gulf South: the first place he mentioned was the Mississippi Territory, which the United States had acquired a few years earlier. But it is also possible that DeLacy meant his words more literally, that U.S. expansion entailed incorporating lands and peoples that had long been part of other empires. DeLacy's reference to "any empire now under [Jefferson's] particular charge" could have included Spain, France, or Britain, all of which had claimed sovereignty over the Mississippi Territory and other parts of the Gulf South before the United States existed and, moreover, continued to compete for the region. As DeLacy recognized, knowing and governing the Gulf South—today's Louisiana, Mississippi, Alabama, and Florida—would be inextricable from encounters with multiple peoples and empires.[1]

1. John Dev[ereux] DeLacy to Thomas Jefferson, Nov. 3, 1801, [1] (quote), TJP. DeLacy was born in Ireland, moved to Philadelphia when he was fourteen, and made his way to the Gulf South around 1800. He presented himself alternately as a planter, medical doctor, and lawyer in Spanish New Orleans before journeying into the Floridas and Creek country. For an overview of DeLacy's life, see John C. Van Horne and Lee W. Formwalt, eds., *The Correspondence and Miscellaneous Papers of Benjamin Henry Latrobe,* 3 vols. (New Haven, Conn., 1984–1988), III, 390n. On DeLacy in the Spanish Gulf provinces,

DeLacy knew that establishing himself as an expert on the Gulf South depended on building connections both in and beyond the region. He thus circulated the knowledge he acquired through exchanges with French, Spanish, and native inhabitants to officials who, he hoped, would reward him financially and endorse his claims to being a man of learning. His two letters to Jefferson in late 1801 described profitable natural productions, explained some of the best trade routes across the Gulf South, and shared ethnographic observations on Indians and Africans. He also tried to persuade Jefferson that southeastern Indians' love of liberty made them natural allies of the United States against the "Monkish Cruelties and Bigottry . . . of Spanish Governors." DeLacy imagined that this combination of expertise and political attachment would win the president's support for his scientific endeavors, including "an excursion to the W. and N.W." But Jefferson neither offered to fund this expedition, which would have prefigured that of Meriwether Lewis and William Clark in 1804, nor bothered to write DeLacy back.[2]

But the plucky adventurer had other irons in the fire. Like other men and women in the Gulf South, DeLacy's loyalties were fluid, and, within months of contacting Jefferson, he was soliciting support from some of the other polities vying for power in the region. In a March 1802 letter to Ventura Perez, a Spanish official in Havana, DeLacy warned that Spain's "Powerful aspiring and covetous neighbours the North Americans . . . sleep but to dream of the possession of South America and its immense riches" and suggested that his own ties with southeastern Indian groups could "bend them to the Spanish interest" and create a buffer against U.S. expansion. DeLacy ensured Perez that he had a "predilection and reverence for Spain" and that his knowledge of "the western Parts of North America and the province of Louisiana" was greater than that of "any other man in existence." DeLacy was, however, simultaneously offering information and loyalty to William Augustus Bowles, the Anglo-American leader of the British-supported State of Muskogee, an envisioned Creek nation that challenged Spanish and U.S. interests in East and West Florida. In December, DeLacy had sent Bowles—who, according to DeLacy, was a learned man of science—a de-

---

see Gilbert C. Din, *War on the Gulf Coast: The Spanish Fight against William Augustus Bowles* (Gainesville, Fla., 2012), 185–195. DeLacy recounted his background in his own words in "Expediente sobre el encarcelamiento de Juan De Lacy," Nueva Orleans, 1802–1803, PC-LLMVC, leg. 220-A, reel 85, Ser. 106, 52–62.

2. John Dev[ereux] DeLacy to Thomas Jefferson, Dec. 18, 1801, [16] ("Monkish"), TJP, DeLacy to Jefferson, Nov. 3, 1801, [17] ("excursion").

tailed list of local natural resources that would help make Muskogee prosperous. Bowles recognized the "advantages that your extensive information has for the commerce of this Country," and, for a short time, DeLacy found himself a patron. Unfortunately for DeLacy, Spanish agents soon caught him smuggling Bahamian trade goods to Muskogee and locked him away in a New Orleans prison.[3]

Yet, even behind bars, DeLacy knew an opportunity when he saw one. Just weeks after news reached New Orleans in October 1803 that the United States had purchased Louisiana from France (Spain had recently announced that it was transferring Louisiana back to France), DeLacy sent a letter from his cell to Secretary of State James Madison in which he outlined how the United States could best "consolidate her empire" through the Louisiana Purchase, a "moment which must be seized by the forelock." He explained that, through his personal influence, the Creeks, Choctaws, Chickasaws, and Cherokees were all eager to sell "the whole of their possessions East of the Mississippi to the U.S. . . . and then go and take possession of the Colony of Louisiana." This independent Indian polity would be a bulwark against "neighbouring powers who might be tempted to be troublesomely encroaching" on the western states. "Nature," he argued, had made the Mississippi River the continent's ideal political boundary because it placed the United States beyond the reach of imperial rivals while also protecting Indians in the West from being "oppressed distressed or harrassed by the [Anglo-American] frontier settlers." DeLacy claimed that his "allegiance [to] the U.S." and his geographic and ethnographic knowledge of North America made him uniquely qualified to chart the course of the continent's political future: "This Sir," he told Madison, "is the light it strikes

3. John Dev[ereux] DeLacy to Ventura Perez, Mar. 23, 1802, PC-LLMVC, leg. 2367, reel 49, Ser. 145, 296 ("western," "any other man"), 297 ("Powerful"), 299 ("predilection"), 300 ("bend"); William Augustus Bowles to DeLacy, Dec. 18, 1801, MPA, SD, VII, reel 61W, 495, PKY ("advantages"); Bowles to Hunter and Waler, Dec. 24 1801, PC-LLMVC, leg. 2372, reel 54, Ser. 145. DeLacy's list of resources emphasized commodities like lumber, medicinal plants, cotton, and black slaves because they "require no extraordinary degree of labor in the procuring and are therefore well adapted to the Indian disposition." See DeLacy to Bowles, Dec. 9, 1801, PC-LLMVC, leg. 2367, reel 50, Ser. 145, 376. On Bowles and the State of Muskogee, see Eliga Gould, "Independence and Interdependence: The American Revolution and the Problem of Postcolonial Nationhood, circa 1802," *WMQ*, 3d Ser., LXXIV (2017), 729–752. On Bowles as a man of science, see DeLacy to Jefferson, Dec. 18, 1801, [23], TJP. On the relationship between DeLacy and Bowles, see J. Leitch Wright, Jr., *William Augustus Bowles: Director General of the Creek Nation* (Athens, Ga., 1967), 142–154.

me in and I am well convinced that I know the Geography of these Countries and the people of them better ... than any other individual in existence." But Madison had little reason to trust DeLacy's loyalty or expertise, so DeLacy remained in prison while Louisiana became part of the United States. According to U.S. territorial governor William C. C. Claiborne, "His confinement ... did not seem to excite a great share of public sympathy." Claiborne eventually released DeLacy in 1804 but expelled him from Louisiana.[4]

By the 1810s, Anglo-Americans controlled the lion's share of wealth and power in the Gulf South, and DeLacy once again reconfigured his intellectual interests and connections to adjust to the region's latest context as a slaveocracy ruled by Anglo citizen-planters. Since steamboats were becoming essential to expanding and enriching the plantation society of the lower Mississippi Valley, the ever-opportunistic DeLacy allied himself with early leaders of the steamboat industry, surveying southern rivers for Robert Fulton and then promoting a rival enterprise that aimed to dominate the transportation of goods and people on the Mississippi River. DeLacy was also designing a steam-powered sugar roller that the master class could use to squeeze every last ounce of wealth out of Louisiana sugarcane, a technology that would complement their goal of squeezing every last ounce of labor out of their slaves.[5]

U.S. expansion into the Gulf South continued to inspire DeLacy's ambitions in 1821 as Spain transferred Florida to the United States. He again sought Jefferson's "patronage," this time with the aim of "obtaining the appointment of Attorney General for East Florida" by emphasizing both his "connections among the Spaniards" and emotional attachment to the United States. As in 1801, Jefferson was unmoved. And even though DeLacy's "habiliments approached what might have been called seediness" by the late

4. John Devereux DeLacy to James Madison, Oct. 14, 1803, Founders Online, http://founders.archives.gov/documents/Madison/02-05-02-0533 ("consolidate"); William C. C. Claiborne to Madison, July 30, 1804, in Dunbar Rowland, ed., *Official Letter Books of W.C.C. Claiborne,* 6 vols. (1917; rpt. New York, 1972), II, 280–281 ("His confinement," 281).

5. Robert Fulton and John D. DeLacy, *Report of the Practicability of Navigating with Steam Boats on the Southern Waters of the United States ...*, 2d ed. (1813; rpt. Philadelphia, 1828); Benjamin Henry Latrobe to Nicholas J. Roosevelt, Feb. 20, 1815, in Van Horne and Formwalt, eds., *Correspondence and Miscellaneous Papers of Benjamin Henry Latrobe,* III, 621, and editors' notes, III, 390, 592, 613, 620. On steamboats, see Walter Johnson, *River of Dark Dreams: Slavery and Empire in the Cotton Kingdom* (Cambridge, Mass., 2013), esp. 73–125.

1820s, he remained as ambitious as ever: he aspired to sue "every steamboat owner in the United States" for stealing a paddle design that he and his partners had patented. But, being "one of those men who are always in trouble," DeLacy could not muster enough money to follow through with the suit.[6]

DeLacy spent most of his adult life trying to profit from knowledge that he thought would benefit the officials, adventurers, and planters competing for power in the Gulf South. But, by his death in 1837, he had achieved little fame or fortune to show for it. Despite his expertise, flexible approach to allegiance, and bold plans to order continental geopolitics from the borderlands outward, DeLacy, like so many people in North America, lost more than he gained during the era of U.S. expansion.

At first blush, DeLacy might seem like an outlandish figure who has no right introducing a history of natural knowledge in early America. But his lifelong effort to navigate and manipulate a world of rapidly shifting power relations and possibilities would have seemed all too common to men and women living in North America from the sixteenth to nineteenth centuries. Far from being peripheral, people like DeLacy—and the Spanish, French, Anglo, creole, black, and native individuals discussed in this book—were central to encounters that made the pursuit of knowledge in early America what it was.

• • •

I first came across copies of DeLacy's letters to Jefferson at the P. K. Yonge Library in Gainesville, Florida, one of the first archives I visited to research a dissertation on the relationship between science and expansion in the early republic. My idea at the time was to find evidence that the scientific practices of the Spanish Empire had influenced the ways science promoted territorial expansion in the United States (I did, in fact, find much evidence for this, and I make a similar argument here in Chapter Three). Since the Gulf South was the first place where the expanding United States collided with Spanish America, Gulf South archives like the P. K. Yonge Library seemed like a good place to start looking for connections between them.

DeLacy did bridge Spanish and Anglo America, but his story caught my

6. John Dev[ereux] DeLacy to Thomas Jefferson, Apr. 16, 1821, [1] ("patronage"), [2] ("connections"), TJP. DeLacy evinced his attachment to the United States by telling Jefferson that he had "been honored when a Boy with the notice of your illustrious freind *[sic]* General Washington" (ibid., [1]). For observations about DeLacy in the 1820s, see John H. B. Latrobe, *A Lost Chapter in the History of the Steamboat* (Baltimore, 1871), 6 ("habiliments"), 11 ("every").

eye primarily because it seemed extraordinary and, at least compared to most stories about early American science, exciting. Here was a frontier naturalist-desperado, a man who hoped to use his wits and learning to build a better life by exploiting competitions among Spain, the United States, and native groups. After discovering DeLacy, I kept coming across more and more extraordinary stories that somehow involved knowledge production in the Gulf South, accounts of spies, mutilation, magic, and monsters. But many of these stories simply seemed too weird to matter. Nevertheless (and maybe against my better judgment), I found myself chasing the most intriguing of them across multiple archives and across national and linguistic boundaries.

It was not immediately apparent to me why these stories, although interesting in and of themselves, were more than the sum of their parts. They all took place in the Gulf South, and they all involved the creation or circulation of knowledge about the natural world. But the range of individuals—African herbalists, European men of science, indigenous shamans—I found striving for knowledge defied easy characterization, as did the kinds of knowledge they sought. The main theme the stories I uncovered seemed to share was that they did not fit within the frameworks scholars have used to study knowledge in early America. Neither the history of American science nor the broader field of American intellectual history seemed able to encompass these stories about knowledge in the Gulf South.

It took me years of accumulating and analyzing these sources before I was able to hear what they had been saying all along: these sources did not make sense within existing frameworks for interpreting the history of knowledge in early America because these frameworks were flawed. Stories like DeLacy's only seemed too weird to matter because we have continued to mischaracterize what was normal and significant about knowledge in America. To put it crudely (I go into more historiographical detail below), the long-standing consensus has been that the important story of knowledge in early America involved Anglos along the Atlantic seaboard, how their modes of producing and interpreting knowledge were molded by contexts created by British colonialism, and how these modes changed after these colonies became the first thirteen United States.

But early America was more than North America's eastern Anglo fringe, whether we use "America" to mean the hemisphere, the continent, the North American spaces that the United States would eventually occupy, or even the early United States' territorial claims. The voices I recovered from the Gulf South demonstrate that the history of American knowledge must involve a more diverse set of places than the Atlantic seaboard, a more diverse set of actors than Anglos, a more diverse set of practices than those that we would comfortably call scientific, and more emphasis on the continuities between the colonial and national eras, periods that seemed less distinct in the rest of the continent than they did in the first thirteen states.

This book analyzes several case studies that, collectively, suggest contours for a new framework for studying the history of knowledge in America, one that spans the years before and after 1776 and is capacious enough to include the various places, actors, and epistemologies of the continent's vast borderlands. This framework, I argue, should emphasize how knowledge developed and circulated amid the ongoing encounters and unequal power relations engendered by imperialism. Imperialism—as practiced first by Europeans and then, in a different form, by the United States—was the common factor that shaped knowledge in all American places (even eastern cities like Philadelphia). By focusing on the seemingly bizarre stories I uncovered about the production and circulation of knowledge in the borderlands, this book begins the work of revealing the lost intellectual world of early America.

This study also explores how this intellectual world became lost. Since the nineteenth century, historians of the early United States have portrayed spaces beyond the Anglo East (and especially the Northeast) as zones of ignorance with no place in America's intellectual history, much less the history of science. But these early scholars did not invent this portrait out of thin air. They found evidence in the writings of Spanish and French observers who mocked the ignorance of creole colonists and in reports by Anglo-Americans that stressed the mental incapacity of the borderlands' nonwhites and the backwardness of the continent's more western and southern spaces. These historical and ethnographic perspectives, like the other forms of knowledge I examine in this book, grew out of encounters spurred by imperialism, and they proved foundational in American historiography. They helped define a fundamental division between the Atlantic-oriented East, a place where science and learning had a history, and the rest of the continent, where—with the telling exception of narratives about the forays of eastern Anglos like Lewis and Clark—they did not. *Frontiers of Science* tells stories about the Gulf South that, I hope, will show that America's borderlands had a vibrant and enduring intellectual life that, far from being separate and irrelevant, was thoroughly connected with the Anglo East and vital to understanding the intellectual history of early America as a whole.

• • •

This is a book about natural knowledge in the Gulf South—also known as the southeast borderlands—from the 1500s to the mid-1800s. By "natural knowledge," I mean knowledge that humans develop about nature, a category that encompasses things like animals, plants, planets, minerals, lands, waters, and peoples. Natural knowledge obviously covers many of the same subjects as scientific knowledge, and there were European (and some non-European) individuals in the Gulf South who studied nature in ways that would, then as now, be readily identifiable as science. But America's borderlands were also home to a culturally diverse array of people who worked to learn about and manipulate nature in ways that could not, either then or now, be easily characterized as science.[7]

My use of "natural knowledge" as a catchall term that embraces science as well as other approaches to studying nature is not meant to imply that every self-proclaimed expert was capable of making equally valid conclusions about the natural world (nor is it to say that white men of science always reached more valid conclusions than other experts). It is instead to assert that all forms of natural knowledge in the southeast borderlands need to be viewed through the same analytical lens. This is, at a general level, because all kinds of natural knowledge are to some degree human constructs susceptible to personal, material, and social influences. It is also because the lines separating science from religion, intrigue, magic, and non-European ways of knowing were porous and unstable in the early modern period, particularly in multicultural borderlands. The cases I examine from the Gulf South indicate that neither natives nor blacks nor Europeans had a set tool kit for understanding and describing nature; rather, place-specific social and political relationships conditioned how individuals acquired knowl-

7. Scholars of frontiers and borderlands have largely ignored intellectual history and the history of natural knowledge. Frederick Jackson Turner did hint in 1893 that "from the conditions of frontier life came intellectual traits of profound importance," but he later argued that science in the United States only began to matter after the supposed closing of the nation's territorial frontiers. His portrayal of twentieth-century scientists as a new generation of pioneers whose explorations would advance American democracy set an influential precedent in American historiography by suggesting that America's physical frontiers and frontiers of knowledge did not overlap or affect each other. See Turner, "The Significance of the Frontier in American History," *Annual Report of the American Historical Association for the Year 1893* (Washington, D.C., 1894), 226 ("conditions"); Turner, *The Frontier in American History* (New York, 1920), 284, 287, 300–301, 331, 357; and Leah Ceccarelli, *On the Frontier of Science: An American Rhetoric of Exploration and Exploitation* (East Lansing, Mich., 2013), esp. 29–51. Far more than Turner, Herbert Eugene Bolton, the founder of borderlands history, recognized the significance of transnational contexts. He encouraged scholarship on the transnational dimensions of science and intellectual life in the Americas, but few have followed up on this suggestion. See Bolton, "The Epic of Greater America," *AHR*, XXXVIII (1933), 449, 473–474.

edge in particular settings, the means of narrating knowledge that could be recognized as legitimate across cultural boundaries, and the itineraries that determined which individuals, institutions, and governments could benefit from it. Lastly, the diverse men and women who sought natural knowledge in the Gulf South can and should be studied as part of the same history because they shared an overarching social and political context that marked their work. This context was a world in which the expansion of imperial powers affected humans' relationships with each other and nature.[8]

I also use the term "local knowledge," but not to refer to some kind of stagnant wisdom that a people possessed simply by dint of sustained experience in a given place. Such a definition is all but meaningless in areas like the Gulf South where imperialism catalyzed a series of population movements and ensured that no group constructed knowledge about the nature, beings, and history of a place in isolation. Local knowledge is better defined as the ever-evolving understandings of a place's nature and inhabitants that individuals and groups elaborated within a shifting matrix of interpersonal and international relations. In other words, encounters could inspire new local knowledge among all of a place's inhabitants and influence how that knowledge moved. Thus, while some knowledge in the Gulf South was local because it depended on encounters specific to the region, it was—like local knowledge in contact zones throughout the world—also global because it emerged from and was part of an international milieu of exchanges, migrations, and competitions. The main thing that differentiated local knowledge from other forms of knowledge is the extent to which those who possessed it or those who described it perceived certain information or practices to be derived from experiences in, and limited to the people of, particular locales.[9]

8. I use "science" to describe the natural knowledge of learned Europeans and Euro-Americans, and I refer to the European and Euro-American men (and they were mostly men) who engaged in the systematic study of nature as "men of science" because these were terms that they used to define themselves and what they did. For an overview of the constructivist position that all natural knowledge (including science) is a human product and, thus, can be analyzed in light of historical and social contexts, see Jan Golinski, *Making Natural Knowledge: Constructivism and the History of Science* (Cambridge, 1998), esp. ix, xi, 6, 9, 17.

9. On the inseparability of the histories of southeastern cultural groups, see Joshua Piker, *The Four Deaths of Acorn Whistler: Telling Stories in Colonial America* (Cambridge, Mass., 2013), esp. 13, 194. On how imperialism could blur boundaries between local knowledge and European science, see Julie Cruikshank, *Do Glaciers Listen?: Local Knowledge, Colonial Encounters, and Social Imagination* (Vancouver, B.C.,

All people in the Gulf South generated natural knowledge through the same basic methods: observation, experimentation, circulation, and inspiration. The boundaries between these methods, just like the boundaries between borderland peoples, were often blurry. The most common method was observation. All peoples used their senses to examine things like stones, stars, and each other, and, while observation was ubiquitous, it was not timeless: observational practices could change when new social relationships, instruments, or ideas appeared in the borderlands. Diverse individuals also practiced experimentation—the application of tests to develop or verify knowledge—to better know or exploit nature. Social factors, especially stratified power relationships, affected the goals, methods, and results of these experiments. Circulation was another means of making knowledge. Circulation entailed the movement of information and things within and beyond the borderlands, and these narratives and objects created and constituted new knowledge as they traveled among places and peoples. Lastly, all the peoples of America formed natural knowledge through inspiration, the understandings that stemmed from divine revelations, creative stories, hypotheses, innovation, and lies. Such inspirations reflected how natural knowledge was rarely isolated from religious beliefs in early America and, more broadly, that even the theoretical and technological inventions of individuals who seemed to work in isolation hinged on social, sacred, and political affairs.[10]

2005); and Neil Safier, "Global Knowledge on the Move: Itineraries, Amerindian Narratives, and Deep Histories of Science," *Isis*, CI (2010), 133–145. On how place and both local and transnational relationships mattered to natural knowledge, see David N. Livingstone, *Putting Science in Its Place: Geographies of Scientific Knowledge* (Chicago, 2003); Fa-ti Fan, "Science in Cultural Borderlands: Methodological Reflections on the Study of Science, European Imperialism, and Cultural Encounter," *East Asian Science, Technology, and Society: An International Journal*, I (2007), 213–231; and Lissa Roberts, "Situating Science in Global History: Local Exchanges and Networks of Circulation," *Itinerario*, XXXIII, no. 1 (March 2009), 9–30.

10. On observation as a knowledge-producing method that could be shaped by personal experience and social context, see Lorraine Daston and Elizabeth Lunbeck, eds., *Histories of Scientific Observation* (Chicago, 2011). The classic work on experimentation as a socially embedded practice is Steven Shapin and Simon Schaffer, *Leviathan and the Air-Pump: Hobbes, Boyle, and the Experimental Life* (Princeton, 1985). On circulation, see James A. Secord, "Knowledge in Transit," *Isis*, XCV (2004), 654–672; and Lissa Roberts, "The Circulation of Knowledge in Early Modern Europe: Embodiment, Mobility, Learning and Knowing," in Ian Inkster, ed., *History of Technology*, XXXI (London, 2012), 47–68.

Ongoing encounters were the very stuff that defined borderlands as borderlands, and such encounters were central to the production, movement, and application of natural knowledge in the Gulf South. The most omnipresent of these interactions were encounters with nature. Humans and environments altered each other, and humans who investigated nature engaged with their environments in physical, historical, and narrative ways: they not only examined and exploited nature but also narrated stories in which relationships between people and nature explained a place's human and environmental histories.[11]

Encounters based on violence and geopolitical competition were similarly critical to intellectual life in the borderlands. Brutality was a basic fact of life in the Gulf South, and warfare, slavery, and abuses against the dead enabled, colored, and circumscribed the study of nature from the 1500s to at least the mid-1800s. Violence was also one of several manifestations of geopolitical competition. Imperial powers (Spain, France, Britain, and the United States), Indian nations and confederacies (Timucuas, Creeks, Choctaws, Chickasaws, and Seminoles, among others), and extra-state polities (including groups of pirates and adventurers) all vied for greater influence in the Gulf South. Competitions for resources, prestige, land, and souls among these groups brought diverse experts—many of whom had fluid allegiances—into contact with each other and motivated them to share their discoveries with officials. Geopolitical competition entailed multiple layers of negotiation because none of the powers competing for the Gulf South had the strength to control or know the region without local cooperation.[12]

11. My definition of borderlands as places of multivalent and ongoing encounter is indebted to David J. Weber's discussion of "zones of interaction . . . places where the cultures of the invader and the invaded contend with one another and with their physical environment to produce a dynamic that is unique to time and place." See Weber, *The Spanish Frontier in North America* (New Haven, Conn., 1992), 11. Historians since Frederick Jackson Turner have noted the importance of encounters between Anglo-Americans and the natural environment to U.S. history. See Turner, "Significance of the Frontier in American History," *Annual Report of the American Historical Association for the Year 1893*, 197–227. On the diversity of engagements with the environment in early America, see Christopher M. Parsons and Cameron B. Strang, "Old Roots, New Shoots: Early American Environmental History," *Early American Studies*, XIII (2015), esp. 280.

12. On the centrality of violence to borderlands and Gulf South slavery, see, respectively, Ned Blackhawk, *Violence over the Land: Indians and Empires in the Early American West* (Cambridge, Mass., 2006); and Edward E. Baptist, *The Half Has Never Been*

Encounters also encouraged exchanges across cultural and political boundaries. Men and women in the Gulf South exchanged things, which ranged from seeds to telescopes to body parts, and information, which usually took the form of oral or written narratives. These objects and narratives determined what individuals in and beyond the borderlands understood about the region and, therefore, often guided policy. Yet stories and things did not simply move across space; they also moved across time. Encounters with history, including engagements with older narratives (oral and written) and objects (ruins, human remains, preserved specimens), left impressions on how all groups in the Gulf South viewed each other and the environment.[13]

Frontiers of Science argues that encounters in America's borderlands shaped the production, circulation, and application of natural knowledge within these contested regions and, more broadly, throughout the empires and nations competing for them. The expansion of European powers and the United States were the primary motors that drove these encounters. Between the 1500s and the mid-1800s, Spanish, British, French, and U.S. imperialism brought hitherto unconnected individuals, nations, and environments into intellectually productive (though often physically destructive) contact. These expansion-instigated encounters, moreover, resulted in new material, social, and political circumstances that influenced how people created and shared natural knowledge.

I examine these processes in the Gulf South, but this was by no means the only American region where the expansion of European empires and the United States led to new connections and conditions that affected natural knowledge. European and U.S. expansion had a similar impact throughout America, and, therefore, focusing on encounters and their outcomes offers a new way to envision the history of natural knowledge in early America on the whole. This approach reveals connectedness across national boundaries and persistence across time.

Several transnational contexts—regional, continental, hemispheric, and

---

Told: Slavery and the Making of American Capitalism (New York, 2014). On geopolitical competition, see Weber, *Spanish Frontier in North America*, 147–203, 236–270; and David Narrett, *Adventurism and Empire: The Struggle for Mastery in the Louisiana-Florida Borderlands, 1762–1803* (Chapel Hill, N.C., 2015).

13. On exchange, see Joseph M. Hall, Jr., *Zamumo's Gifts: Indian-European Exchange in the Colonial Southeast* (Philadelphia, 2009). On some of the ways engagement with the past mattered in borderlands, see Samuel Truett, "The Borderlands and Lost Worlds of Early America," in Juliana Barr and Edward Countryman, eds., *Contested Spaces of Early America* (Philadelphia, 2014), 300–324.

Atlantic—framed social and intellectual developments in the Gulf South. At the regional level, the Gulf South (some two hundred thousand square miles of Louisiana, Mississippi, Alabama, and Florida) was a borderland, and one of this book's most basic goals is to begin to recover the richness of intellectual life within America's borderlands. Knowledge production in the Gulf South also mattered at the continental level. Networks of power, information, and human trafficking meant that events in the Chesapeake, the far West, and other parts of North America could impinge on, and be changed by, local knowledge in the Gulf South. At the hemispheric level, the Gulf South was simultaneously part of the Caribbean and North America. The region was integral to circum-Caribbean networks of exchange and embroiled in imperial competitions involving Caribbean islands and shipping routes. Moreover, both Europeans and Anglo-Americans in the Gulf South tried to imitate the hierarchies, agronomic success, and technologies of domination in Caribbean plantation societies. As part of Anglo, French, and Spanish America, the Gulf South also exemplified how the Americas had a common hemispheric history in which the similarities and connections between North and South America outweighed the distinctions between them. More broadly still, the Gulf South was thoroughly connected with the Atlantic world. Free and unfree emigrants brought epistemologies from Europe and Africa to the Gulf South, and natural knowledge did not merely flow from the Old World to the Americas; insights from the New World altered Europeans' scientific theories and practices.[14]

14. There is a wealth of excellent scholarship on the social and cultural dynamics of borderlands. An overview is Pekka Hämäläinen and Samuel Truett, "On Borderlands," *JAH*, XCVIII (2011), 338–361. On the importance of borderland encounters beyond borderlands themselves, see François Furstenberg, "The Significance of the Trans-Appalachian Frontier in Atlantic History," *AHR*, CXIII (2008), 647–677; and Deborah A. Rosen, *Border Law: The First Seminole War and American Nationhood* (Cambridge, Mass., 2015). On how knowledge production in the borderlands mattered to European geopolitics, see Paul W. Mapp, *The Elusive West and the Contest for Empire, 1713–1763* (Chapel Hill, N.C., 2011). On the Gulf South as part of the Caribbean, see Amy Turner Bushnell, *Situado and Sabana: Spain's Support System for the Presidio and Mission Provinces of Florida* (Athens, Ga., 1994), esp. 27; and Rebecca J. Scott, *Degrees of Freedom: Louisiana and Cuba After Slavery* (Cambridge, Mass., 2005). On hemispheric history and connections between Latin America and Anglo America, see Bolton, "Epic of Greater America," *AHR*, XXXVIII (1933), 448–474; Eliga H. Gould, "Entangled Histories, Entangled Worlds: The English-Speaking Atlantic as a Spanish Periphery," *AHR*, CXII (2007), 764–786; and Daniel B. Rood, *The Reinvention of Atlantic Slavery: Technology, Labor, Race, and Capitalism in the Greater Caribbean* (New York, 2017). On the Gulf South as part of the Atlantic world, see Cécile Vidal, ed., *Louisiana: Crossroads of*

From the first Spanish entradas in the early 1500s to the age of U.S. domination in the mid-1800s, the history of natural knowledge in the Gulf South was a story of persistence because imperialism conditioned social and political relationships throughout this era. The imperial powers that expanded into the region were Spain, France, Britain, and, starting in the 1790s, the United States. Unlike Spain, France, and Britain, the United States was a republic in which citizens, not a monarch, were sovereign. But republican political organization in no way prevented the early United States from doing the imperial things that other imperial powers did. Far from marking a clean break between colonial and national periods, inhabitants of the Gulf South would have found that U.S. rule perpetuated, and sometimes intensified, many of the same sorts of practices, relationships, and hierarchies previously introduced by European empires.[15]

For one, U.S. officials, like those in European empires, employed diplomacy and military force to support the spread of their nation's people and territory. As a commercial hub, strategic frontier, farmland to expand freedom, plantation land to expand slavery, native homeland, and haven for runaway slaves, the Gulf South was the most desirable part of the continent for many Anglo-Americans, and the federal government took decisive action to help ensure that this region became part of the United States. Federal negotiators finalized land purchases, trade deals, and treaties to secure land for, and the loyalty of, citizens who moved into Gulf South spaces claimed by native groups, Spain, and France. Federal officials then deployed the army to clear Indians from these lands and enforce slavery within them.[16]

U.S. rule also prolonged an imperial context in which officials relied on

---

the Atlantic World (Philadelphia, 2014). On how imperial expansion into the Americas shaped the pursuit of science among Europeans, see Antonio Barrera-Osorio, *Experiencing Nature: The Spanish American Empire and the Early Scientific Revolution* (Austin, Tex., 2006); and James Delbourgo and Nicholas Dew, eds., *Science and Empire in the Atlantic World* (New York, 2008).

15. On the importance of studying how the effects of U.S. expansion were imperial instead of fretting over whether the United States was an "empire," see Paul A. Kramer, "Power and Connection: Imperial Histories of the United States in the World," *AHR*, CXVI (2011), 1348–1391.

16. Adam Rothman, *Slave Country: American Expansion and the Origins of the Deep South* (Cambridge, Mass., 2005); Eliga H. Gould, *Among the Powers of the Earth: The American Revolution and the Making of a New World Empire* (Cambridge, Mass., 2012); Alan Taylor, *American Revolutions: A Continental History, 1750–1804* (New York, 2016), esp. 403–408, 430, 479.

local cooperation to exercise governance. Federal officials hoped to deal with the difficulties of governing an enormous territory in a manner that was at once revolutionary and a throwback to the decentralized imperialism that had made Europe's early modern empires possible: they believed in diffusing power across a vast territory in which authority was, necessarily, negotiated. This was Jefferson's ostensibly centerless and peripheryless "empire for liberty," an expansive state in which a collective sense of nationhood would bind far-flung Americans together in lieu of metropolitan coercion. This mode of governance offered Gulf South inhabitants the opportunity to challenge and shape the still inchoate practices of U.S. imperialism and the still emerging contours of Anglo-American natural knowledge because officials and men of science in the United States, like European ones before them, relied on local power brokers and experts to understand and rule these territories. Moreover, the very way that the United States expanded its influence by incorporating local knowledge and negotiating with prominent inhabitants was largely a continuation of Spain's approach to controlling its Gulf South colonies, an approach that power brokers in the region obliged U.S. officials to co-opt. U.S. officials and men of science were, for their part, eager for the cooperation of individuals with experience as agents of the Spanish Empire: eastern Anglos remained as greedy for information, expansion-promoting methods, and (of course) territory from the Spanish world as Anglos in sixteenth-century England. Federal power in the Gulf South even relied on local cooperation throughout the antebellum era. When the master class ceased to support federal authority, as they did during the 1860s, U.S. governance faltered.[17]

The most significant way that U.S. rule intensified the social and political conditions first imposed by European empires was by enforcing hierarchies based on racial difference. Spanish, French, and British colonial societies were certainly highly unequal, but blacks, natives, and whites in these colonies could all share a similar status as subjects subordinate to a

17. Peter S. Onuf, *Jefferson's Empire: The Language of American Nationhood* (Charlottesville, Va., 2000); Jack P. Greene, "Colonial History and National History: Reflections on a Continuing Problem," *WMQ*, LXIV, (2007), 235–250. For entanglements between the Spanish and Anglo worlds, see Ralph Bauer, *The Cultural Geography of Colonial American Literatures: Empire, Travel, Modernity* (Cambridge, 2003); Jorge Cañizares-Esguerra, *Puritan Conquistadors: Iberianizing the Atlantic, 1550–1700* (Stanford, Calif., 2006); and Gould, "Entangled Histories, Entangled Worlds," *AHR*, CXII (2007), 764–786. On imperialism and natural knowledge in the Spanish Atlantic, see Daniela Bleichmar et al., eds., *Science in the Spanish and Portuguese Empires, 1500–1800* (Stanford, Calif., 2009).

monarch and, at least in theory, enjoy the monarch's protection. The racial hierarchy that the United States introduced was more clearly stratified and more effectively implemented. White citizens shared sovereignty and, in most cases, permanently denied the rights of citizenship to blacks and natives. Anglos in the Gulf South not only chose to perpetuate the old European practice of asserting racial hierarchies but, by the 1810s, could marshal overwhelming demographic and military power to uphold them. In short, the United States, like earlier imperial powers, was a collection of distant places and unequal peoples that came under one government through a combination of conquest, cooperation, and coercion.[18]

Imperialism was as vital to natural knowledge in the United States as it was in the Spanish, French, and British empires, and, as in these powers, natural knowledge and imperialism developed together in the United States. Atlantic historians have made convincing arguments that science, technology, and medicine helped establish, sustain, and challenge Europe's overseas empires. They have also debunked diffusionist notions that scientific knowledge traveled mono-directionally from European centers to colonial peripheries and, instead, have revealed polycentric networks of overlapping trajectories and diverse agents. Surprisingly, these perspectives have had little impact on U.S. history. The legacy of American exceptionalism has continued to convince us that the history of knowledge and expansion in the early United States fit the same diffusion model so widely criticized by scholars of European imperialism. However, U.S. imperialism and American science were both things in the making, and the experience of incorporating ethnically diverse borderlands was essential to their evolution.[19]

18. Kathleen DuVal, *Independence Lost: Lives on the Edge of the American Revolution* (New York, 2015). On how expansion into the Gulf South influenced the boundaries of citizenship and nationhood, see Peter J. Kastor, *The Nation's Crucible: The Louisiana Purchase and the Creation of America* (New Haven, Conn., 2004).

19. On natural knowledge in Europe's early modern empires, see Londa Schiebinger, *Plants and Empire: Colonial Bioprospecting in the Atlantic World* (Cambridge, Mass., 2004); Jorge Cañizares-Esguerra, *Nature, Empire, and Nation: Explorations of the History of Science in the Iberian World* (Stanford, Calif., 2006); and Delbourgo and Dew, eds., *Science and Empire in the Atlantic World.* On how Afro-Atlantic healers and ritual practitioners challenged imperialism, see James H. Sweet, *Domingos Álvares, African Healing, and the Intellectual History of the Atlantic World* (Chapel Hill, N.C., 2011); and Pablo F. Gómez, *The Experiential Caribbean: Creating Knowledge and Healing in the Early Modern Atlantic* (Chapel Hill, N.C., 2017). Historians of science and U.S. expansion have argued that exploration, imperatives to "civilize" other peoples, and geographic

As an expanding power, the United States prolonged, and built off of, a nearly three-hundred-year-old history of colonialism in the Southeast. Nevertheless, studies of natural knowledge in early America have largely ignored imperial continuities in favor of elaborating ways that independence, democracy, postcoloniality, and sectionalism made intellectual life in the United States exceptional and novel. Nationalism and democratization did put their stamp on how at least some Anglo-Americans produced knowledge, but, as in the years before 1776, imperialism and inequality remained fundamental contexts for both the hundreds of thousands of people living in America's borderlands and the eastern Anglo-Americans who sought to know and rule their continent. The study of nature in the United States did not cohere around patriotism, democratic politics, or postcolonial attachments. It was, as in European empires, divided by multiple loyalties and identities, organized through contested hierarchies of ethnicity and place, and reliant on violence.[20]

---

conceptions of imminent continental dominion motivated Anglo-American expansion and diffused Anglo-Atlantic intellectual culture into the West and the wider world. See William H. Goetzmann, *Exploration and Empire: The Explorer and the Scientist in the Winning of the American West* (New York, 1966); Michael Adas, *Dominance by Design: Technological Imperatives and America's Civilizing Mission* (Cambridge, Mass., 2006); Martin Brückner, *The Geographic Revolution in Early America: Maps, Literacy, and National Identity* (Chapel Hill, N.C., 2006), 204–263; and James D. Drake, *The Nation's Nature: How Continental Presumptions Gave Rise to the United States of America* (Charlottesville, Va., 2011). On how perceptions of exceptionalism have resulted from an overly narrow vision of America and American science, see Joyce E. Chaplin, "Expansion and Exceptionalism in Early American History," *JAH*, LXXXIX (2003), 1431–1455; and Laura Dassow Walls, *The Passage to Cosmos: Alexander von Humboldt and the Shaping of America* (Chicago, 2009), esp. x.

20. Alexis de Tocqueville (a French observer in 1830s America) set an influential precedent by striving to identify the social and intellectual conditions that made the United States exceptional as a nation. He argued that "a democratic state of society and democratic institutions" distinguished the pursuit of knowledge in the United States. See De Tocqueville, *Democracy in America . . .* , trans. Henry Reeve, II (New York, 1840), 41. Studies of the national and Anglo-Atlantic aspects of early American science include Brooke Hindle, *The Pursuit of Science in Revolutionary America, 1735–1789* (New York, 1974), 248–385; John C. Greene, *American Science in the Age of Jefferson* (Ames, Iowa, 1984); James Delbourgo, *A Most Amazing Scene of Wonders: Electricity and Enlightenment in Early America* (Cambridge, Mass., 2006); and Andrew J. Lewis, *A Democracy of Facts: Natural History in the Early Republic* (Philadelphia, 2011). Works that define American science primarily through ongoing dependence on Europe include I. Bernard Cohen, *Science and American Society in the First Century of the Republic* (Columbus, Oh., 1961); George Basalla, "The Spread of Western Science," *Science*, CLVI (1967), 614,

The persistence of imperialism ensured that cross-cultural negotiation and brokerage remained integral to Euro-American science in the nineteenth century. Historians of science have argued that the rise of nation states and a greater emphasis on utilizing precision instruments combined to limit the importance of go-betweens and local knowledge to European science during the late eighteenth and early nineteenth centuries. Yet it was largely through engagements with each other that Anglo-Americans, Indians, and other peoples constructed the ethnographic interpretations through which they differentiated themselves in terms of both epistemology and biology. Anglo-Americans' sense that their nation, ethnicity, and capacity for generating legitimate natural knowledge were distinct and exceptional arose because of ongoing competition, violence, and exchange with other ethnicities and empires, not because encounters with these peoples ceased to affect Euro-American science after U.S. independence.[21]

Although imperialism consistently influenced observation, experimentation, circulation, and inspiration in the Gulf South from the 1500s to the 1800s, intellectual and social life did change dramatically during these years. A few of the catalysts for change included Old World diseases, new methods and theories for understanding nature, and the territorial reshufflings through which parts of the Gulf South shifted between imperial powers with little or no say from the peoples living there. Yet two eras, 1670 to 1715 and 1812 to 1821, were particularly revolutionary because they marked critical turning points in the history of geopolitical competition, one of the primary forms of encounter that shaped the region. The decades between 1670 and 1715 witnessed the beginning of a new phase of competition as

---

617; and Kariann Akemi Yokota, *Unbecoming British: How Revolutionary America Became a Postcolonial Nation* (Oxford, 2011). As for sectionalism, historians have tended to relegate knowledge production in the area from Maryland to Texas as southern science, a topic that has more often been contrasted with an American science located in the Northeast than analyzed as constitutive of it. See Thomas Cary Johnson, *Scientific Interests in the Old South* (New York, 1936); and Ronald L. Numbers and Todd L. Savitt, eds., *Science and Medicine in the Old South* (Baton Rouge, La., 1989). However, as one nuanced intellectual history has argued, "Southerners were national, postcolonial, and imperial, all at once, and partly invented their culture in the tense encounters among these conditions." See Michael O'Brien, *Conjectures of Order: Intellectual Life and the American South, 1810–1860*, 2 vols. (Chapel Hill, N.C., 2004), I, 2.

21. On how cross-cultural exchanges remained vital to natural knowledge despite the rise of nation states and the proliferation of precision instruments, see Simon Schaffer et al., eds., *The Brokered World: Go-Betweens and Global Intelligence, 1770–1820* (Sagamore Beach, Mass., 2009), esp. xxi–xxx.

France and Britain founded colonies that encroached on Spanish Florida and as formidable native confederacies, including the Creeks, Chickasaws, and Choctaws, emerged amid the demographic catastrophes initiated by disease and the Indian slave trade. This period of geopolitical competition and mutual weakness endured until the years between 1812 and 1821, when France, Britain, and Spain ceased to contend for the Gulf South and the United States achieved unmatched political dominance. Local blacks, Spanish and French creoles, and natives enjoyed far less leverage and independence after the 1810s than they had when imperial rivalry had ensured that they could secure powerful allies against an unwanted government, and Anglo-American planters acquired unprecedented, but by no means absolute, power in their relations with the Gulf South's other peoples.[22]

Racial hierarchies became more entrenched by the 1820s, yet the same sorts of interactions that had permeated the study of nature in the colonial period continued to do so in the antebellum era. These ongoing encounters included local relationships between Anglos and enslaved blacks that were rooted in violence and continent-spanning relationships between southern and northern Anglos that were usually based on exchange and competition. Sectional tension over whether Anglos would turn Indian lands in the far West into free soil or slave soil was the latest phase in a long history of geopolitical rivalry for North America (a rivalry that enslaved blacks exploited by running away to the North, encouraging abolitionism, and, eventually, pushing an emancipationist agenda during the Civil War). In short, political and demographic changes between the 1500s and the mid-1800s mattered enormously to the history of natural knowledge in America, but such changes mattered inasmuch as they affected *how*—and not the fact *that*—encounters inspired by imperialism generated natural knowledge and structured its movement.[23]

22. Many histories have argued that U.S. expansion fundamentally changed social and political dynamics throughout the continent by transforming fluid borderland worlds of negotiated power into places where Anglo males severely limited the options and independence of other peoples. See Richard White, *The Middle Ground: Indians, Empires, and Republics in the Great Lakes Region, 1650–1815* (Cambridge, 1991); Jeremy Adelman and Stephen Aron, "From Borderlands to Borders: Empires, Nation-States, and the Peoples in between in North American History," *AHR*, CIV (1999), 814–841; and Anne F. Hyde, *Empires, Nations, and Families: A History of the North American West, 1800–1860* (Lincoln, Neb., 2011).

23. On how sectional rivalry over the expansion of slavery was similar to older imperial competitions for the borderlands, see John Craig Hammond, "Slavery, Sovereignty,

• • •

Instead of seeking to impose a clean narrative on a messy history, my method has been to embrace the complexity of the sources and organize this book around detailed case studies from the eras before, during, and after the shift to U.S. governance. This method has several advantages. Case studies can show how, precisely, personal and social circumstances influenced the pursuit of natural knowledge as well as the impact that this knowledge had on geopolitics and peoples' lives. Analyzing case studies can also lay bare commonalities shared across national and temporal boundaries, including the enduring importance of imperialism. Moreover, the episodes in this book make clear how varied and, I think, interesting natural knowledge was in the southeast borderlands, a region where whites, blacks, and natives studied plants, peoples, animals, the cosmos, and each other. Last but not least, case studies offer the opportunity to engage with more specific historical fields than the book's overarching topic (the history of natural knowledge in early America). I hope the analyses presented here will improve historians' understandings of a range of related subjects, including the histories of various branches of natural knowledge (such as astronomy and geology), the scope of the United States' scientific community, and the ways natural knowledge did (and did not) support imperialism in Spain and the United States.

Frontiers of Science follows a roughly chronological arc. The first two chapters consider the Spanish, French, and British colonies of the southeast borderlands from the early 1500s to the late 1700s. These chapters introduce some of the many ways that local, regional, and international relationships catalyzed by imperialism affected observation, experimentation, circulation, and inspiration among the region's inhabitants. The Spanish colonial context was particularly important because Spain had been involved in the Gulf South longer than any other empire and because it claimed sovereignty from Louisiana to East Florida in the decades immediately preceding U.S. expansion.

Chapters Three and Four cover U.S. expansion into the Gulf South between the 1790s and 1810s. Chapter Three concentrates on how the practices, methods, and goals of astronomy in the United States developed hand in hand with expansion into the lower Mississippi Valley and encounters with the Spanish Empire. Chapter Four zeroes in on the experiences of three Euro-American men of science—one Spanish, one Anglo, one French—to

---

and Empires: North American Borderlands and the American Civil War, 1660–1860," *Journal of the Civil War Era,* IV (2014), 264–298.

explore the ways individuals in the borderlands blurred the boundaries between national scientific communities as they sought support and recognition from multiple polities.

Chapter Five, which examines ethnographic discourses about the mental capacities of Gulf South ethnic groups from 1800 to the 1830s, marks a turning point in the book. The United States ousted its imperial rivals during these years, and Anglo-American planters, riding the tide of the cotton and sugar booms, achieved enormous wealth and power. Chapter Six looks particularly at how Anglo masters' relationships with enslaved blacks and northeastern men of science affected geological research and theories during the 1830s. Chapter Seven analyzes violent encounters among Anglos and Seminoles in Florida, particularly violence against the dead, that generated new knowledge and identities during the Second Seminole War (1835–1842). Just as bloody clashes between Spaniards and Indians had initiated the colonial era in the Gulf South, so, too, did warfare between Anglos and Indians mark the effective end of the Gulf South as a borderland.

Yet the conclusion of the Second Seminole War did not mean that Anglos and the other peoples in the Gulf South ceased to engage with each other in ways that engendered natural knowledge. Nor did it mean that natural knowledge in the United States was no longer shaped by transnational encounters in borderlands. Only four years after the end of the war in Florida, Anglo-Americans launched another imperial war, this time into the southwest borderlands, which added new territories and ethnicities to the United States. The Epilogue considers how U.S. expansion into the Southwest and the rise of the Smithsonian Institution were not unprecedented developments in the histories of U.S. expansion and science but rather reflected and perpetuated a much older history in which natural knowledge and imperialism evolved together. Encounters inspired by imperialism remained as inextricable from natural knowledge in the nineteenth-century United States as they had been in sixteenth-century Florida.

{ 1 }

VIOLENCE, COMPETITION, AND EXCHANGE IN THE EARLY COLONIAL ERA

• • •

Between the 1500s and mid-1700s, the Gulf South became a new world as natives, Europeans, Africans, and (eventually) creoles exchanged biota, technologies, cosmological concepts, and specimens. For many—the hundreds of Catholic conquistadors and martyrs who died trying to save and subject the region's natives, the thousands of African and creole blacks who grappled with life as commodities, and the hundreds of thousands of Indians who died from disease and enslavement—life in this new world was a brutish and short nightmare. For others, the new Gulf South was a place of opportunity. These luckier few included native chiefs who looked to European goods and experts to sustain their power and Euro-American naturalists who garnered prestige by circulating local specimens to Europe.

Natives and newcomers valued natural knowledge in surprisingly similar ways. Gulf South natives and Europeans both considered access to difficult-to-acquire information and objects to be a source and reflection of power; therefore, they prized specialized knowledge and supported the individuals who seemed able to obtain and interpret it. The encounters initiated by European imperialism offered Europeans and Indians alike an opportunity to investigate previously hidden aspects of nature (though at enormous cost for natives). For Europeans, laying hands on power-promoting knowledge in the Gulf South could mean relying on armies and slave raiders to explore the region, interrogating native neophytes, pumping onsite informants for geographic details, and trading with natives for plants and animals. Natives tapped into similar forms of knowledge by promoting the pursuits of learned Indian and European priests, accepting gifts of foreign animals and apparatus, and visiting distant lands as explorers, emissaries, and captives. Encounters involving violence, geopolitical competition, and exchange generated new materials and methods for understanding nature and, largely because of the shared value that natives and newcomers placed on goods and information from distant realms, intertwined native and European intellectual networks.

The colonial Gulf South was a complex and constantly changing place in which seeking natural knowledge could be a viable path to prestige and power but just as often reflected an individual's or group's weakness. The first wave of European invaders, Spanish adelantados (advancers), used violence to advance knowledge as part of their larger goals of advancing their own wealth, the glory of the crown, and the Kingdom of Christ. Spain eventually won a foothold in 1565 and, for more than a century, colonized Florida by establishing Catholic missions and sharing power with native chiefs. Yet, while Florida chiefs did maintain considerable influence by allying with Spanish leaders, these alliances obliged many chiefs to begin supporting learned Catholic priests through the same forms of patronage they offered to learned indigenous priests; there was, for natives and Europeans alike, no clear boundary between natural and spiritual expertise. The patronage networks that sustained native priests changed beyond Florida as well during the epidemics and slave raids of the 1670s to 1710s, an era with far-reaching consequences for the organization of Indian communities. The slave raids were one manifestation of a broader era of international competition during which France and England challenged Spanish rule. This competition drove men in all three empires and natives caught in the middle of their power struggle to generate and circulate maps as part of their efforts to claim, or simply survive in, the Gulf South.

By the 1700s, native American networks of information, alliance, and prestige goods were fully intermeshed with networks of science and commodities oriented toward the Atlantic. The result was a polycentric web of intellectual and material exchange that connected the southeast borderlands, the Caribbean, Africa, and Europe. This web was neither indigenous nor European; rather, it blurred both the channels through which items and ideas flowed and the logics on which such exchanges were based. Cross-cultural exchanges enabled European technologies and creatures to bolster the prestige of natives and facilitated the circulation of specimens and information to Europe. In short, natives and Europeans valued natural knowledge, European expansion led to encounters that enabled natives and Europeans to incorporate hitherto unavailable information and materials, and these encounters engendered new networks that transcended regional and national boundaries. The most intellectually productive of these encounters centered around violence, geopolitical competition, and exchange.[1]

1. On how the colonial Southeast was an uncomfortably shared new world tied to the Atlantic, see James H. Merrell, *The Indians' New World: Catawbas and Their Neighbors from European Contact through the Era of Removal* (Chapel Hill, N.C., 1989); Daniel H.

ADVANCERS OF KNOWLEDGE

Many of the same adventurers and entrepreneurs who sought to advance Spain's empire and their own glory by conquering parts of the Americas were also dedicated (sometimes contractually) to producing knowledge about their would-be domains. Several adelantados signed contracts with the Spanish crown to conquer Florida at their own expense in exchange for seigniorial rights to the land and the labor of its inhabitants. Adelantados were, in effect, a heavily armed subset of a larger group of sixteenth-century Spanish entrepreneurs, men who relied on firsthand observation, experimentation, Indian informants, and state-of-the-art instruments to profit from the Indies. The adelantados in Florida and the men who accompanied them were not simply slave-raiding gold seekers (though they certainly were these things). They were also geographic explorers, observant naturalists, and agricultural experimenters who built on the hard-won experience of earlier expeditions. Most importantly, the case of Florida's adelantados reveals that violence—particularly against Indians and enslaved blacks—was at the heart of European efforts to know and order the Gulf South from the very beginning.[2]

Juan Ponce de León, the former governor of Puerto Rico and the first conquistador to invade Florida, combined military conquest and exploration as a matter of contractual obligation. The contract through which the Spanish crown granted him a license to "discover and settle the Island of Binini"—the fabulous land he was looking for when he chanced on Florida in 1513—offered Ponce de León governance of the island, the bulk of its

---

Usner, Jr., *Indians, Settlers, and Slaves in a Frontier Exchange Economy: The Lower Mississippi Valley Before 1783* (Chapel Hill, N.C., 1992); Joseph M. Hall, Jr., *Zamumo's Gifts: Indian-European Exchange in the Colonial Southeast* (Philadelphia, 2009); Joshua Piker, *The Four Deaths of Acorn Whistler: Telling Stories in Colonial America* (Cambridge, Mass., 2013); Cécile Vidal, ed., *Louisiana: Crossroads of the Atlantic World* (Philadelphia, 2014); and Alejandra Dubcovsky, *Informed Power: Communication in the Early American South* (Cambridge, Mass., 2016).

2. On entrepreneurs in the sixteenth-century Iberian Atlantic and their role in developing empirical science, see Antonio Barrera-Osorio, *Experiencing Nature: The Spanish American Empire and the Early Scientific Revolution* (Austin, Tex., 2006). On how adelantados merged exploration and conquest, see Charles Hudson, *Knights of Spain, Warriors of the Sun: Hernando De Soto and the South's Ancient Chiefdoms* (Athens, Ga., 1997), 2–3; and David J. Weber, *The Spanish Frontier in North America* (New Haven, Conn., 1992), 29, 64. On how adelantados aimed to become "neo-feudal" lords in the Americas, see Daniel K. Richter, *Before the Revolution: America's Ancient Pasts* (Cambridge, Mass., 2011), 67–87 (quote, 72).

gold, and the power to extract labor from its inhabitants. In return, Ponce de León had to finance the conquest and exploration of Binini (also called Bimini) and pen reports about the land and its resources. At least during his second expedition to Florida in 1521, Ponce de León made a good-faith effort to fulfill this part of his contract. He assured the king that he would "discover more of the coast" and "endeavor to find out everything I can" so as to "make a relation to Your Majesty of whatever is done or seen in those parts." Ponce de León and his men attempted to develop a clearer picture of Florida's geography through a combination of research and violence: for instance, they conducted astronomical observations to determine latitude and also captured a Calusa Indian "as a navigator." But the Calusas, who had been warned about Spaniards by refugees from the Caribbean, proved more than a match for Spain's soldier-explorers, repelling Ponce de León's men in 1513 and, in 1521, laying him low with a mortal wound. Ponce de León's experiences were the first of many hard lessons for Spaniards in the land he named La Florida. Despite their military technology, eye for discovery, and eagerness to learn from native informants, colonizing and knowing North America would not be easy.[3]

Subsequent expeditions were even more oriented toward studying Florida's geography and nature but fared no better than Ponce de León's. In 1523, sugar-planter-turned-adelantado Lucas Vázquez de Ayllón signed a contract to colonize Florida in which he agreed to acquire information on the territory's resources and inhabitants. Ayllón based his plans on Indian stories, particularly the lavish accounts that Francisco de Chicora, who Ayllón had captured during a slave raid in present-day South Carolina in 1521, gave of the region's valuable plants and agricultural potential. Ayllón's attempt to found a colony near Chicora's homeland in 1525–1526 had much

3. "1512 Royal Contract with Juan Ponce de León," in John E. Worth, ed. and trans., *Discovering Florida: First-Contact Narratives from Spanish Expeditions along the Lower Gulf Coast* (Gainesville, Fla., 2014), 73 ("discover and settle"), 76, Juan Ponce de León to King of Spain, Feb. 10, 1521, 83–85 ("discover more," 83, "endeavor," 84), "The 1610 Antonio Herrera Account," 50 ("navigator"), 51. Although less precise than the famous *Relaciones geográficas* (Geographic relations) of the 1570s, adelantado contracts like Ponce de León's were early examples of the kinds of scientific instructions through which metropolitan bureaucrats tried to order the production and circulation of knowledge in Atlantic empires. See María M. Portuondo, *Secret Science: Spanish Cosmography and the New World* (Chicago, 2009). On scientific instructions more broadly, see Daniel Carey, "Inquiries, Heads, and Directions: Orienting Early Modern Travel," in Judy A. Hayden, ed., *Travel Narratives, the New Science, and Literary Discourse, 1569–1750* (Burlington, Vt., 2012), 25–51.

in common with later English colonies in the same region, especially in his focus on using black and Indian slaves to harvest agricultural commodities for export. But internal strife and less-than-cooperative Guale Indians meant that 450 of the original 600 colonists, including Ayllón, died during this experimental effort to establish a new Andalusia. Two years later, the adelantado Pánfilo de Narváez also failed to satisfy his contractual obligation "to explore, conquer, and settle" Florida. Apalachee archers forced Narváez's men to flee by sea, and most of the 242 remaining Spaniards perished in the Gulf of Mexico. A few survivors washed ashore in Texas, including Alvar Nuñez Cabeza de Vaca who, after eight years living and traveling in the Southwest, found his way to Mexico City and added his observations to Spain's painfully growing body of knowledge about Florida.[4]

The adelantado Hernando de Soto, a veteran of Peru, led the largest yet expedition of discovery and conquest into the Gulf South from 1539 until his death in 1542. After arriving in Florida, de Soto tried to make use of a mass of information he had accumulated from kidnapped Indians, older maps, and tight-lipped veterans of previous entradas. But neither his careful planning, six-hundred-strong army, nor constant interrogations of local informants enabled him to fulfill his dreams of finding a new Peru—that is, a centralized Indian empire to make his own—and precious metals. Instead, he had to settle for indulging his passion for hurting Indians. De Soto and his men maimed, killed, raped, enslaved, and warred from Florida to Tennessee to the Mississippi River, where de Soto perished from disease before actually conquering anything.[5]

4. Pánfilo de Narváez, *Capitulación,* Dec. 11, 1526, quoted in Weber, *Spanish Frontier in North America,* 42–43 (quote); Paul E. Hoffman, *A New Andalucia and a Way to the Orient: The American Southeast during the Sixteenth Century* (Baton Rouge, La., 2004), esp. 35, 60.

5. De Soto had acquired a map of the Gulf coast and, in 1538, sent a reconnaissance mission to Florida that returned to Cuba with a report on the coastline and four captive Indian guides. But knowledge did not always flow freely, even among Spaniards, and entrepreneurial conquistadors could be reluctant to share details about the lands they coveted. Before departing for Florida, de Soto met Cabeza de Vaca—who envied de Soto's new title of governor of Florida—in Madrid. Cabeza de Vaca had "a relation of what he had seen in Florida" but refused "to divulge certain things" to de Soto. See "The Account by a Gentleman from Elvas," trans. and ed. James Alexander Robertson, in Lawrence A. Clayton, Vernon James Knight, Jr., and Edward C. Moore, eds., *The De Soto Chronicles: The Expedition of Hernando De Soto to North America in 1539–1543,* 2 vols. (Tuscaloosa, Ala., 1993), I, 48–49 (quotes, 48). See also Garcilaso de la Vega, "La Florida, by the Inca," trans. Charmion Shelby, ed. David Bost, ibid., II, 89; Robert S. Weddle, "Soto's Problems of Orientation: Maps, Navigation, and Instruments in the Florida Expedition," in

The violence at the heart of de Soto's entrada was by no means distinct from the expedition members' efforts to learn about geography, nature, and resources. This was apparent in de Soto's eagerness to acquire and validate the knowledge of Indian informants. Time and time again, de Soto seized natives and forced them to lead the Spanish to population centers and rumored mineral deposits. Violence also gave de Soto some standard by which to assess the credibility of informants from cultures about which he knew almost nothing. After one captive supposedly reported that the far-off town of Ocale had an "abundance of gold and silver and many pearls," de Soto mused, "would to God that it be so, for I do not believe of these Indians only what I see ... although they know and consider it for a fact that if they lie to me, it must be at the cost of their life." De Soto, like other Spanish entrepreneurs throughout the Atlantic, yearned for reliable eyewitness evidence, and threatening Indian informants with a painful death was the closest he could get to verifying the truth of their reports. But de Soto's reputation for violence also had the contradictory effect of inspiring some captive guides to mislead the Spaniards to keep them away from population centers. After concluding that one Indian was guiding his men to nowhere, de Soto had him torn apart by dogs.[6]

De Soto intended his entrada to be both a military and exploratory expedition and, to at least some extent, succeeded in this goal. For one, de Soto's army included Juan de Añasco who "prided himself on being a cosmographer." Añasco had reconnoitered the expedition's landing spot, brought astronomical instruments to observe latitudes throughout Florida, and used his geographical and navigational training to help de Soto and his

Patricia Galloway, ed., *The Hernando De Soto Expedition: History, Historiography, and "Discovery" in the Southeast* (Lincoln, Neb., 1997), 219–220; and Juan Gaytan, Jno. de An Asco, and Luis Fernandez de Biedma, "Letter to the King of Spain from Officers at Havana in the Army of De Soto," [ed. and] trans. Buc[k]ingham Smith, *FHQ*, XVI (1938), 222. On de Soto's use of knowledge accumulated during earlier invasions, see Paul E. Hoffman, "Introduction: The De Soto Expedition, a Cultural Crossroads," in Clayton, Knight, and Moore, eds., *De Soto Chronicles*, I, 1–17. For an overview of de Soto's entrada, see Hudson, *Knights of Spain, Warriors of the Sun*.

6. Hernando de Soto, "Letter Written to the Secular Cabildo of Santiago De Cuba ..., Espiritu Santo, Florida, July 9, 1539," [ed. and] trans. James A. Robertson, *FHQ*, XVI (1938), 176 (quotes); De la Vega, "La Florida," trans. Shelby, ed. Bost, in Clayton, Knight, and Moore, eds., *De Soto Chronicles*, II, 199–202. Some of the many examples of de Soto using violence to extract Indian knowledge include "Account by a Gentleman from Elvas," trans. and ed. Robertson, in Clayton, Knight, and Moore, eds., *De Soto Chronicles*, I, 74, 80, 82, 84, and "Relation of the Island of Florida, by Luys Hernández de Biedma," trans. and ed. John E. Worth, I, 226.

men get where they wanted to go. His skills proved most valuable in 1543 as the Spaniards who outlived de Soto tried to retreat by sea. Añasco used an astrolabe (a navigational instrument that he had salvaged and repaired after a disastrous battle with the chiefdom of Mabila), a jerry-rigged fore-staff, and a sea chart he drew on a deerskin to help the survivors straggle to safety in Mexico.[7]

De Soto's army also included men dedicated to natural history and ethnography. The Gentleman of Elvas, who penned a chronicle of the expedition, included a chapter on "Certain diversities and peculiarities of the Land of Florida; and the Products and Birds and Animals of that Land" in which he listed potentially valuable botanical commodities and compared the region's fauna with those of Africa and Spain. Although the natural history writings of another observer, Rodrigo Rangel, have not survived, the chronicler Gonzalo Fernández de Oviedo summarized that Rangel had drafted a chapter on "the animals of that land" as well as the "fruits of that land and trees of sweetgum and sables and many other particulars." One of the friars who accompanied de Soto, Sebastián de Cañete, recorded notes on botanical resources and native peoples, whom he considered "very bright and well featured and of acute judgment." The Incan chronicler Garcilaso de la Vega was not present during the expedition but nevertheless backed up Cañete's argument that Florida Indians were just as capable of reason as Spaniards. These ethnographic arguments were part of a broader debate about Indians' mental capacity that would go on for centuries. But they also suggest that violence against Indians, including both living natives whom Spaniards tortured and dead natives whose graves de Soto's men ransacked, did not always lead to negative assessments of Indians' character and abilities.[8]

7. Not everyone respected Añasco's navigational expertise. One survivor recalled: "The mariners ... who knew that [Añasco] was not a seaman ... ridiculed him, and when he learned how they were jeering at him he threw all the instruments except the astrolabe into the sea. Another brigantine that was coming behind picked them up .... Thus we traveled, or rather navigated, seven or eight days ...." See De la Vega, "La Florida," trans. Shelby, ed. Bost, in Clayton, Knight, and Moore, eds., *De Soto Chronicles,* II, 403, 455 ("prided"), 527 ("mariners"); and Weddle, "Soto's Problems of Orientation," in Galloway, ed., *Hernando De Soto Expedition,* 220–221, 224.

8. "Account by a Gentleman from Elvas," trans. and ed. Robertson, in Clayton, Knight, and Moore, eds., *De Soto Chronicles,* I, 168–170 ("Certain," I, 168), "Account of the Northern Conquest and Discovery of Hernando De Soto, by Rodrigo Rangel (Drawn from *Historia general y natural de las Indias* by Gonzalo Fernández de Oviedo y Valdés)," ed. and trans. John E. Worth, I, 306 ("animals"), "The Cañete Fragment: Another Nar-

The knowledge that de Soto's followers collected through violence in Florida circulated throughout the Atlantic world. Only one of the expedition's four chroniclers published their accounts in Spain; two had their narratives published in Portugal, and one of these, the chronicle by the Gentleman of Elvas, was translated and published in England by Richard Hakluyt in 1609 to promote English colonialism. These chronicles crossed international boundaries because Portugal and England leapt at opportunities to co-opt Spanish knowledge and challenge Spanish power. The expedition's survivors also circulated ethnographic details and natural history specimens within the Americas. They told the viceroy of New Spain about the "monstrous deformity" that Gulf coast natives "gave their heads and faces by artificial means," but the viceroy was especially interested in learning about lucrative resources and "the nature of the lands of La Florida." What he heard "increased [his] desire to make the conquest." Additionally, the survivors provided geographic details to the institutional center of Spanish cosmography, the Casa de Contratación in Seville. Royal Cosmographer Juan López de Velasco incorporated the so-called De Soto Map (circa 1544)—which included reports from the de Soto, Ayllón, and Narváez entradas—as part of his effort to construct a more complete map of the world by synthesizing data from throughout the Spanish Empire.[9]

De Soto and his predecessors failed to conquer the Gulf South, but, by the 1560s, their work had encouraged officials and men of science in Spain,

---

rative of Hernando De Soto," [ed. and trans.] Eugene Lyon, I, 308–309 ("very bright," I, 308). Garcilaso couched his claim about natives' capacity for reason in defensive language to stave off accusations that he lauded Florida Indians merely as a way "to praise my own nation because I am an Indian." See De la Vega, "La Florida," trans. Shelby, ed. Bost, ibid., II, 175–176 (quote, 176). For de Soto and Indian graves, see "Relation of the Island of Florida, by Luys Hernández de Biedma," trans. and ed. Worth, ibid., I, 230–231. On how Europeans debated Indians' rationality in the 1500s, see Anthony Pagden, *The Fall of Natural Man: The American Indian and the Origins of Comparative Ethnology* (Cambridge, 1982).

9. De la Vega, "La Florida," trans. Shelby, ed. Bost, in Clayton, Knight, and Moore, eds., *De Soto Chronicles,* II, 545–549 ("monstrous," II, 548, "nature," II, 549). On the circulation of the chronicles from De Soto's expedition, see Galloway, ed., *Hernando De Soto Expedition.* On Hakluyt's publication of Elvas, see Peter C. Mancall, *Hakluyt's Promise: An Elizabethan's Obsession for an English America* (New Haven, Conn., 2007), 270–272. On the Casa de Contratación and its royal cosmographers, see Portuondo, *Secret Science.* For the De Soto map, see Barbara Boston, "The 'De Soto Map,'" *Mid-America,* XXIII (1941), 236–250; William P[atterson] Cumming, *The Southeast in Early Maps,* rev. and enl. by Louis De Vorsey, Jr., 3d ed. (Chapel Hill, N.C., 1998), 108–111; and Weddle, "Soto's Problems of Orientation," in Galloway, ed., *Hernando De Soto Expedition,* 229–230.

FIGURE 1. *Dibujo del Golfo de México y costa de Nueva España* ("De Soto Map"). [Circa 1544]. Ministerio de Educación, Cultura y Deporte. Archivo General de Indias. AGI, MP-México, 1

tortugas
los martires
florida
agnada
farellones
capalar
finar

England, and, especially, France to begin incorporating firsthand information about the region into their imperial imaginations. Between 1562 and 1565, French Protestants allied with France's Catholic government established a settlement on Florida's Atlantic coast and, in 1564, built Fort Caroline near the mouth of the Saint Johns' River. This colony initiated what would prove to be a 250-year era of imperial competition for the Gulf South.[10] The French Protestants produced a striking volume of writings and illustrations about Florida's plants, animals, agricultural potential, precious metals, and Timucuan peoples. One French officer explained that such depictions were invaluable because "public opinion is formed and instructed with new knowledge, and the uncertainty of things imagined becomes an assurance by beholding with the eye things as marvelous as prodigious land and sea monsters." France's expedition was, moreover, staffed with self-proclaimed experts, including an astronomer, a physician, an apothecary, an artist / cartographer, and a soldier who claimed to be "a great Magician [who] by the secrets of magic . . . had discovered a mine of gold or silver." But the main reason the French learned so much about Florida was that they approached knowledge gathering as part of a broader practice of gift exchange. They offered exotic objects to Timucuas that, in turn, inspired them to present the French with specimens and grant them access to lands and resources. Gift exchange undergirded French exploration from the beginning: on arriving on the Florida coast in 1562, expedition leader Jean Ribaut presented "some loking glases and other prety thinges" to an Indian who showed them "the easiest landing place." The French then offered gifts that built sufficient trust for the Timucuas to assist their efforts "to veue and knowe those Indians" and observe plants and animals.[11] The Timucuas

10. On how Spanish reports inspired French efforts to settle Florida, see Jean Ribaut, "The True Discouerie of Terra Florrida, from a Manuscript Copy in the British Museum," in Ribaut, *The Whole and True Discouerye of Terra Florida: A Facsimile Reprint of the London Edition of 1563 . . .*, [ed.] David L. Dowd and Jeannette Thurber Connor (Gainesville, Fla., 1964), 75; and Paul E. Hoffman, "The Chicora Legend and Franco-Spanish Rivalry in *La Florida*," *FHQ*, LII (1984), 419–438.

11. Charles E. Bennett, ed., *Laudonniere and Fort Caroline: History and Documents* (Gainesville, Fla., 1964), 6–16, Giles de Pysière, "Discourse on the Enterprise and Pillage That the Criminals in the Isle of Florida Did to Their Captains and Governors . . . with a Description of the Wild Beasts . . . ," 1565, ibid., 74 ("public opinion"); René [Goulaine de] Laudonnière, *Three Voyages*, ed. and trans. Charles E. Bennett (Gainesville, Fla., 1975), 92 ("great Magician"); Ribaut, "True Discouerie of Terra Florrida," in Ribaut, *Whole and True Discouerye of Terra Florida*, [ed.] Dowd and Connor, 67–89 ("some loking," 67,

seem to have appreciated the Frenchmen's interest in natural history, and one chief gifted Ribaut a deerskin covered with drawings of several animal species. Indigenous networks of gift exchange, Atlantic scientific networks, and global commodity networks were, by the late 1500s, already becoming intertwined.[12]

The man most responsible for destroying the French settlement and establishing Spanish colonialism in Florida was Pedro Menéndez de Avilés. Not only was Menéndez a knight and the latest Spaniard to be named governor and adelantado of Florida, he was also among the Spanish explorers who combined a crusading spirit, advanced technology, and scientific pursuits to establish and expand Spain's Atlantic empire. Menéndez was both an aspiring neo-feudal lord who dreamed of acquiring estates and laborers in the New World and an entrepreneur who developed new techniques and technologies through firsthand experience: after years at sea fighting "pirates and Lutherans," for example, he designed models for faster ships and a new instrument for measuring longitude at sea based on deviations in the compass. Yet Menéndez's scientific mindset in no way offset his belief that conquering Florida also demanded the conversion of its natives, a

---

"veue," 71); "The Narrative of Jacques le Moyne de Morgues, an Artist Who Accompanied the French Expedition to Florida under René de Laudonnière in the Year 1564," in Stefan Lorant, ed., *The New World: The First Pictures of America* ... (New York, 1946), 33–87, "Nicolas Le Challeux's Narrative of Captain Jean Ribaut's Last Voyage in 1565, Undertaken at the King's Command, to an Island in the Indies Commonly Called Florida," esp. 94, 96; "Copy of a Letter Coming from Florida, Sent to Rouen and Then to M. D'Everon ...," 1564, in Bennett, ed., *Laudonniere and Fort Caroline,* 67, 68, 70, Bennett, "The First Part—The History," 20; Lorant, ed., *New World,* 12. The distribution of prestige goods was at the center of southeastern Indians' political economy, a legacy from their Mississippian ancestors. Europeans likewise understood that gifts had power to convey prestige and establish mutually beneficial alliances. See Hall, *Zamumo's Gifts,* esp. 7, 40.

12. John H. Hann, *A History of the Timucua Indians and Missions* (Gainesville, Fla., 1996), 36. Decades after Fort Caroline met its doom, French corsairs continued trading with Florida Indians for ambergris and medicinal herbs. See Amy Turner Bushnell, *Situado and Sabana: Spain's Support System for the Presidio and Mission Provinces of Florida* (Athens, Ga., 1994), 64, 68. As with de Soto's expedition, the failure of the French colonists did not prevent the knowledge they collected from circulating across imperial boundaries. See, for example, H. P. Biggar, "Jean Ribaut's Discoverye of Terra Florida," in Ribaut, *Whole and True Discouerye of Terra Florida,* [ed.] Dowd and Connor, 48–49; "Narrative of Jacques le Moyne de Morgues," in Lorant, ed., *New World,* 30–31; and Nicholas Monardes, *Joyful Newes out of the Newe-Founde Worlde* (London, 1596), quoted in Bennett, ed., *Laudonniere and Fort Caroline,* 186.

spiritual crusade that inspired him to invite the first Jesuit missionaries to Florida in 1566 and the first Franciscans in 1573.[13]

Menéndez outdid his predecessors in colonizing and knowing Florida by combining an agenda of exploration grounded in military might with the administrative apparatus of a colonial bureaucracy. Like Ponce de León, Menéndez's contract with the king stipulated that he combine conquest and exploration, using his five hundred men "supplied and prepared for war" for protection while he reconnoitered the coast and determined the latitude of strategic points. And, as a 1571 epic poem about his invasion made clear, Spaniards observed the topography, "types of trees," and peoples of Florida as part of their bloody campaigns against French "Lutherans" (actually Calvinist Huguenots) and enemy Indians. Since he aspired to spread Spanish rule across North America, Menéndez turned Florida's capital—which he founded at Saint Augustine in 1565 and moved to Santa Elena in 1566 (before it returned to Saint Augustine in 1587)—into one of the administrative nodes that ordered the accumulation of knowledge in Spanish America. He organized expeditions up the Atlantic coast and into the interior, most notably Juan Pardo's expedition through present-day South Carolina and Tennessee from 1566 to 1568. In accordance with Menéndez's instructions, Pardo's team searched for a land route to New Mexico, studied the region's soils, plants, and animals, and employed an alchemist to assay the minerals they found in the Appalachian mountains. Beyond being a provincial node in a Euro-centric scientific network, Menéndez also turned Santa Elena into the focal point of a continental web of alliance and exchange through which the Spanish acquired sassafras, sarsaparilla, and other botanical commodities for export across the Atlantic.[14]

13. "Order Granting Pedro Menéndez a Royal Patent for His Device to Determine Longitude," trans. Eugene Lyon, in Lyon, ed., *Pedro Menéndez de Avilés* (New York, 1995), 57 (quote). On Menéndez as an innovator and religious crusader, see Lyon, "Aspects of Pedro Menéndez the Man," ibid., 17–18, 21. On Menéndez as a knight, see "The Application of Pedro Menéndez for the Grade of Cavalier in the Religious-Military Order of Santiago," trans. Lyon, ibid., 41–47. On other knight-cosmographers in the Spanish Atlantic, see Jorge Cañizares-Esguerra, *Nature, Empire, and Nation: Explorations of the History of Science in the Iberian World* (Stanford, Calif., 2006), 10. On entrepreneurs and innovators in the Spanish Atlantic, see Barrera-Osorio, *Experiencing Nature.* On missionaries, see Weber, *Spanish Frontier in North America,* 69, 100.

14. "Pedro Menéndez' Contract for the Settlement and Pacification of Florida," Mar. 20, 1565, trans. Lyon, in Lyon, ed., *Pedro Menéndez de Avilés,* 79–86 ("supplied," 80); Bartolomé de Flores, *La felice victoria* (1571), [ed. and trans.] E. Thomson Shields, [Jr.], and Thomas Hallock, *Common-Place,* XV, no. 4.5 (Summer 2015), http://common-place

Yet Menéndez's most ambitious project was to develop Florida into a colony that would thrive through experimental agricultural and the exploitation of enslaved laborers. In this sense, Menéndez combined the neofeudal dream that had driven adelantados since the fifteenth century with the innovation and empiricism of Spanish Atlantic entrepreneurs. His goal was to test commodities like grapes, silkworms, and, especially, sugar, which was only just beginning to emerge as the plantation commodity par excellence of the circum-Caribbean. The establishment of an experimental sugar colony worked by African slaves was included in Menéndez's instructions from the crown in his adelantado contract, in which Menéndez pledged "to import ... five hundred [African] slaves ... for planting sugar cane for the sugar mills that may be built, and for building the said sugar mills." Philip II, it seems, saw the rising economic importance of sugar plantations and hoped that a proven experimenter like Menéndez would put Spain on the cutting edge of this new wave of colonial exploitation. Menéndez tried and failed to realize this dream, and it would be one hundred years before a southeastern colony—English Carolina—successfully combined agricultural experimentation and violence against slaves to form a viable plantation society.[15]

Nevertheless, Spanish explorers and experimenters already had decades of experience in the Gulf South before the first Englishmen established toeholds on the continent. These experiences confirmed their expectations that violence was an effective means of acquiring knowledge in colonial frontiers, spaces where brutality allowed Europeans to forego niceties like gentlemanly collaboration and the appearance of disinterestedness that

---

.org/book/la-felice-victoria/ ("types of trees"); Eugene Lyon, "Introduction," in Lyon, ed., *Pedro Menéndez de Avilés,* xx, "Pedro Menéndez Marqués' Florida Voyages: From Newfoundland to the Gulf of Mexico," Jan. 29, 1573, trans. Lyon, 277, "Exploration of Juan Pardo," Jan. 23, 1569, trans. Lyon, 280, Lyon, "Aspects of Pedro Menéndez the Man," 20; Chester B. DePratter, Charles M. Hudson, and Marvin T. Smith, "The Route of Juan Pardo's Explorations in the Interior Southeast, 1566–1568," *FHQ,* LXII (1983), 125–158; Weber, *Spanish Frontier in North America,* 70–71. On how Menéndez depended on Indian alliances and gifts to secure his position in Florida, see Hall, *Zamumo's Gifts,* 40–43.

15. "Pedro Menéndez' Contract for the Settlement and Pacification of Florida," Mar. 20, 1565, trans. Lyon, in Lyon, ed., *Pedro Menéndez de Avilés,* 82 (quote), "Pedro Menéndez' Letter to Phillip II of October 15, 1565," trans. Lyon, 169, "Pedro Menéndez' Plan for the Economic Development of the Florida Colony," [1569], trans. Lyon, 464, Lyon, "Aspects of Pedro Menéndez the Man," 20, Lyon, "Introduction," xxi; S. Max Edelson, *Plantation Enterprise in Colonial South Carolina* (Cambridge, Mass., 2006).

were increasingly expected of men of science working in Europe. The fact that force also enabled Europeans to establish and maintain colonial societies ordered through hierarchies of ethnic difference helped ensure that violence, particularly against natives and blacks, would remain a viable method for learning about the nature and peoples of the Gulf South.[16]

NATIVE PATRONS AND LEARNED PRIESTS

The Spaniards who invaded the Gulf South entered an Indian world, one with rich intellectual traditions that had existed long before contact and endured for centuries after. But imperialism—and the violence and exchanges that accompanied it—also introduced ideas and goods that altered many Indians' views of the world. One reason why some European practices and beliefs took hold among many Florida Indians between the late 1500s and the early 1700s was that learned Europeans won the patronage of several of the chiefs who had, before the colonial era, supported natives.

Patronage is not a word that comes up often in histories of native America. Historians of science, however, have long insisted that patronage relationships—which, in the early modern era, often entailed powerful individuals backing experts by providing them with material support, status, and social connections—affected the intellectual pursuits and self-presentation of men of science. Patronage relationships similarly structured connections between political leaders and learned men in Gulf South native societies. These relationships influenced how native priests, who claimed to be specialists in both the spiritual and natural worlds, pursued knowledge and fashioned their identities. Chiefs in the Gulf South had long promoted the work of priests whose expertise seemed likely to augment their influence. In the colonial era, some chiefs in Florida decided that European priests—who had extraordinary access to otherwise-unobtainable knowledge, goods, and allies—were more likely to bolster their power than native ones. The patronage relationships between chiefs and Europeans were not limited to top-down hierarchies: a web of dependencies connected native chiefs, Spanish officials, and European priests. Still, the credibility and aid that chiefs provided to missionaries helped Europeans make significant inroads into the social and intellectual lives of Florida Indians. New diseases and the slave trade, however, led to the collapse of Florida's mission communities in the early 1700s and also reconfigured native societies throughout the

16. For how forms of violence unacceptable in Europe could be acceptable in colonial frontiers, see Eliga H. Gould, "Zones of Law, Zones of Violence: The Legal Geography of the British Atlantic, circa 1772," *WMQ*, 3d Ser., LX (2003), 471–510.

Gulf South. These new societies—including confederacies like the Creeks, Choctaws, and Chickasaws—were less hierarchical than earlier chiefdoms, and some Indian priests in these confederacies partially refashioned their identities in hopes of earning support and recognition from a wide swath of society (and not, as in earlier centuries, from individual chiefs). Nevertheless, despite adapting to new social and political circumstances, native priests in the eighteenth-century Gulf South maintained many of the same practices and faced many of the same challenges as their sixteenth-century predecessors.[17]

Political power and specialized knowledge of the natural and spiritual worlds had been directly linked in southern chiefdoms since at least the heyday of the Mississippian era (circa 900 to 1350). The full flowering of Mississippian culture was centered at Cahokia, the Mound-Builder metropolis of forty thousand inhabitants near present-day Saint Louis. But, even after Cahokia fell in the fourteenth century, Mississippian culture remained foundational to the smaller, conflict-ridden chiefdoms stretching from northern Florida to Oklahoma. These chiefdoms were highly stratified societies, and the families with the most power were those with the greatest access to exotic prestige goods and hidden knowledge. Prestige goods proved elites' connections and ability to procure resources from beyond their own vicinities and, when distributed as gifts, could foster dependence and loyalty among social inferiors and strengthen political and diplomatic alliances. Elites' influence also stemmed from the possession of specialized knowledge about agriculture, death, and war that elites and priests acquired through extraordinary connections with the spirits, monsters, and birds that inhabited parts of the cosmos beyond the terrestrial sphere. The so-called warfare / cosmogony complex, for example, associated violence and monstrous beings—often portrayed on exotic materials—with chiefs

17. The literature on patronage in the history of science is extensive, and numerous case studies have demonstrated the basic point that patronage relationships and funding mattered to how Europeans pursued science. Still, there is no one-size-fits-all definition that encompasses the variety of patronage relationships through which individuals with power or resources supported men of science. The very range of patronage relationships in early modern European science (and politics and art) makes me comfortable referring to relations between native chiefs and the experts they supported as patronage. See Bruce T. Moran, ed., *Patronage and Institutions: Science, Technology, and Medicine at the European Court, 1500–1750* (Rochester, N.Y., 1991); E. C. Spary, *Utopia's Garden: French Natural History from Old Regime to Revolution* (Chicago, 2000); and Casper Andersen, Jakob Bek-Thomsen, and Peter C. Kjærgaard, "The Money Trail: A New Historiography for Networks, Patronage, and Scientific Careers," *Isis*, CIII (2012), 310–315.

to underscore that military, spiritual, and economic power flowed through these leaders. The pursuit and demonstration of knowledge in southeastern chiefdoms was thus focused on emphasizing chiefs' power. Experts in natural and spiritual matters—often labeled as priests, shamans, or medicine men—made it their business to uphold chiefly power, and, in return, chiefs rewarded these learned priests with prestige goods and agricultural surplus.[18]

The learned natives of southeastern chiefdoms might have sometimes specialized in particular branches of knowledge, but natural knowledge and the occult were not mutually exclusive. This was apparent in the questions that Franciscan missionary Francisco Pareja posed in his 1631 *Confessionario,* which included a bilingual Spanish / Timucuan questionnaire meant to gauge the persistence of heathen beliefs. Pareja included distinct sets of questions for sorcerers, doctors, herbalists, and midwives, but he also listed questions that applied to "Doctors, Herbalists, and Sorcerers," a hint that even an experienced observer like Pareja struggled to distinguish between various experts.[19]

The Franciscans who began missionizing Florida in 1573—the same year the Spanish crown made missionaries the primary agents for exploring and pacifying frontiers—also blurred the lines between religious and natural

18. Richter, *Before the Revolution,* 22–36; Neal Salisbury, "The Indians' Old World: Native Americans and the Coming of Europeans," *WMQ,* LIII (1996), 439, 444–449; Robbie Ethridge, *From Chicaza to Chickasaw: The European Invasion and the Transformation of the Mississippian World, 1540–1715* (Chapel Hill, N.C., 2010), 12–22, 78; Randolph J. Widmer, "The Structure of Southeastern Chiefdoms," in Charles Hudson and Carmen Chaves Tesser, ed., *The Forgotten Centuries: Indians and Europeans in the American South, 1521–1704* (Athens, Ga., 1994), 147–148, John F. Scarry, "The Late Prehistoric Southeast," 28–30; Hudson, *Knights of Spain, Warriors of the Sun,* 13–23; Vernon James Knight, Jr., "Some Speculations on Mississippian Monsters," in Patricia Galloway, ed., *The Southeastern Ceremonial Complex: Artifacts and Analysis: The Cottonlandia Conference* (Lincoln, Neb., 1989), 206, 209; John E. Worth, *The Timucuan Chiefdoms of Spanish Florida,* 2 vols. (Gainesville, Fla., 1998), I, 5, 13; Robert C. Galgano, *Feast of Souls: Indians and Spaniards in the Seventeenth-Century Missions of Florida and New Mexico* (Albuquerque, N. Mex., 2005), 24–25; Hall, *Zamumo's Gifts,* esp. 24; Hann, *History of the Timucua Indians and Missions,* 25, 115.

19. The Timucuan translation of the *Confessionario* included only two categories, herbalists / doctors and sorcerers. See Jerald T. Milanich and William C. Sturtevant, eds., *Francisco Pareja's 1613 Confessionario: A Documentary Source for Timucuan Ethnography,* trans. Emilio F. Moran (Tallahassee, Fla., 1972), 28–36, 47n ("Doctors," 30). See also Laudonnière, *Three Voyages,* ed. and trans. Bennett, 13; Hann, *History of the Timucua Indians and Missions,* 155; and Galgano, *Feast of Souls,* 24–26.

knowledge. One Franciscan praised a colleague in Florida for "his devotion to God at all times" while also noting that "there is not a better man in scientific matters in [Spain's] University of Santiago de Compostela." This same friar, like many contemporary Europeans, believed that wonder was the proper response when observing nature, and he tried to convince natives of God's glory by using language borrowed from natural theology: "Oh, my Indians," he rhapsodized, "if you look at the stars, you will see wonders unequalled. Look at the waters and the fish therein. Look at the many animals on Earth. Observe the sun and the moon and the beautiful planets. Look at the mortals of this world and try to place in your mind the eternal happiness of glory." Spanish priests in Florida observed and described many of these same aspects of God's nature as part of their larger project of saving Indians. The bishop of Cuba, for instance, took time away from a grueling eight-month tour of Florida's missions in 1676 to map the province, describe its resources, and observe Indians' customs and appearance (taking special care to count each of the 4,081 bare-breasted women he saw). The bishop managed to send this information to the Spanish court before dying of exhaustion and, perhaps, longing.[20]

Franciscans aimed to Hispanicize groups like the Timucuas, Guales, and Apalachees and thus worked to teach them (and, with the help of Spanish soldiers, sometimes force them) to pray, dress, make love, speak, read, and farm like Spaniards. It is thus not surprising that missionaries were particularly dedicated to ethnography—especially as it concerned Indians' cosmology and capacity for reason—because it overlapped directly with their efforts to convert and Hispanicize Indians. Pareja's *Confessionario* grilled Timucuas about their "Ceremonies, Omens, and Superstitions" while also recommending how they should change their behavior. Pareja did believe that ethnographic research of Indians promoted proselytization, yet the religious aspect of his work also gave him some perspective on ethnogra-

20. Alonso Gregorio de Escobedo, *Pirates, Indians, and Spaniards: Father Escobedo's "La Florida,"* ed. James W. Covington, trans. A. F. Falcones (Saint Petersburg, Fla., [1963]), 126 ("his devotion"), 130 ("Oh, my"); Gabriel Díaz Vara Calderón, *A 17th Century Letter of Gabriel Diaz Vara Calderón, Bishop of Cuba, Describing the Indians and Indian Missions of Florida,* transcr. and trans. Lucy L. Wenhold (Washington, D.C., 1936), 3, 7–14. On the religious practices of Franciscans, see Galgano, *Feast of Souls,* 19–24. The Jesuits who preceded the Franciscans in Florida also combined scientific study and proselytizing. See Antonio Sedeño to Francisco Borgia, Mar. 6, 1570, in Felix Zubillaga, "Selected Letters from *Monumenta Antiquae Florida, 1566–1572 (1946),*" trans. Margaret Dembo, in David Hurst Thomas, ed., *Ethnology of the Indians of Spanish Florida* (New York, 1991), 4.

phy's inherent limits. He admitted that God was the only true judge of faith while "we can judge only by a person's exterior manifestations."[21]

Beyond observing natives' beliefs, missionaries participated in the debate over whether Indians had the same capacity for comprehending natural and supernatural phenomena as Europeans. One Jesuit lamented in 1570 that proselytization had stalled because "the devil has put it into [the Indians'] hearts that we are sorcerers," and he feared that "there is not much hope [of Christianizing them] if God does not change them into different men, because being by nature like beasts ... they believe nothing but what they see." This Jesuit asserted that there was some inherent difference between Indians' and Europeans' mental abilities that, ironically enough, led natives to rely too heavily on firsthand observation. A Franciscan who came to Florida seventeen years later also believed that Indians "are without natural reasonable thinking." Yet he then went on to list several ways that Florida Indians equaled or surpassed Europeans in applying reason to nature, including "the science of catching the whale." Like European observers for centuries to come, missionaries struggled to reconcile obvious evidence of Indians' rationality with their ostensibly irrational desire to continue being Indians.[22]

All of Spain's goals—frontier defense, conversion, procuring a labor force—depended on the cooperation of, and thus negotiations with, Indian chiefs. These alliances were mutually beneficial: chiefs gained access to exotic prestige goods from Spain that sustained their influence among their own people, and Spaniards secured laborers and tribute that sustained their military presence and proselytizing. For the Franciscans, Indian cooperation meant that a small corps of missionaries (about seventy at their peak in the mid-1600s) established nearly forty missions that ministered to some twenty-six thousand natives.[23]

21. Milanich and Sturtevant, eds., *Francisco Pareja's 1613 Confessionario,* trans. Moran, 23 ("Ceremonies"), 36–39; Luís Gerónimo de Oré, *The Martyrs of Florida (1513–1616),* [ed. and] trans. Maynard Geiger (New York, 1936), 103–106 ("we can judge," 106).

22. Sedeño to Borgia, Mar. 6, 1570, in Zubillaga, "Selected Letters from *Monumenta Antiquae Florida,"* trans. Dembo, in Thomas, ed., *Ethnology of the Indians of Spanish Florida,* 4 ("devil"); Escobedo, *Pirates, Indians, and Spaniards,* 142 ("are without"), 145 ("science").

23. These missions hardly ran as smoothly as the Franciscans wished, but, by the mid-1600s, even rebellions against Spanish rule tended to focus more on renegotiating the terms of cooperation between Indian and Spanish leaders than on fully ousting Spanish colonists and priests. See Ethridge, *From Chicaza to Chickasaw,* 78; Galgano, *Feast of Souls,* 1–2, 7–10, 41, 44, 65; Hall, *Zamumo's Gifts,* 34, 44–45, 51–54; Weber, *Spanish*

Even though some Indian leaders allied with Spanish officials, Franciscans still had to compete with natives who asserted the superiority of their own ways of understanding the cosmos. In his questions for "Sorcerers," Pareja recognized the potential efficacy of Indians' occult knowledge, asking them if they had "conjured a rain storm" and if "the spell work[ed]." He also agreed with Indian herbalists' conclusions that various plants were vehicles of divine power but struggled to convince them that this power flowed from the Catholics' God, insisting that they should "heal only with herbs and medicines, since God made them for our health ... without mixing them with useless words and words from the Devil." When persuasion failed, however, friars were not above employing soldiers to flog shamans and the commoners who sought their expertise. Indians also challenged the authority and beliefs of Franciscans. After seizing Francisco de Ávila during the Guale uprising of 1597, Ávila's captors mocked Catholic rituals, threatened to burn him on a cross, and beat him with a book. Apparently impressed with his fortitude, it seems that the Guales tried to co-opt his spiritual power. They made Ávila custodian of a temple in which other "wizards" served and offered him a "beautiful" wife in hopes that he would "give up the things you teach us for they are foolish" and "leave your law and become an Indian." Missionaries were not the only Floridians welcoming converts.[24]

Both native and Spanish priests knew that winning a chief's material and social backing was critical to achieving influence throughout an Indian community. Before and after the arrival of the Spanish, chiefs patronized native priests directly by providing them with prestige goods and agricultural plots worked by common men and, especially, women. This support gave shamans status and the time needed to cultivate knowledge, while shamans reinforced the association between a chief and extranatural forces. Franciscan missionaries recognized this relationship between chiefs and

---

Frontier in North America, 95–96, 100, 105–108, 115; and Michael Gannon, *The Cross in the Sand: The Early Catholic Church in Florida, 1513–1870* (Gainesville, Fla., 1965), 36–57.

24. Milanich and Sturtevant, eds., *Francisco Pareja's 1613 Confessionario,* trans. Moran, 30–32 ("Sorcerers," 30, "conjured," 31, "heal only," 32); "Narrative of the Great Hardships Which Father Avila Endured during the Year and a Half in Which He Was Captive among the Rebel Indians ...," in Oré, *Martyrs of Florida,* ed. Geiger, 87–93 ("wizards," 92). See also Weber, *Spanish Frontier in North America,* 114; and Galgano, *Feast of Souls,* 11, 53. On punishing Indians, see John H. Hann, *Apalachee: The Land between the Rivers* (Gainesville, Fla., 1988), 92; and Jerald T. Milanich, *Florida Indians and the Invasion from Europe* (Gainesville, Fla., 1995), 204. On Indian efforts to integrate Catholic priests' power, see Galgano, *Feast of Souls,* 63, 66, 68.

native experts and, as part of their larger goals of extirpating idolatry and converting chiefs (with the idea that their people would follow), focused on identifying and undercutting shamans' chiefly patronage. Pareja's *Confessionario* asked chiefs if they distributed meat, fish, and grain to sorcerers and pressed sorcerers to confess that they extorted commoners by telling them "that they must give you something, otherwise you will kill all of them." More importantly, Franciscans distributed European goods to chiefs, items that—thanks to their exotic provenance and associations with religious power—could be displayed and redistributed to add to chiefs' influence among their own people and, thus, strengthen chiefs' alliances with Spanish religious and military leaders and make them more willing to provide Spaniards with food and laborers.[25]

A crucial reason that Franciscans managed to supplant many native priests in Florida during the mission era was that several chiefs believed that it was in their own interest to patronize missionaries instead of shamans. Chiefs funded missionaries in part through the fields and laborers that, before a chief's conversion to Catholicism, had supported native shamans. The bishop of Cuba observed in Apalachee that "all in common cultivate and sow the lands of the caciques [chiefs]" including fields of wheat "as alms for the missionaries." Chiefs also patronized missionaries through prestigious gifts, including churches and convents that chiefs forced common laborers to construct. One Franciscan recalled how he "received their gifts with pleasure" and even chided Spaniards for their comparative stinginess, exclaiming "Oh! how many Christians there are who give with regret; but the cacique gives happily." No wonder chiefs found joy in giving: it was how they demonstrated their prestige, secured the attachment of their followers, and built relationships with powerful foreigners. Many Franciscans, in turn, bolstered chiefs' influence and sometimes even sided with their chiefly patrons against Spanish officials.[26]

25. Milanich and Sturtevant, eds., *Francisco Pareja's 1613 Confessionario,* trans. Moran, 26, 31 (quote), 36; Hann, *Apalachee,* 3, 92; Galgano, *Feast of Souls,* 25, 45, 65; Worth, *Timucuan Chiefdoms of Spanish Florida,* I, 38, 137, 167; Sarah M. S. Pearsall, "'Having Many Wives' in Two American Rebellions: The Politics of Households and the Radically Conservative," *AHR,* CXVIII (2013), 1006, 1013; Amy Turner Bushnell, "Ruling 'the Republic of Indians' in Seventeenth-Century Florida," in Gregory A. Waselkov, Peter H. Wood, and Tom Hatley, eds., *Powhatan's Mantle: Indians in the Colonial Southeast,* rev. and expanded ed. (Lincoln, Neb., 2006), 195–200; Hall, *Zamumo's Gifts,* 51–54; Milanich, *Florida Indians and the Invasion from Europe,* 199.

26. Calderón, *A 17th Century Letter of Gabriel Diaz Vara Calderón,* transcr. and trans. Wenhold, 13 ("all in common"); Escobedo, *Pirates, Indians, and Spaniards,* 129 ("re-

Franciscans, moreover, enjoyed support from both the crown and Florida chiefs. This system of mixed funding reflected how Florida on the whole achieved some stability in the seventeenth century through a combination of imported Spanish cash and extorted Indian labor. As these Franciscans realized, and as many aspiring experts in the Gulf South would discover during centuries of competition for the region, getting paid in the borderlands often hinged on securing multiple patrons across cultural and political boundaries.[27]

Missionaries never succeeded in totally replacing native priests in Florida, nor was the success they achieved entirely of their own making: Spanish colonialism introduced Old World diseases that killed many shamans, made their cures and prayers seem less effective, and pushed chiefs and commoners alike to seek out the expertise of European priests who seemed less susceptible to new scourges. Moreover, the places where Spanish missionaries did thrive—especially the Timucuan chiefdoms of northern Florida—were as much on the periphery of the Mississippian world as they were on the periphery of the Spanish Empire. Thus, neither missionaries themselves nor outbreaks of disease within Florida's mission communities immediately transformed the chiefdoms in the rest of the Gulf South. Rather, it was the same forces that ultimately destroyed the Franciscan missions—slave raids by Indians allied with English Carolina and the diseases they introduced—that reduced and reordered the native communities of the Gulf South between the 1670s and 1710s. The combined

ceived"), 148 ("Oh! how many"); Worth, *Timucuan Chiefdoms of Spanish Florida,* I, 167–169; Oré, *Martyrs of Florida,* ed. Geiger, 107; Galgano, *Feast of Souls,* 63, 68, 73. On how Franciscans sometimes sided with chiefs, see Worth, *Timucuan Chiefdoms of Spanish Florida,* I, xix, 80–81. The significance of chiefly patronage to missionaries' ability to establish their credibility and convert natives was also clear in cases where missionaries failed because chiefs refused to support them. From the 1500s to mid-1700s, for example, missionaries made several efforts to convert the Calusas of southwestern Florida by offering gifts and the promise of political alliance to chiefs. But Calusa chiefs usually did not see any inherent superiority in Christian beliefs, and, more importantly, the ties between chiefs' power and the esoteric knowledge of Calusa priests seem to have been so strong that chiefs refused to convert because it would have meant the end of their legitimacy and influence. See Feliciano López to Pedro Taybo, 1697, in John H. Hann, ed. and trans., *Missions to the Calusa* (Gainesville, Fla., 1991), 159–160, Juan Rogel to Jerónimo Ruiz del Portillo, Apr. 25, 1568, 247, 262, Statement of Reverend Father Fray Feliciano López, Feb. 27, 1698, 166; John H. Hann, *Indians of Central and South Florida, 1513–1763* (Gainesville, Fla., 2003), 164–170, 188, 198–199.

27. Bushnell, *Situado and Sabana,* esp. 210–211; Worth, *Timucuan Chiefdoms of Spanish Florida,* I, 126–130.

demographic impact of these slave wars and the diseases they fostered decreased the native population of the region by around 55 percent. Although several aspects of Mississippian culture lived on among the survivors, the strict social and political hierarchies that had ordered Mississippian chiefdoms did not endure. Those who experienced and survived the slave wars of the 1670s to 1710s reorganized primarily as political confederacies of allied towns in which no individuals commanded the same degree of influence as the leaders of earlier chiefdoms. These confederacies included the Creeks, Chickasaws, and Choctaws, all of which became major players in geopolitical competitions for the Gulf South.[28]

Political power in these new confederacies was distributed among several village chiefs, war leaders, and town councils; therefore, Indians who claimed expertise in natural and spiritual matters had fewer opportunities to acquire sustenance and prestige from a single powerful chief. By the 1700s, some shamans had fashioned identities meant to ensure that a town's leaders and commoners still valued their role in society enough to support them. For example, some priests sought to achieve broad appeal by presenting themselves as servants of the people: naturalist William Bartram described shamans distributing food leftover from feasts to the poor, a far cry from the days when priests had eaten food that head chiefs coerced common men and women to produce for them. Self-proclaimed experts in the eighteenth century also continued to blur the lines between natural knowledge and the occult and, perhaps, might have also become more numerous and more specialized. One British observer found that "many" Choctaws "are well acquainted with [medicinal] plants," that "juggling Quacks" were all too common, and that "they have a number of people in their nation called rainmakers." This plurality of specialists might have appealed to eighteenth-century southeastern Indians who disdained centralized power.[29]

28. Hann, *Apalachee,* 92; Weber, *Spanish Frontier in North America,* 115; Worth, *Timucuan Chiefdoms of Spanish Florida,* I, 81; Galgano, *Feast of Souls,* 62, 105; Ethridge, *From Chicaza to Chickasaw,* esp. 2, 86–90, 237, 251–254; Paul Kelton, *Epidemics and Enslavement: Biological Catastrophe in the Native Southeast, 1492–1715* (Lincoln, Neb., 2009), esp. xviii–xx, 82–99; Alan Gallay, *The Indian Slave Trade: The Rise of the English Empire in the American South, 1670–1717* (New Haven, Conn., 2002).

29. Bernard Romans, *A Concise Natural History of East and West-Florida . . .* (New York, 1776), 85 ("many," "are well," "they have a"), 88 ("juggling"); William Bartram, *Travels through North and South Carolina, Georgia, East and West Florida . . .* (Philadelphia, 1791), 236; Ethridge, *From Chicaza to Chickasaw,* 252; Kathleen DuVal, "Inter-

Although learned natives did seem to adjust their identities to the new social circumstances initiated by imperialism and disease, they also maintained roles that priests had fulfilled since the Mississippian era. For one, priests continued to assert their access to esoteric knowledge applicable to war. In 1564, a French artist illustrated how a centenarian "sorcerer" held an "animated conversation" with invisible beings before he "saluted the chief, and revealed to him the number of the enemy and the place where they were to fight." Some two hundred years later, Bartram observed that every Creek town had "a high priest, usually called by the white people jugglers, or conjurers," who "maintains and exercises great influence in the state; particularly in military affairs, the senate never determine on an expedition against their enemy without his counsel and assistance." The southeast borderlands remained a world of competition and conflict in the eighteenth century, so priests who could offer military advantages to a town's headmen continued to find support.[30]

Shamans in the eighteenth-century Gulf South also had to appeal to common townspeople. As Bartram put it, "priests or Doctors" had to "make the people believe by their cunning or craft, that they have a supernatural spiritual communion with invisible spirits or powers." The audience that these doctors had to impress was far broader than in earlier centuries, but the work of actually convincing would-be supporters was as risky in the 1700s as it had been in the 1500s. A Franciscan who came to Florida in 1587 observed how "there are many shamans who attempt to demonstrate their powers before special fires . . . ." "Whoever comes from this ritual without burning his hands is believed holy; given great honor, and considered to be a nobleman," he continued, but, "if any of them is burned, he loses his position, fame, credit and perhaps his life." In the mid-eighteenth century, trader James Adair noted that Creek shamans claimed they could make it rain but tried to avoid doing so "as long as they possibly can, till the murmurs of the people force them to the sacred attempt." "If he fails," he solemnly observed, "the prophet is shot dead." A Creek rainmaker could probably expect fewer rewards for his work than a sixteenth-century shaman, but the dangers of overstating one's ability to manipulate nature remained just as great.[31]

---

connectedness and Diversity in 'French Louisiana,'" in Waselkov, Wood, and Hatley, eds., *Powhatan's Mantle,* 135.

30. "Narrative of Jacques le Moyne de Morgues," in Lorant, ed., *New World,* 59 ("sorcerer"); Bartram, *Travels,* 236, 497 ("high priest").

31. William Bartram, "Answers to Benjamin Smith Barton's Queries about Indians,"

In sum, patronage relationships had long been vital to determining how men of learning and political elites supported each other in the Gulf South's Mississippian chiefdoms. In Florida, Franciscans sometimes succeeded in supplanting learned native priests and developing ties with chiefs that helped missionaries spread Catholicism and Spanish influence. By the early eighteenth century, disease and slave raids had caused the mission system to collapse in Florida and made political hierarchies throughout the Gulf South less centralized. This new context—which developed amid violence and exchanges initiated by European imperialism—prompted some learned natives to adjust their identities but did not mark the end of a native intellectual culture that reached back to the Mississippian era. At the same time that native patronage was becoming more diffuse, however, European monarchies were working to further centralize the sponsorship and circulation of knowledge in order to give their empires an edge in international competitions. This renewed push to acquire empire-promoting information and accumulate it in metropolitan centers was especially apparent in the case of cartography.

COMPETITION, CAPTIVITY, AND CARTOGRAPHY

Between the 1670s and 1710s, imperial competition over Spain's territorial claims in the Gulf South—from England on the Atlantic coast and France along the Gulf coast—catalyzed an explosion of cartographic production among mapmakers, traders, missionaries, and officials in all three empires. Officials valued maps as tools to stake territorial claims, improve commerce and navigation, enhance their military effectiveness, and dream imperial dreams. The importance of cartography to claiming territory, moreover, motivated a frenetic interimperial circulation of maps as officials in each empire sought to counter each other's claims and co-opt each other's hard-won data. But this cartographic competition did not simply involve contests and exchanges among Europeans. The Spanish, French, and English were competing to claim Indian country, secure Indian allies, dominate the Indian trade, and—especially in the case of South Carolina—enslave Indi-

---

1789, Bartram Family Papers, Collection 36, 15, HSP ("priests"); Escobedo, *Pirates, Indians, and Spaniards,* 142 ("there are many"); James Adair, *The History of the American Indians; Particularly Those Nations Adjoining to the Mississippi, East and West Florida, Georgia, South and North Carolina, and Virginia* ... (London, 1775), 85–86 ("as long as," 85). See also John R. Swanton, *Social Organization and Social Usages of the Indians of the Creek Confederacy* (Washington, D.C., 1928), 583–584; and Paul Kelton, "Avoiding the Smallpox Spirits: Colonial Epidemics and Southeastern Indian Survival," *Ethnohistory,* LI (2004), 47.

ans as commodities. English cartographers even used slave raids as opportunities to make geographic observations in the Gulf South. Natives, too, valued cartography as a means of portraying distant places and geopolitical relationships. Yet maps drawn by enslaved natives evince that cartography was not always a means of representing or effecting power. Instead, such maps were products of weakness that depicted a landscape of suffering. Case studies of three mapmakers—François Le Maire, Lamhatty, and Thomas Nairne—reveal the extent to which imperial competition and the violence that accompanied it motivated and influenced the work of European and native cartographers.[32]

Interimperial competition to settle the Gulf coast began in earnest in 1682 when René Robert Cavelier, sieur de La Salle, explored the lower Mississippi Valley and claimed it for France as Louisiana. When La Salle and his men returned in 1684, they wound up settling a doomed colony on the Texas coast before La Salle's own men murdered him. Word of La Salle's venture nevertheless reignited Spain's long-dormant interest in the Gulf coast, and Spain sent five expeditions by sea and six by land that reconnoitered the coast from Texas to Florida with the aim of finding and eliminating the French settlement. For Spain, maps and reports proved to be the real payoff of the four-year hunt for La Salle, particularly the enticing descriptions of Pensacola Bay that circulated among officials in Mexico City. One of the naval officers who searched for La Salle brought news of Pensacola's harbor and strategic significance to the viceroy and creole cosmographer Carlos de Sigüenza y Góngora, one of New Spain's leading men of science. Convinced of the strategic importance of Pensacola for keeping the French and English out of the Gulf, the viceroy, with the king's consent, organized an expedition to explore and map Pensacola Bay in 1693. He ordered the expedition—which included Sigüenza y Góngora—to take soundings of the bay and note "the fruits and trees, the nature and disposition of the Indians, and everything else that multiplies and grows." He also instructed Sigüenza y Góngora to produce a map and expected the explorers to win the friendship of local Indians with gifts. Sigüenza y Góngora surveyed the bay, described its naval stores and natural history, and proudly

32. On maps as expressions of power, see J. B. Harley, *The New Nature of Maps: Essays in the History of Cartography*, ed. Paul Laxton (Baltimore, 2001). On the role of maps in imperial competition, see Paul W. Mapp, *The Elusive West and the Contest for Empire, 1713–1763* (Chapel Hill, N.C., 2011); and Cumming, *Southeast in Early Maps*, rev. and enl. De Vorsey. On how suffering could be central to the experience of Indians in American borderlands, see Ned Blackhawk, *Violence over the Land: Indians and Empires in the Early American West* (Cambridge, Mass., 2006).

offered the viceroy "the map requested." He argued that Pensacola was "the finest jewel possessed by His Majesty ... not only here in America but in all his kingdom," and the king—inspired more by the need to prevent French and English incursions than the expedition's glowing reports—ordered the occupation of Pensacola.[33]

But an actual European presence on the Gulf coast only came in 1698 when England, France, and Spain all raced to settle Pensacola. Spain won the dash and fortified Pensacola Bay in November 1698, but this was a pyrrhic victory. In 1699, French colonizers led by Pierre LeMoyne, sieur d'Iberville, found Pensacola occupied, so they continued on to establish French Louisiana in the Mississippi delta, which proved to be far more valuable strategically than Pensacola. After 1699, then, France had a foothold in the Gulf South, but Louisiana never attracted enough French settlers or investment to overwhelm its Spanish and English competitors. Instead, France relied on Indian alliances to exercise influence in the region and relied on maps to announce and exaggerate its claims to European diplomats. The most important of these French maps depended on Father François Le Maire, a French missionary stationed in Spanish Pensacola and French Mobile from 1706 to 1720 who drew maps based on data from Indian, French, and Spanish sources and sent these to Spanish and French officials.[34]

33. "Orders and Instructions Issued to Don Andrés de Pez for the Expedition," Jan. 12, 1693, in Irving A. Leonard, [ed.] and trans., *Spanish Approach to Pensacola, 1689–1693* (Albuquerque, N.Mex., 1939), 145–148 ("fruits," 145), "Description of the Bay of Santa María de Galve (Formerly Pensacola) ... by Don Carlos de Sigüenza y Góngora, Cosmographer of the King ...," June 1, 1693, 152–192 ("map," 185), "Report of Don Carlos de Sigüenza," June 1, 1693, 193 ("finest jewel"), Leonard, "Introduction," 7–12, 19; Weber, *Spanish Frontier in North America,* 148–152, 155–156.

34. Weber, *Spanish Frontier in North America,* 156–160; Mathé Allain, *"Not Worth a Straw": French Colonial Policy and the Early Years of Louisiana* (Lafayette, La., 1988); Christine Marie Petto, *When France Was King of Cartography: The Patronage and Production of Maps in Early Modern France* (Lanham, Md., 2007), 100–106; Gordon Sayre, "How to Succeed in Exploration without Really Discovering Anything: Four French Travelers in Colonial Louisiana, 1714–63," *Atlantic Studies,* X (2013), 51–68; Conrad E. Heidenreich and Edward H. Dahl, "The French Mapping of North America, 1600–1760," *Map Collector,* XIX (June 1982), 2–7; Usner, *Indians, Settlers, and Slaves.* On Le Maire and his scientific endeavors, see "M. Le Maire on Louisiana," in Mildred Mott Wedel, ed., *A Jean Delanglez, S.J., Anthology: Selections Useful for Mississippi Valley and Trans-Mississippi American Indian Studies* (New York, 1985), 124–154; Jack Jackson and Winston DeVille, "Le Maire and the 'Mother Map' of Delisle," in Jackson, Robert S. Weddle, and DeVille, [eds.], *Mapping Texas and the Gulf Coast: The Contributions of Saint-Denis,*

Imperial competition inspired Le Maire's cartographic efforts, provided a multinational source base from which he could draw geographic information, and, by putting Le Maire in contact with both Spanish and French officials, fostered the interimperial circulation of maps. Part of the reason why Le Maire had such an outsized influence on North American cartography was that he siphoned information contained in the manuscript maps of Spanish explorers to leading French cartographers. Le Maire learned Spanish while missionizing in Pensacola and seems to have gained access to Spanish maps, including a copy of Sigüenza y Góngora's 1693 map of Pensacola Bay, by befriending the garrison's Spanish officers. Beyond simply extracting Spanish knowledge for French benefit, Le Maire was unapologetic about producing maps directly for Spain and, perhaps, considered it a way to improve his own condition and connections while living in Spanish territory. Le Maire was one of the earliest Gulf South cartographers to realize that maps made valuable gifts for winning official patronage.[35]

Le Maire's most significant cartographic connection in France was Guillaume Delisle, and Le Maire and Delisle produced their maps in a transatlantic dialogue. Delisle was a scion of a renowned map-making family and eventually became royal geographer of France. He discovered Le Maire's maps through a mutual acquaintance, Versailles chaplain Jean Bobé, who both solicited information from Le Maire through questionnaires and had the ear of leading French ministers involved with Louisiana. Delisle had long since been in the habit of accumulating present and past French, Spanish, and English explorers' observations and maps, many of which were, in turn, derived from Indian informants. For example, Delisle's 1703 *Carte de Mexique et de la Floride* integrated information from de Soto's chroniclers, the writings of French colonists in sixteenth-century Florida, and interviews in which Iberville related estimates of distances "he had learned from savages."[36]

Oliván, and Le Maire (College Station, Tex., 1990), esp. 22–30; and Shannon Lee Dawdy, *Building the Devil's Empire: French Colonial New Orleans* (Chicago, 2008), 25–26, 35.

35. "M. Le Maire on Louisiana," in Wedel, ed., *A Jean Delanglez, S.J., Anthology*, 126; Jackson and DeVille, "Le Maire and the 'Mother Map' of Delisle," in Jackson, Weddle, and DeVille, [eds.], *Mapping Texas and the Gulf Coast*, 31, 34, 37–39, 48, 49, 59; Robert S. Weddle, *The French Thorn: Rival Explorers in the Spanish Sea, 1682–1762* (College Station, Tex., 1991), 346; [François] Le Maire to [Jean Bobé], Jan. 15, 1714, in "M. Le Maire on Louisiana," in Wedel, ed., *A Jean Delanglez, S.J., Anthology*, 137.

36. "Livres mémoires cartes imprimées ou M.S. dont on s'est servi pour dresser la carte de ... Canada du Mississipi et de la Floride qui a été presenté à mgr. le cte de Pont-

The collaboration between Delisle and Le Maire was not simply that of a metropolitan expert compiling and publishing observations from an eyewitness in the colonies. Le Maire did send Delisle data, but his most significant contribution was as an editor, improving Delisle's maps and those of other French mapmakers by incorporating the latest information he picked up from his Gulf South informants. For example, Le Maire "scribble[d] a small plan" in 1714 that he sent to Bobé, noting that he had placed several Indian groups "differently from their location on the map of M. Delisle" with the explanation that "you must know that these changes and migrations have taken place within the last four or five years." Le Maire, in effect, updated Delisle's maps to include the enormous demographic shifts that the Indian slave trade had initiated across the region. After consulting Le Maire's sketch, Delisle could re-place or delete Indian towns in later maps.[37]

Le Maire's most complete manuscript map, his 1716 *Carte nouvelle de la Louisiane et pays circonvoisins,* both built off of Delisle's 1703 map and became the main source for Delisle's 1718 masterpiece, the *Carte de la Louisiane et du cours du Mississipi, dressée sur un grand nombre de mémoires entrautres sur ceux de Mr. le Maire.* Delisle's *Carte de la Louisiane* mapped both the geography and history of imperial competition for the Gulf South, including the route of de Soto's entrada and the position of France's 1564 Fort Caroline. But Delisle's map also became a key text in

---

chartrain . . . ," in Wedel, ed., *A Jean Delanglez, S.J., Anthology,* 279–298 (quote, 291); Petto, *When France Was King of Cartography,* 100, 103; Jackson and DeVille, "Le Maire and the 'Mother Map' of Delisle," in Jackson, Weddle, and DeVille, [eds.], *Mapping Texas and the Gulf Coast,* 28, 31; Paul E. Hoffman, "Discovery and Early Cartography of the Northern Gulf Coast," in Alfred E. Lemmon, John T. Magill, and Jason R. Wiese, eds., *Charting Louisiana: Five Hundred Years of Maps* (New Orleans, 2003), 9, 18–19; Patricia Galloway, "Debriefing Explorers: Amerindian Information in the Delisles' Mapping of the Southeast," in G. Malcolm Lewis, ed., *Cartographic Encounters: Perspectives on Native American Mapmaking and Map Use* (Chicago, 1998), 223–240; Weddle, *French Thorn,* 346.

37. Le Maire to [Bobé], Jan. 15, 1714, in "M. Le Maire on Louisiana," in Wedel, ed., *A Jean Delanglez, S.J., Anthology,* 137 ("scribble[d]"), 147 ("differently"); "M. Le Maire on Louisiana," in Wedel, ed., *A Jean Delanglez, S.J., Anthology,* 127; Jackson and DeVille, "Le Maire and the 'Mother Map' of Delisle," in Jackson, Weddle, and DeVille, [eds.], *Mapping Texas and the Gulf Coast,* 37. For a similar example of how information coming from the Gulf South during the Indian slave trade led Delisle to alter the locations of Indian groups, see Patricia Galloway, "Henry de Tonti du village des Chacta, 1702: The Beginning of the French Alliance," in Galloway, ed., *La Salle and His Legacy: Frenchmen and Indians in the Lower Mississippi Valley* (Jackson, Miss., 1982), 146, 150, 154–155.

FIGURE 2.
Carte nouvelle de la Louisiane et pays circonvoisins . . . . By F[rançois] Le Maire. 1716. Bibliothèque nationale de France, Cartes et plans, CPL GE D-7883

FIGURE 3. *Carte de la Louisiane et du cours du Mississippi: Dressée sur un grand nombre de mémoires entrautres sur ceux de Mr. le Maire.* By Guillaume Delisle. 1718. Library of Congress, Geography and Map Division, G3700 1718 .L5

future arguments among France, Spain, and Britain about which empire had sovereignty over various parts of the region. The map pushed French Louisiana east into Carolina and southwest into New Spain, prompting both Spain and Britain to produce maps contesting Delisle's cartographic claims.[38]

Far more than in French Louisiana, English efforts to dominate the Southeast from the 1670s to 1710s centered on the Indian slave trade, and two maps, both originally drawn in 1708, made this region-shattering eco-

38. Jackson and DeVille, "Le Maire and the 'Mother Map' of Delisle," in Jackson, Weddle, and DeVille, [eds.], *Mapping Texas and the Gulf Coast,* 34–36, 42–43, 45, 52–54; Petto, *When France Was King of Cartography,* 100, 106; Weddle, *French Thorn,* 317–323; [William Patterson Cumming], "The Early Maps of Southeastern North America: An Introductory Essay," in Cumming, *Southeast in Early Maps,* rev. and enl. De Vorsey, 21, 23, 34.

nomic and political enterprise visible. The first was by Lamhatty, an enslaved Tawasa man from Florida's Gulf coast. Although Lamhatty's map is unusual as one of the few surviving examples of southeastern Indian cartography, his enslavement was all too typical of the experience of Gulf South natives. From the 1670s to the 1710s, English marauders from Carolina and their Indian allies enslaved at least thirty thousand to fifty thousand Indians, many of whom ended up on plantations in Carolina, Virginia, and Barbados. These raids destroyed the missions of northern Spanish Florida—effectively ending, among many other things, the chiefly patronage of Spanish priests—and reached across the Gulf South. Those who survived often left their homes to join or build new communities and, it seems, Lamhatty's town was composed largely of refugees from Timucuan and Apalachee missions that Yamasee, Creek, and English invaders had wiped out between 1702 and 1705.[39]

Violence and slavery saturated Lamhatty's map: he based it on observations he made as a captive to Creeks and Shawnees who bound him with ropes and drew it at the behest of Virginian masters who bound him with chains. Creek raiders took Lamhatty captive in the spring of 1707 and transported him to Creek towns on the Coosa and Tallapoosa Rivers "where they made him worke in ye Ground." Agricultural labor was women's work and probably added an element of humiliation to the usual isolation and pain of captivity. His Creek captors then took him east, probably with the idea of selling him in Charleston, but they sold him instead to Shawnees they met en route who took Lamhatty north. He broke free from his captors and headed east where, in early January, he stumbled "into one of ye houses of the upper Inhabitants" of Virginia who "tied him by ye arm" despite the way Lamhatty "Shed tears and Shewed them how his hands were galled and Swelled by being tyed before." This Virginian then took Lamhatty to Lieutenant Colonel John Walker who "put him in Irons." It seems that Lamhatty's fate was not unique among Tawasas, for, "after some of his Country folks were found servants" in other Virginian households, Lamhatty "became verry melancholly often fasting and crying Several days together Sometimes using little Conjurations." These conjurations suggest that Lamhatty might have been a shaman, and, just maybe, his esoteric knowledge—

39. On the impact of the Indian slave trade, see Gallay, *Indian Slave Trade,* esp. 143–148, 299; and Ethridge, *From Chicaza to Chickasaw,* esp. 209–211. On the Tawasas, see Gregory A. Waselkov, "Indian Maps of the Colonial Southeast," in Waselkov, Wood, and Hatley, eds., *Powhatan's Mantle,* 467–468; and John Reed Swanton, *Early History of the Creek Indians and Their Neighbors* (Washington, D.C., 1922), 137–143.

FIGURE 4. Map Drawn by Lamhatty. [1708]. Virginia Historical Society, Richmond. MSS1 L51 fol. 676–682 Manuscripts; Lee Family Papers, 1638–1867. Section 163

or his observations of the surrounding landscape—gave Lamhatty the edge he needed when he escaped the next spring.[40]

Walker probably compelled Lamhatty to draw a map of the contested lands between the Gulf coast and Virginia so that Walker could present it as a gift to Virginia governor Edmund Jennings and, thus, earn the governor's favor. Yet, while Europeans sometimes did extract information from native maps, it is unlikely that Jennings found much of geopolitical value in Walker's gift. For one, Lamhatty's geographic descriptions suffered from inadequate and, it seems, intentionally biased translation: the Tuscarora interpreter not only admitted that he could not speak Lamhatty's language but told the Virginians that his own people—and not the Creeks—had raided the Tawasas, probably in order to highlight the Tuscarora's alliance with the English. Additionally, although it seems that Lamhatty made a good faith effort to depict the terrain through which he passed, the map nevertheless reflected his own geographic perspectives and priorities. For example, the area around Tawasa was larger and more detailed on his map than places farther from Lamhatty's home, and his captivity limited the map's scope to the places where he was transported and forced to labor. His shame and pain while enslaved by natives and his desire to please or misdirect his Anglo master probably affected his observations and illustrations as well.[41]

40. This account of Lamhatty's captivity is based on two contemporary descriptions, Lieutenant Colonel John Walker's letter to Virginia's governor, in which he included Lamhatty's original drawing of his map, and Robert Beverley's account, which he penned on the back of the extant copy of Lamhatty's map. See John Walker to Edmund Jennings, Jan. 16, 1708, in Waselkov, "Indian Maps of the Colonial Southeast," in Waselkov, Wood, and Hatley, eds., *Powhatan's Mantle,* 462 ("into one of ye houses," "tied"), 463 ("put him in Irons"); and "Mr. Robert Beverley's Acct. of Lamhatty," 1708, in Waselkov, "Indian Maps of the Colonial Southeast," ibid., 465 ("where they made him worke," "Shed tears," "after some of his Country folks," "became very melancholly"). On Lamhatty's agricultural labor as women's work, see Denise I. Bossy, "Indian Slavery in Southeastern Indian and British Societies, 1670–1730," in Alan Gallay, ed., *Indian Slavery in Colonial America* (Lincoln, Neb., 2009), 214–215.

41. Waselkov, "Indian Maps of the Colonial Southeast," in Waselkov, Wood, and Hatley, eds., *Powhatan's Mantle,* 440, 443, 463, 466, 468–469. On how Europeans struggled to make sense of Indian maps and how the experience of captivity could limit and skew natives' geographic knowledge, see Mapp, *Elusive West,* 70–84, 200–202, 251–254. On how enslaved natives carried information across the Southeast, see Dubcovsky, *Informed Power,* 126. Although Virginia was not among the immediate contestants for the Gulf South, it is significant that the networks of exchange were already connecting powerful officials in the Chesapeake to sources of geographic information in the Gulf South.

Lamhatty's map does have some features typical of native cartography in the Gulf South, yet comparing his map to those made by more powerful southeastern groups reveals how, far from being an instrument of power, maps could also be windows into weakness and pain. The most obvious resemblance to contemporary southeastern native maps was Lamhatty's use of circles to indicate towns and, in the case of some of the Creek locales he identified, clusters of allied villages. Contemporary maps—including two that Chickasaw cartographers gave to the French in 1737—similarly used circles to represent towns and even entire confederacies. The 1737 Chickasaw maps were declarations of power: the cartographers who drew them put their own groups at the center of their maps and, by using lines to illustrate the bonds and paths that linked them with Indian and European allies, a web of political connections. For Indians and Europeans alike, exercising power in the Gulf South was all about being part of larger networks of alliance and exchange, and, according to these Chickasaw maps, the Chickasaws saw themselves as key to determining whether Britain or France would gain influence among the satellite groups surrounding the Chickasaws.[42]

Lamhatty's map, however, depicted his town as one of several unconnected Gulf coast communities on the very edge of his map. The map's center concentrated instead on the spaces of his captivity—Creek towns, slave-trading paths, Virginia farms—and the only line that connected the Indian and English communities was the one that traced his own middle passage. Lamhatty was not a confident native cartographer who used maps to stake claims to territory or boast about the extent of his peoples' geopolitical in-

These networks had begun to emerge in the 1690s—Governor Francis Nicholson had collected Indian and European maps of the lower Mississippi Valley—and would reach a new height in the early 1800s as Thomas Jefferson and other officials in Washington, D.C., accumulated maps from diverse Gulf South cartographers. See G. Malcolm Lewis, "Frontier Encounters in the Field: 1511–1925," in Lewis, ed., *Cartographic Encounters,* 21; and Waselkov, "Indian Maps of the Colonial Southeast," in Waselkov, Wood, and Hatley, eds., *Powhatan's Mantle,* 436, 457–462.

42. On the use of circles in southeastern Indian maps, see Galloway, "Debriefing Explorers," in Lewis, ed., *Cartographic Encounters,* 224–227; Waselkov, "Indian Maps of the Colonial Southeast," in Waselkov, Wood, and Hatley, eds., *Powhatan's Mantle,* 443–445, 447–448, 463, 466. On Indian maps as sources of power and resistance, see J. Brian Harley, "Rereading the Maps of the Columbian Encounter," *Annals of the Association of American Geographers,* LXXII (1992), 527. On how Indians sometimes "had the power to draw borders in early America," see Juliana Barr, "Geographies of Power: Mapping Indian Borders in the 'Borderlands' of the Early Southwest," *WMQ,* LXVIII (2011), 43.

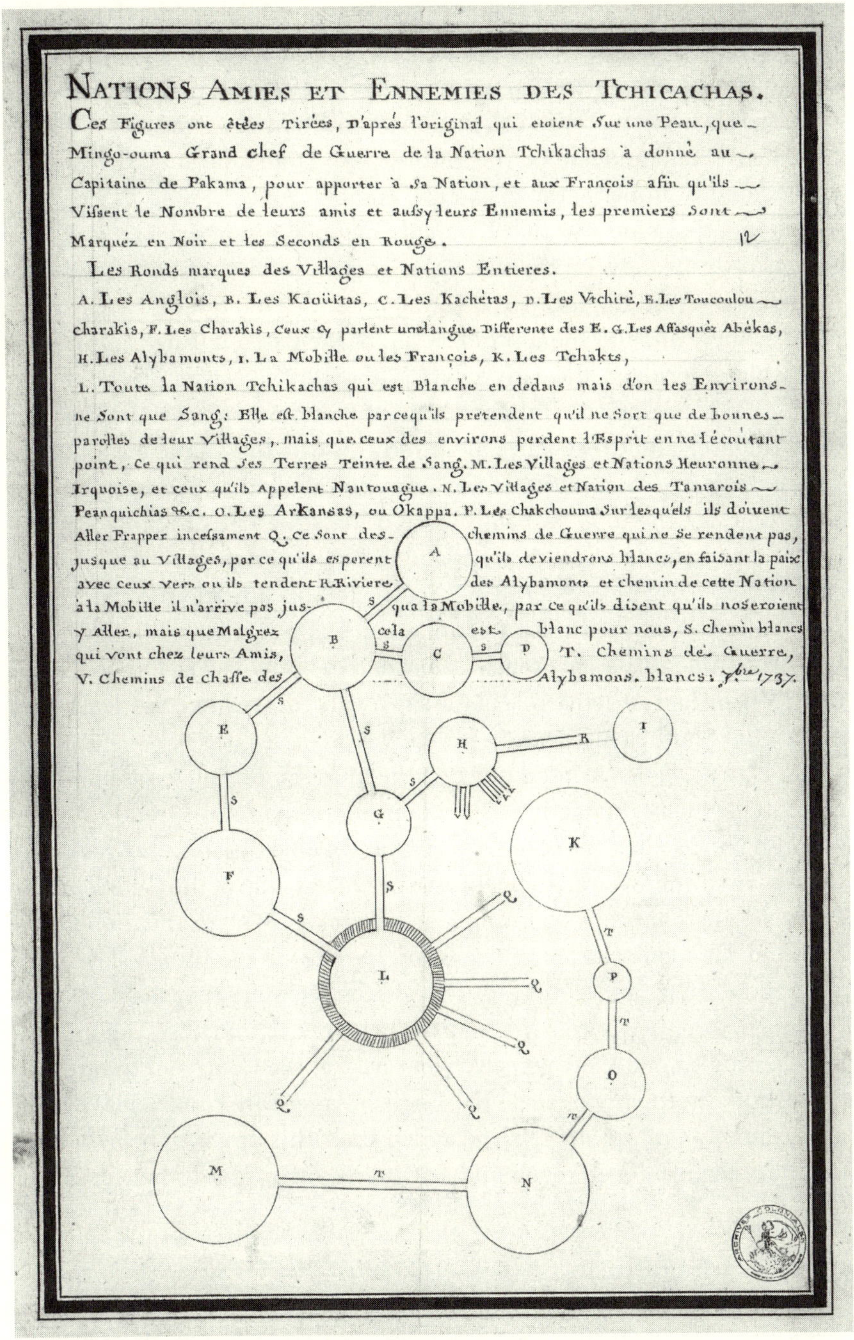

NATIONS AMIES ET ENNEMIES DES TCHICACHAS.

Ces Figures ont êtées Tirées, D'aprés l'original qui etoient Sur une Peau, que Mingo-ouma Grand chef de Guerre de la Nation Tchikachas a donné au Capitaine de Pakama, pour apporter à Sa Nation, et aux François afin qu'ils Vissent le Nombre de leurs amis et aussy leurs Ennemis, les premiers Sont Marquéz en Noir et les Seconds en Rouge. 12

Les Ronds marques des Villages et Nations Entieres.

A. Les Anglois, B. Les Kaoüitas, C. Les Kachetas, D. Les Vtchirè, E. Les Toucoulou charakis, F. Les Charakis, Ceux cy parlent une langue Differente des E. G. Les Affasquez Abekas, H. Les Alybamonts, I. La Mobille ou les François, K. Les Tchakts, L. Toute la Nation Tchikachas qui est Blanche en dedans mais don les Environs ne Sont que Sang: Elle est blanche parcequ'ils pretendent qu'il ne Sort que de bonnes parolles de leur Villages, mais que ceux des environs perdent l'Esprit en ne l'écoutant point, Ce qui rend ses Terres Teinte de Sang. M. Les Villages et Nations Heuronne Irquoise, et ceux qu'ils Appelent Nantouague. N. Les Villages et Nation des Tamarois Peanquichias &c. O. Les Arkansas, ou Okappa. P. Les Chakchouma Sur lesquels ils doivent Aller Frapper incessament Q. Ce Sont des chemins de Guerre qui ne se rendent pas jusque au Villages, par ce qu'ils esperent qu'ils deviendrons blancs, en faisant la paix avec ceux vers ou ils tendent R. Riviere des Alybamonts et chemin de cette Nation à la Mobille il n'arrive pas jusqua la Mobille, par ce qu'ils disent qu'ils noseroient y Aller, mais que Malgrez cela est blanc pour nous, S. Chemin blancs qui vont chez leurs Amis, T. Chemins de Guerre, V. Chemins de chasse des Alybamons. blancs: 7bre 1737.

FIGURE 5. *Nations amies et ennemies de Tchicachas*. 1737. Archives Nationales d'Outre-Mer, Aix-en-Provence. Fr Anom F3/290/12. All rights reserved

fluence. He was a traumatized man, and his map, which included the hometown he had seen destroyed and the Indian and Anglo settlements where he was abused, was a portrait of weakness and suffering. Lamhatty's map, in other words, was an accurate depiction of how tens of thousands of natives experienced and understood the southeastern landscape amid the violence engendered by imperial expansion and competition.

In 1708, the same year that Lamhatty mapped his experience as a slave, Thomas Nairne drew on his own experience as a slave raider to produce a map that asserted English power throughout the Gulf South. Nairne was a planter, surveyor, and Indian trader who participated in English invasions of Spanish Florida in the early 1700s. Although the English failed to take Saint Augustine or push the Spanish out of Florida, English marauders and their Yamasee and Creek allies did utterly destroy the Timucuan and Apalachee mission communities of northern Florida and raided deep into the peninsula. Nairne took part in these Florida slave raids to both accumulate human commodities for profit and to further English power at Spain's expense. By early 1708, Nairne had redirected his attention to the Mississippi Valley where he worked to strengthen English alliances with Chickasaws who raided slaves in French Louisiana. In both Florida and Louisiana, then, Nairne had an active role in the slave wars that marked the death knell of the Gulf South's chiefdoms.[43]

For Nairne, violence in Florida and the Mississippi Valley was politically and intellectually productive. His role in the Indian slave trade strengthened England's alliances with powerful native allies, especially Creeks and Chickasaws, while also enabling him to produce a map of the Gulf South. In 1708, Nairne sent this map to England's secretary of state and, in both the map and the memorial accompanying it, portrayed the Indian slave trade as integral to England's interests. He told the secretary that the purpose of "Laying before your Lordship a map of Such Travells and observations" was to help him "at one View perceive what part of the Continent we are now possest off" in the hopes that "the English American Empire may not be unreasonably Crampt up." Nairne added that "this province being a frontier, both against The French and Spaniards, ought not to be Neglected"

43. Alexander Moore, "Introduction," in Moore, ed., *Nairne's Muskhogean Journals: The 1708 Expedition to the Mississippi River* (Jackson, Miss., 1988), 7–9; Verner W. Crane, *The Southern Frontier, 1670–1732* (Durham, N.C., 1928), esp. 81; [Cumming], "Early Maps of Southeastern North America," in Cumming, *Southeast in Early Maps,* rev. and enl. De Vorsey, 22, 54n; Gallay, *Indian Slave Trade,* 135–136, 145, 147–148, 154–155, 164, 168; Ethridge, *From Chicaza to Chickasaw,* 209–210.

and emphasized how Florida's "Indian towns [were] all Consumed Either by us ... or by our Indian Subjects" while further west "Our friend the Talapoosies [Upper Creeks] and Chicasas Imploy themselves in making Slaves of such Indians ... as are now Subject to the french." Indian slavery, Nairne argued, had augmented English power at the expense of its Spanish and French rivals.[44]

Although Nairne's 1708 map is now lost, it was the basis for a series of printed maps through which England staked its territorial claims and, with apparent pride, charted the slave trade. On the version of Nairne's map published in 1711 by South Carolina merchant Edward Crisp, the lines that connected peoples and places traced—as on Lamhatty's map—the routes of Indian slavery. Nairne's map marked the north-south "road of the Ochese [Lower Creeks] going to War with the Florideans" and the east-west paths through which slaves captured in the Mississippi Valley were transported to Carolina and Virginia. Yet, where Lamhatty had included Tawasa and several other villages along the Florida panhandle, Nairne's map simply marked this region as having "no Inhabitants" without mentioning that this desertion was a recent result of slave raids. Nairne also noted the spot where England's Indian allies left their canoes "when they goe to War against ye Florideans," information that—like other details on the map about peninsular Florida—he had observed as a participant in Yamasee slave raids. The 1711 map was an artifact of the slave trade, but it was also a geopolitical statement. Once published by Crisp, Nairne's map became an argument that South Carolina's boundaries reached below Saint Augustine and west to the Mississippi River. In stark contrast to Lamhatty's map, Nairne's map reflected and projected power.[45]

Nairne was tortured to death by his former Yamasee allies in 1715, but his map lived on and evolved into a clearer declaration about the significance of the slave trade to southeastern geopolitics and a bolder exaggeration of British territorial possessions. Herman Moll's 1720 *New Map of*

44. Thomas Nairne's Memorial to Charles Spencer, earl of Sunderland, July 10, 1708, in Moore, ed., *Nairne's Muskhogean Journals*, 73–79 ("Laying before," 73, "this province," 74, "Indian towns," 75).

45. Edw[ard] Crisp, "A Compleat Description of the Province of Carolina in 3 Parts; 1st. the Improved Part from the Surveys of Maurice Mathews and Mr. John Love; 21y. the West Part by Capt. Tho. Nairn; 31y. a Chart of the Coast from Virginia to Cape Florida," (London, [1711]) (quotes); [Cumming], "Early Maps of Southeastern North America," in Cumming, *Southeast in Early Maps*, rev. an enl. De Vorsey, 21–22. On European maps as tools of empire, see Harley, "Rereading the Maps of the Columbian Encounter," *Annals of the Association of American Geographers*, LXXII (1992), 523, 528.

FIGURE 6. *A Compleat Description of the Province of Carolina in 3 Parts; 1st the Improved Part from the Surveys of Maurice Mathews and Mr. John Love; 2ly the West Part by Capt. Tho. Nairn; 3ly a Chart of the Coast from Virginia to Cape Florida*. By Edw[ard] Crisp. [1711?]. Detail. Library of Congress, Geography and Map Division, G3870 1711 .C6

the North Parts of America claimed by France, a pro-British response to Delisle's 1718 *Carte de la Louisiane,* incorporated Nairne's geographic data and narratives. Like Nairne, Moll extended Carolina's boundaries south of Saint Augustine, depicted the paths that carried English trade goods and Indian slaves across the Gulf South, and, in far more detail than Crisp's map, described the path that Nairne and thirty-three Yamasees took "to go a Slave-Hunting" in southern Florida. Moll used Nairne's path to tell a story that interwove details of Florida geography—including the location of landmarks and (former) Indian towns—with the history of Indian slavery. Networks of knowledge, power, and human trafficking were entwined in early America.[46]

Imperial competition was the driving force that led Le Maire, Lamhatty, and Nairne to produce maps, shaped the content of their maps, and determined the routes through which their maps circulated. Le Maire drew on his connections with diverse local inhabitants to satisfy officials' desire for

46. [Cumming], "Early Maps of Southeastern North America," in Cumming, *Southeast in Early Maps,* rev. and enl. De Vorsey, 23, ibid., 205–206; H[erman] Moll, "A New Map of the North Parts of America Claimed by France . . ." ([Cornhill], 1720) (quote).

FIGURE 7. *A New Map of the North Parts of America Claimed by France* .... By H[erman] Moll. 1720. Library of Congress, Geography and Map Division, G3300 1720 .M6

geographic information that would bolster their claims to a contested region. Lamhatty's and Nairne's maps reflected that the violence catalyzed by imperial competition could also promote knowledge production. Lamhatty's map, born from desperation, circulated geographical information but did little or nothing to improve his own condition or affect regional geopolitics. Nairne's map, born from aggression and observations made while enslaving people like Lamhatty, supported British arguments about the extent of their empire. Knowledge was not power for everyone: both Lamhatty and Nairne knew about the lands along southern slave-trading paths. Rather, powerful people valued knowledge because it helped make them even more powerful.

NATURAL HISTORY AND GIFTS IN THE BORDERLANDS AND ATLANTIC WORLD

In the wake of the slave wars, no nation dominated the southeast borderlands: British, Spanish, and French colonists as well as Creeks, Chickasaws, and Choctaws all depended on one or more other groups to achieve profit and influence. This mutual dependence made Indians and Europeans alike eager to exchange goods and—since natives and whites continued to value exotic knowledge—information. Animals, plants, and ethnographic accounts circulated among Indians and Europeans in the eighteenth-century Gulf South because a Mississippian political tradition in which gift exchange and access to specialized knowledge were sources of prestige became entangled with an Atlantic world in which commodity exchange and access to American natural knowledge heightened social status. By the early 1700s, then, overlapping regional and Atlantic relationships of competition, violence, and exchange fully contextualized the production and circulation of knowledge in and beyond the Gulf South.[47]

Indian and Atlantic networks of exchange penetrated each other so thoroughly because of fundamental congruencies between them: the presentation and reception of exotic goods, many of which doubled as natural history specimens, strengthened relationships of patronage, alliance, and reciprocal obligation that were as much a path to prestige for Indians as they were for white naturalists. For southeastern natives who inherited the Mississippian tradition of exchange, distributing gifts, particularly those associated with knowledge from distant realms, proved one's capacity to obtain extraordinary items and formed the basis of unequal relationships with people who wanted those same items but had no other means of getting them. Some of the most highly prized gifts were European manufactures, especially things like guns that were not made in Indian country and necessitated European expertise. Similarly, the naturalists who gathered flora and fauna in the Gulf South and sent these specimens across the Atlantic did so largely to build relationships with prominent men in Europe and to achieve personal prestige. Just as Indians could accrue influence by tapping into Atlantic trade networks for exotic goods and redistributing them, so, too, could whites engage with Indian networks of gift exchange to acquire specimens and stories that, once recirculated to Europe, increased their own authority as men of science.[48]

47. On the weakness of Indians and Europeans and how the circulation of stories between them "show us a world that cannot be divided into tidy categories" like white and Indian, see Piker, *Four Deaths of Acorn Whistler,* 13 (quote). On the interconnectedness of Europeans and Indians and their dependence on each other for survival and information, see Dubcovsky, *Informed Power.* On how networks of science and commerce overlapped, see Londa Schiebinger and Claudia Swan, eds., *Colonial Botany: Science, Commerce, and Politics in the Early Modern World* (Philadelphia, 2005); and Harold J. Cook, *Matters of Exchange: Commerce, Medicine, and Science in the Dutch Golden Age* (New Haven, Conn., 2007).

48. Hall, *Zamumo's Gifts,* esp. 5, 7–9, 13, 118, 129; Salisbury, "Indians' Old World," *WMQ,* LIII (1996), 436–437, 454, 457–458; Richter, *Before the Revolution,* 26–29, 123,

There was, moreover, considerable overlap between what constituted a natural history specimen, a gift, and a commodity. Deerskins—the central trade good of the eighteenth-century Gulf South—could be sent to Europe as zoological specimens or processed and packaged as commodities, while Indians also presented painted deerskins to each other and Europeans as gifts that, in European hands, sometimes transformed into ethnographic artifacts. Feathers had value for Indians and Europeans alike as fashion accessories and embodiments of authority: eagles, for example, signified power among European emperors and Gulf South headmen. European colonists thus sent southeastern feathers to Europe as both natural history specimens and commodities destined for ladies' hats while also distributing feathers as presents to Indians in hopes of keeping them as allies and trading partners. And, although natives continued to associate feathers with particular birds' spiritual and political power—by attaching eagle feathers to calumets, for instance—they sometimes bought and sold feathers as commodities. Natives and Europeans also prized and circulated medicinal plants for the same set of reasons. They esteemed the chemical and spiritual qualities that made them effective medicines, they studied their physical and pharmaceutical properties, and they sought them out and exchanged them as both marketable commodities and gifts through which they could build relationships.[49]

128; Usner, *Indians, Settlers, and Slaves,* 63–66, 95; Gallay, *Indian Slave Trade,* 123–125; James Taylor Carson, "Sacred Circles and Dangerous People: Native American Cosmology and the French Settlement of Louisiana," in Bradley G. Bond, ed., *French Colonial Louisiana and the Atlantic World* (Baton Rouge, La., 2005), 65–82. On how Anglo naturalists in colonial America circulated specimens to Europe, and couched these specimens as gifts, to build relationships, see Susan Scott Parrish, *American Curiosity: Cultures of Natural History in the Colonial British Atlantic World* (Chapel Hill, N.C., 2006), 22, 107, 136, 169–173.

49. On how knowledge was embedded in, and circulated through, objects and specimens, see Lissa Roberts, "The Circulation of Knowledge in Early Modern Europe: Embodiment, Mobility, Learning and Knowing," in Ian Inkster, ed., *History of Technology,* XXXI (London, 2012), 47–68. On deerskins, see Jean-Françoise-Benjamin Dumont de Montigny, *The Memoir of Lieutenant Dumont, 1715–1747: A Sojourner in the French Atlantic,* ed. Gordon M. Sayre and Carla Zecher, trans. Sayre (Chapel Hill, N.C., 2012), 339–342; Kathryn E. Holland Braund, *Deerskins and Duffels: The Creek Indian Trade with Anglo-America, 1685–1815* (Lincoln, Neb., 1993); and Hall, *Zamumo's Gifts,* 52, 67. On feathers, see [Antoine-Simon] Le Page Du Pratz, *Histoire de la Louisiane . . . ,* 3 vols. (Paris, 1758), II, 109; Dumont de Montigny, *Memoir of Lieutenant Dumont,* ed. Sayre and Zecher, trans. Sayre, 342–343; Shepard Krech, III, *Spirits of the Air: Birds and American Indians in the South* (Athens, Ga., 2009), esp. 63, 69, 71, 77–78, 104, 111–113, 133; Dawdy,

Rare animals, alive and dead, were valuable in both Atlantic scientific networks and borderland gift networks, and the exchange of animals often hinged on gender roles. Women in Europe were among the main consumers for exotic birds, and their demands could determine the supply from the Gulf South. One naturalist in Louisiana observed that local parrots' beautiful plumage did not compensate for their taciturnity "because a silent Parrot would not make a fortune among our Ladies, [and therefore] we do not see these in France." Europeans also distributed exotic animals as gifts to Indian women: an English trader observed "a young Chicasaw princess who was carrying from the English setlement 2 young catts, to her country as a great rarity." Leaving aside that there is "an indefinable *je ne sais quoi* about cats," it is possible that Chickasaw women, like native women in more southerly parts of the circum-Caribbean, were responsible for adopting and taming animals and exchanging them with other groups to build ties of friendship. Gift giving and zoological specimen exchange overlapped in male-dominated spheres as well, including tobacco culture. During a visit to a Creek town, William Bartram offered an elderly chief "presents," including "a twist of choice Tobacco," and the old man "thanked me, returning the favour immediately with his own stone pipe and cat skin of Tobacco." Animal skin tobacco pouches were prized gifts between native men and became a way for European naturalists to get their hands on the pelt of a rare animal. Naturalist Mark Catesby, for one, relied on such a gift for his account and illustration of a whooping crane. The sharing of tobacco among men both strengthened homosocial bonds of alliance and exchange while also supplying European naturalists with ready-preserved specimens.[50]

Building the Devil's Empire, 17; "Statement of the Requests to the King from the Colony, Louisiana," 1759, in Dunbar Rowland and A. G. Sanders, eds., *Mississippi Provincial Archives,* V, *French Dominion, 1749–1763,* rev. and ed. Patricia Kay Galloway (Baton Rouge, La., 1984), 228–229. On medicinal plants, see Régis du Roullet to Périer, Mar. 16, 1731, in Rowland and Sanders, eds., *Mississippi Provincial Archives,* IV, *French Dominion, 1729–1748,* rev. and ed. Galloway (Baton Rouge, La., 1984), 69; Bartram, *Travels,* 185; Mark Catesby, *The Natural History of Carolina, Florida, and the Bahama Islands . . . ,* I (London, 1731), 35; Hall, *Zamumo's Gifts,* 95–96; Robbie Ethridge, *Creek Country: The Creek Indians and Their World* (Chapel Hill, N.C., 2003), 179; and Rudolph Matas, *The Rudolph Matas History of Medicine in Louisiana,* ed. John Duffy, 2 vols. ([Baton Rouge, La.], 1958–1962), I, 29, 112.

50. Le Page Du Pratz, *Histoire de la Louisiane,* II, 128 ("because a silent"); [Thomas Nairne] to Robert Fenwick, Apr. 13, 1708, in Moore, ed., *Nairne's Muskhogean Journals,* 51 ("young Chicasaw"); Robert Darnton, *The Great Cat Massacre and Other Episodes in French Cultural History* (New York, 1999), 89 ("indefinable"); Bartram, *Travels,* 453, 499

For European naturalists, tapping into Indian exchange networks—many of which connected the Gulf South to distant parts of North America—could be an effective way to acquire specimens for shipment to Europe. Catesby described how he obtained "the largest white-bill Wood-pecker" from southern Indians already in the habit of selling the bills of these birds to *"Canada Indians,* who make Coronets of 'em ... at the price of two, and sometimes three Buck-skins a Bill." More impressively, the planter and naturalist Antoine-Simon Le Page du Pratz, who lived in Louisiana from 1718 to 1734, managed to acquire a huge stock of medicinal plants by gifting the Great Sun, the Natchez chief, a silver-bedecked calumet. After receiving the pipe, the Great Sun "named me his true friend" and, in return, offered the services of his personal "Physician" to treat Le Page's swollen eye. The treatment worked, and Le Page praised the doctor's knowledge of herbal remedies. When Le Page learned that the directors of the West India Company desired Frenchmen privy to "the secrets of the Natives" to "research the Simples used for Medicine and dyeing," he rushed to supply European chemists and capitalists with local plants by exploiting his connections with native experts and chiefs. Presumably with the assistance of the Natchez physician, Le Page gathered and shipped "more than three hundred Simples" and penned "a Memoire that detailed their qualities, and taught how to use them." These transplants eventually made it to a "botanical Garden" in France. It was Le Page's ties of friendship with the Great Sun, initiated by the gift of a calumet, that gave him access to medicinal plants that he sent to France in order to build bonds with prominent officials.[51]

Native leaders not only introduced Europeans to experts in return for gifts but also offered to pay Europeans in exchange for knowledge and technologies. Le Page claimed to have astonished the Great Sun by using a magnifying glass to rekindle the Natchez's sacred fire with sunlight, adding a religious aura to the show by pronouncing "the word *Caheuch,* which means 'come,'" as he lit the blaze. Le Page then had the Great Sun "perform the experiment for himself," after which the chief decided to purchase the instrument at any price. The Great Sun's offer to buy the glass suggests that

("presents"); Catesby, *Natural History of Carolina, Florida, and the Bahama Islands,* I, 38, 75; Marcy Norton, "The Chicken or the *Iegue:* Human-Animal Relationships and the Columbian Exchange," *AHR,* CXX (2015), 40, 43, 46–47, 51–53. All translations of foreign language sources are my own unless otherwise noted.

51. Catesby, *Natural History of Carolina, Florida, and the Bahama Islands,* I, 16 ("largest"); Le Page Du Pratz, *Histoire de la Louisiane,* I, 206 ("named"), 209 ("Physician"), 212 ("the secrets," "research," "more than," "a Memoire," "botanical").

the Natchez quickly adapted to European notions of buying and selling, but Le Page also recognized the advantages of completing the transaction as an exchange of gifts, telling the Great Sun that he gave the glass freely as a token of affection and "asked for nothing more in return but was necessary to my subsistence." The Natchez, usually considered the last of the Gulf South's chiefdoms, maintained the Mississippian tradition of grounding the chief's authority in his extraordinary access to spiritual power, and once the Great Sun and his retinue acquired Le Page's glass—an exotic gift through which they could tap into the otherwise remote power of the sun—they immediately conspired to keep their new technology a state secret. Le Page reported that "the common people, always interested in penetrating the secrets of the court," assembled to watch the Great Sun kindle a flame from sunlight. When the glass produced fire "their curiosity became even greater but was not further satisfied" for the Great Sun told them merely that Le Page "had given it to him, guided more by my friendship with him than by my own interests," and that "the Natchez had a great obligation to me." By presenting the Natchez chief with an exotic gift associated with power, Le Page acquired influence among, and material rewards from, an entire chiefdom.[52]

Obtaining natural history specimens through native gift exchange networks was far safer for European naturalists than exploring the Gulf South themselves. From the 1710s to the 1780s, the region experienced periodic violence between natives and whites as well as several imperial wars, most notably the Seven Years' War (which made western Louisiana Spanish and Florida and eastern Louisiana British) and the American Revolution (which made Florida and eastern Louisiana Spanish again and introduced a new nation, the United States, into ongoing competitions for the Gulf South). Scientific work in the region could, therefore, be deadly business. The French missionary François Le Maire told a correspondent at Versailles that "the savages perform every day marvelous cures with the herbs that grow in this country" and that he had received requests for specimens of

52. Antoine-Simon Le Page Du Pratz, *Histoire de la Louisiane,* 3 vols. (Paris, 1758), trans. Gordon Sayre, II, [342–351] ("word," [342], "to perform," [343], "asked for," [349], "the common," "their curiosity," "had given," [350], "the Natchez," [351]), https://blogs.uoregon.edu/lpdpanddumont/. Le Page was so aware of the effectiveness of gifts as a means of securing specimens that he once forbade some Indian companions from collecting crystals because he "feared that some Frenchman seeing the stones would persuade the Naturals, by the force of gifts, to reveal the location" of minerals that Le Page wanted to reserve for himself. See ibid., I, [238].

these plants from "the chair of Botany of the Royal Garden." But Le Maire regretted that collecting such plants was "impossible in Pensacola, where to botanize is a question of life and death" owing to "the risk of being caught by the savages at war with the Spaniards." Philadelphian naturalist John Bartram recognized that safety during scientific expeditions depended on imperial domination: challenges from competing empires and less-than-complete subjugations of indigenous populations made doing science in the colonies hazardous. As he wrote while planning an expedition in 1764: "I should be exceedingly pleased . . . to make A thorough search not only at pensacola but ye coast of florida[,] alabama, Georgia and ye banks of ye Misisipi[.] I make no difference who got it, if I could but travail safely in it." His son William experienced firsthand how colonial unrest could limit the collection of specimens. William had planned to explore Creek country in 1775, but "the alarm from the Frontiers of hostilities commencing between the Indians and whites put a stop to that scheme." Without the armies that had made Spanish exploration of the Gulf South possible in the sixteenth century, his best hope for collecting specimens was to gain entrée to native exchange networks and secure the protection of native headmen.[53]

Indians contributed significantly to the ongoing flow of gifts, specimens, and contraband between the Gulf South and the rest of the circum-Caribbean. For one, Indians sometimes collected the specimens that Europeans exchanged across imperial boundaries. Catesby painted "The Parrot of Paradise of Cuba" based on a specimen "shot by an *Indian,* on the Island Cuba," who "carried it to the Governour of the *Havana,* who presented it to a Gentlewoman of *Carolina.*" Indians were also go-betweens in the gift

53. Le Maire to [Bobé], Jan. 15, 1714, in "M. Le Maire on Louisiana," in Wedel, ed., *A Jean Delanglez, S.J., Anthology,* 140 ("savages"); John Bartram to Peter Collinson, Mar. 4, 1764, in Edmund Berkeley and Dorothy Smith Berkeley, eds., *The Correspondence of John Bartram, 1734–1777* (Gainesville, Fla., 1992), 622 ("I should be exceedingly"), William Bartram to [John Bartram], Mar. 27, 1775, 769 ("alarm"). For how the Revolutionary War contextualized William Bartram's expedition in the Gulf South, see Edward J. Cashin, *William Bartram and the American Revolution on the Southern Frontier* (Columbia, S.C., 2000). William Bartram's botanical and zoological research among the Creeks and Seminoles during the Revolution sometimes depended on the consent of Indian headmen. The Seminole Cowkeeper gave Bartram the name "PUC PUGGY or the Flower hunter" and offered him protection as he botanized in Seminole territory. But this protection also seemed to entail a reciprocal obligation of friendship that required Bartram to use his expertise to protect the Seminoles, such as when three Seminoles demanded that he kill a rattlesnake that had invaded their town. See Bartram, *Travels,* 185 (quote), 238, 260–262.

and commodity networks connecting the Gulf South and the West Indies, such as one Florida Seminole who gave William Bartram "a choice piece of Tobacco, which he ... had received from the governor of Cuba." And Indians could thwart the circulation of commercially valuable specimens around the Gulf of Mexico. Officials in Louisiana were eager to establish illicit trade with Mexico, but, when they asked a Spanish smuggler from Campeche "to try to bring us a few cochineal eggs and a memorandum about it," they were disappointed to learn that the smuggler "could not get the eggs because the Indians [in Mexico] are not willing to let them leave." Indians were not involved with every transfer of specimens around the Caribbean—whites and blacks also circulated flora and fauna—but they were nevertheless active in the distribution of curiosities and commodities.[54]

Natives were also directly involved in circulating specimen gifts across the Atlantic, such as when Tomochichi, a headman of the Lower Creek town of Yamacraw, gave George II "Feathers of the Eagle" during a trip to England in 1734. The express purpose of the visit was to build an "everlasting Peace" between the Creeks and the English and address trade grievances, but it seems that Tomochichi's gift to the king was part of a broader effort to acquire knowledge from England. As Tomochichi put it, "I am come for the Good of the Children of the Nations of the Upper and of the Lower *Creeks,* that they may be instructed in the Knowledge of the *English.*" Part of this knowledge was religious: Tomochichi aimed to bring missionaries to his town who would instruct his people, especially his heir, in spiritual mysteries. But he also sought natural and medical expertise in England. He visited the Royal Society, where he watched natural philosophers perform experiments, and toured the Royal Hospital. Tomochichi, like several other Creek headmen, exaggerated his own importance by claiming to be the principal leader of the still protean Creek nation. But Tomochichi's presentation of prestige goods in England in hopes of receiving knowledge in return suggests that he fancied himself as perpetuating the role of earlier

54. Catesby, *Natural History of Carolina, Florida, and the Bahama Islands,* I, 10 ("Parrot"), emphasis in original; Bartram, *Travels,* 227 ("choice"); Perier and De La Chaise to the Directors of the Company of the Indies, July 31, 1728, in Dunbar Rowland and Albert Godfrey Sanders, eds., *Mississippi Provincial Archives,* II, *French Dominion 1701–1729* (Jackson, Miss., 1929), 577 ("try to bring"); Le Maire to [Bobé], Jan. 15, 1714, in "M. Le Maire on Louisiana," in Wedel, ed., *A Jean Delanglez, S.J., Anthology,* 140; Le Page Du Pratz, *Histoire de la Louisiane,* III, 390; Romans, *Concise Natural History of East and West Florida,* 246. On Caribbean smuggling networks and how these circuits followed paths pioneered by natives, see Dawdy, *Building the Devil's Empire,* 3–4, 102.

Mississippian chiefs who patronized learned priests with gifts in order to bolster their own influence.[55]

European colonists, however, supplied the majority of natural history specimens that reached Europe from the Gulf South, and collectors sent gifts of plants, animals, and ethnographic artifacts as a way to amplify their own prestige as naturalists and build closer relationships with European patrons and men of science. The presentation of specimens as gifts was not divorced from the circulation of specimens as potentially valuable commodities: these items were often one and the same, but the rhetoric of gift giving allowed collectors and patrons alike to present themselves as driven by friendship and disinterested curiosity instead of a crass desire for profit. As with other items that traveled between metropoles and colonies, the meaning and uses of specimen gifts often transformed as they circulated around the Atlantic world. The Spanish governor of Pensacola, for example, sent the viceroy of New Spain some enormous acorns he collected during an excursion to the Texas coast, "acorns so big that their cups, after adding a stem, could be used as chocolate cups; and such was the use which the ... Viceroy of Mexico, to whom they were presented, made of them, and he sent half a dozen such cups to the late King of Spain." The viceroy remade a curiosity from the fringes of New Spain into an exotic royal gift that, not coincidentally, kept the chocolate and timber resources of New Spain in the king's hand, mouth, and mind.[56]

This system of transatlantic specimen exchange worked because the support that European patrons gave collectors in America was itself a gift, one that added to the prestige of patrons ranging from metropolitan naturalists to royalty. The king of England named John Bartram as king's botanist in the Floridas despite that, as jealous South Carolina naturalist Alexan-

55. *Gentleman's Monthly Intelligencer* (August 1734), 446, quoted in Julie Anne Sweet, "Bearing Feathers of the Eagle: Tomochichi's Trip to England," *Georgia Historical Quarterly,* LXXXVI (2002), 353 (quotes), 361; Krech, *Spirits of the Air,* 107–113; Coll Thrush, *Indigenous London: Native Travelers at the Heart of Empire* (New Haven, Conn., 2016), 83–88. Tomochichi expected George II to reciprocate his present with gift specimens of his own, particularly two live swans, but the king never delivered the birds. For how Creek headmen exaggerated their own power and the Creeks' degree of nationhood, see Piker, *Four Deaths of Acorn Whistler,* 78–131.

56. Le Maire to [Bobé], Jan. 15, 1714, in "M. Le Maire on Louisiana," in Wedel, ed., *A Jean Delanglez, S.J., Anthology,* 139 (quote); Parrish, *American Curiosity,* 136, 141, 169–173, 177, 193; Nicholas Thomas, *Entangled Objects: Exchange, Material Culture, and Colonialism in the Pacific* (Cambridge, Mass., 1991).

der Garden put it, Bartram could "scarcely spell, much less make out the characters of any one genus of plants." But Garden nevertheless praised the king for patronizing a naturalist dedicated to "searching out the plants of East and West Florida." "Surely our King is a great King!" he exclaimed, "The very idea of ordering such a search is noble, grand, royal." The governor of British East Florida, for his part, knew that he could add to his own prestige by sending Florida seeds to the king's exotic garden at Kew, where the king and his horticulturally inclined ministers would prize the seeds as a means of displaying the fruits of British overseas expansion. Like Mississippian chiefs, European monarchs recognized that demonstrating their exceptional access to exotic specimens, particularly those that embodied knowledge from distant realms, was an effective way to convey their own centrality to webs of exchange, information, and power.[57]

Natural history in the eighteenth century also encompassed ethnography. As French missionary Le Maire put it, the study of "nature" embraced "plants, minerals[,] and animals" as well as "habitants [and] their customs." Collectors in the Gulf South did send ethnographic artifacts to Europe, but most of what Europeans learned about Indians circulated in narrative form. Some ethnographic observers told stories about Indians that seemed to support grand theories about Indians' culture and origins. Trader James Adair studied contemporary Creeks, Choctaws, and Chickasaws as evidence for an "INDIAN SYSTEM" that used linguistic and cultural evidence to argue that Indians had descended from Hebrews. Yet Adair and other ethnographers were careful to explain that the stories they told about Indians were not simply curious accounts but were directly useful to officials. Le Page, for instance, was explicit that his "design is only to make known, from the general character of these people, what course we ought to observe in order to draw advantage from our intercourse with them." Like the specimens that collec-

57. Dr. [Alexander] Garden to John Ellis, July 15, 1765, in John Bartram, "Diary of a Journey through the Carolinas, Georgia, and Florida, from July 1, 1765, to April 10, 1766," [ed.] Francis Harper, American Philosophical Society, *Transactions*, XXXIII, New Ser., part I (1942), 56 (quotes); David R. Brigham, "Mark Catesby and the Patronage of Natural History in the First Half of the Eighteenth Century," in Amy R. W. Meyers and Margaret Beck Pritchard, eds., *Empire's Nature: Mark Catesby's New World Vision* (Chapel Hill, N.C., 1998), 91–146; Robert Olwell, "Seeds of Empire: Florida, Kew, and the British Imperial Meridian in the 1760s," in Elizabeth Mancke and Carole Shammas, eds., *The Creation of the British Atlantic World* (Baltimore, 2005), 263–282. See also Kathleen S. Murphy, "To Make Florida Answer to Its Name: John Ellis, Bernard Romans, and the Atlantic Science of British West Florida," *British Journal for the History of Science*, XLVII (2014), 43–65.

tors sent to Europe, ethnographic stories blurred the line between curiosities that demonstrated an observer's ability to obtain unfamiliar knowledge and commodities that promoted imperial wealth and power.[58]

Europeans were by no means the only peoples in the Gulf South interested in the history and origins of Indians. Natives also pursued and narrated ethnographic knowledge about their own and other nations. Similarly to the Europeans who described Indians, natives were not ethnographic specialists but rather studied the traditions and customs of Indian groups as one of many intellectual interests. For example, a Mustee Creek with a Choctaw mother and a Creek-English father accompanied William Bartram as part of a trading party to Choctaw country, and, according to Bartram, his "designs [were] in the highest degree commendable, nothing less than to inform himself of every species of arts and sciences, that might be of use and advantage, when introduced into his own country, but more particularly music and poetry." This same goal—the accumulation of nationally beneficial knowledge among other peoples—also inspired many experts from Britain, France, Spain, and, eventually, the United States. The Choctaws, however, grew wary of this "young philosopher" and denounced him as a "spy," forcing him to flee for his life. As with Europeans who went on scientific expeditions in the contested Gulf South, suspicion of international intrigue and the threat of violence could circumscribe Indians' research.[59]

The Yazoo Indian explorer Moncacht-apé sought ethnographic knowledge across a far wider field as he traveled from his home in the lower Mississippi Valley to the northeastern Atlantic coast and far into the West around the year 1700. According to Le Page, who gave him a mirror and other gifts in exchange for narrating his stories, Moncacht-apé was inter-

58. Le Maire to [Bobé], Jan. 15, 1714, in "M. Le Maire on Louisiana," in Wedel, ed., *A Jean Delanglez, S.J., Anthology,* 134 ("nature"); Adair, *History of the American Indians,* 1, 10 ("INDIAN"), 261, 281; Le Page Du Pratz, *Histoire de la Louisiane,* trans. Sayre, II, [307] ("design"); Thomas Nairne, "An Account of the Customs, Humers, and Present State of the Chicasaws," Apr. 12, 1708, in Moore, ed., *Nairne's Muskhogean Journals,* 36–50. On how "stories themselves were products of exchange," see Hall, *Zamumo's Gifts,* 29. Several other Gulf South authors included extensive narratives about natives and their cultural practices as part of their broader efforts to present themselves as regional experts. See Bartram, *Travels,* esp. 481–522; Pierre F. X. de Charlevoix, *Charlevoix's Louisiana: Selections from the History and the Journal,* ed. Charles E. O'Neill (Baton Rouge, La., 1977), esp. 139–149; and Dumont de Montigny, *Memoir of Lieutenant Dumont,* ed. Sayre and Zecher, trans. Sayre, 36–37, 333–365.

59. Bartram, *Travels,* 440, 506–507 ("designs," 506, "young philosopher," 507).

ested in "discovering the origin of his people" and spent five years traveling the West to find "the country from which his ancestors emigrated." When one western chief asked him "What do you want here with your short hair[?]," Moncacht-apé replied: "I am in search of knowledge, and I come to see you so that you might provide it. My hair is short so that it may not bother me, but my courage is good . . . ." "During winter, like the bear," he explained, "I seek a covert, and in summer I imitate the eagle, who moves about to satisfy his curiosity." Yet Moncacht-apé did not merely satisfy his curiosity about the Indian peoples he met. He also joined western groups in studying the light-skinned, bearded men who sometimes appeared among them and, in one instance, based his ethnographic analysis of white men on their bodily remains. After joining a western nation in an ambush against a company of Europeans, Moncacht-apé recalled how "we then went to examine the dead" who were "much smaller than we were, and very white" with "large heads." As early as 1700, Indians had already begun studying bodies to understand ethnic difference.[60]

By the 1720s, however, Gulf South Indians did not have to travel across the continent to learn about human difference and their own origins; instead, they simply had to listen to the new stories about the creations of Indians, blacks, and whites circulating around their own region. One of the earliest records of such stories came from a Taensa Indian in Louisiana, who told how "long ago . . . there were three men in a cave, one white, one red[,] and one black." The white man left the cave first and claimed the best hunting grounds, the Indian came out second and had to settle for inferior lands, and "the black man, who is the negro, having been the third to go out, got entirely lost in a very bad country." The Jesuit who recorded this story concluded that the white man's primacy in emerging meant that Indians "drew the inference . . . that they must listen to us as to men who have more intelligence than they." Adair heard similar accounts among the Choctaws and Chickasaws, but he argued that far from reflecting Indian traditions (which, in his mind, should have demonstrated their descent from Israelites), stories of separate creations had been inspired by "the innovating superstitious ignorance of the popish priests, to the south-west of us." For

60. Le Page Du Pratz, *Histoire de la Louisiane*, trans. Sayre, III, [102] ("discovering"), [112] ("What do you"), [113] ("During winter"), [125] ("we then went"), [130]. Le Page is the only author who mentions Moncacht-Apé, and it is possible that he invented this story to entertain his readers. See Gordon M. Sayre, "A Native American Scoops Lewis and Clark: The Voyage of Moncacht-Apé," *Common-Place*, V, no. 4 (July 2005), http://www.common-place-archives.org/vol-05/no-04/sayre/index.shtml.

Adair, Catholic proselytizers, and not Indian theorizers, were responsible for introducing ideas of racial difference among Gulf South natives, but it seems unlikely that Catholic missionaries would have intentionally spread the notion of separate creations because, after all, this theory contradicts scripture. These stories were, therefore, probably the work of natives trying to make sense of the new ethnic complexity of the colonial era. More importantly, these myths were precocious articulations of the theory of polygenesis, the idea that whites, blacks, and Indians had been created separately and were, therefore, essentially different peoples. In the Gulf South, Indians were pioneers in the ethnographic practice of differentiating racial groups based on supposedly inborn and immutable traits.[61]

Europeans wrote these stories down and shared them with other Europeans at least in part because doing so demonstrated that they possessed specialized ethnographic knowledge that many European readers valued (either as a curiosity or, perhaps, as an insight that could help in future conversion efforts). In short, the circulation of stories and specimens in the 1700s suggests that Gulf South natives and Europeans considered difficult-to-access knowledge a source of prestige and power, and this led individuals in both groups to obtain information and materials from exotic places and cultures. By the 1700s, this shared ambition culminated in a new web of exchange that blurred Indian and Atlantic networks of knowledge and power.

• • •

Europeans and natives valued knowledge and the experts who produced it as sources of power and, from the 1500s through the 1700s, learned about their mutually new world during encounters involving violence, geopolitical competition, and exchange. These same encounters would continue to contextualize and shape knowledge production as Spain struggled to maintain control over the entire Gulf South in the late eighteenth century and as the United States vied to dominate the region in the nineteenth century. And since information and texts traveled across time (not just space), nineteenth-century officials and men of science from President Thomas Jefferson to craniologist Samuel George Morton based much of what they thought they knew about the Gulf South and its inhabitants on colonial

61. Rapha[ë]l [de Luxembourg] to [Jean Baptiste] Raguet, May 15, 1725, in Rowland and Sanders, eds., *Mississippi Provincial Archives,* II, 485–486 ("long ago," II, 485–486, "black man," II, 486); Adair, *History of the American Indians,* 194–195 ("innovating"); Carson, "Sacred Circles and Dangerous People," in Bond, ed., *French Colonial Louisiana and the Atlantic World,* esp. 73; Nancy Shoemaker, "How Indians Got to Be Red," *AHR,* CII (1997), 625–644.

authors like de Soto's chroniclers, Le Page du Pratz, and William Bartram. Both the knowledge that individuals in the Gulf South produced in the colonial era and the context of imperialism in which they produced it would continue to influence what diverse men and women knew about Gulf South nature and peoples for generations to come.[62]

62. [Thomas Jefferson, ed.], *An Account of Louisiana* (Newbern, [N.C.], 1804); Shannon Lee Dawdy, "Proper Caresses and Prudent Distance: A How-To Manual from Colonial Louisiana," in Ann Laura Stoler, ed., *Haunted by Empire: Geographies of Intimacy in North American History* (Durham, N.C., 2006), 152–155; Samuel George Morton, *Crania Americana; or, A Comparative View of the Skulls of Various Aboriginal Nations of North and South America* ... (Philadelphia, 1839), 158–165.

{ 2 }

KNOWLEDGE, WEAKNESS, AND NARRATIVE IN THE LATE EIGHTEENTH CENTURY

• • •

In the late 1700s, Spanish dominion reached from East Florida to Louisiana and, indeed, across the southern part of North America all the way to the Pacific Ocean. Spain acquired western Louisiana from France in 1763 as part of the aftermath of the Seven Years' War and, after losing Florida to Britain that same year, regained sovereignty over the provinces of East and West Florida (Britain had divided them) in 1783 as part of the fallout from the American Revolution. Yet, despite Spanish claims to rule the entire Gulf South, imperial competition for the region persisted and remained tied to global contests for territory and wealth. Spanish officials thus had hemispheric security and the global economy in mind as they tried to develop the Gulf South into both a strategic buffer protecting the rich territories of Mexico and a profitable plantation region that filled a niche in Spain's imperial economy. But Spain lacked the regional strength needed to defend its borders or coerce inhabitants into complying with imperial economic schemes. Spanish officials, therefore, had to depend on the cooperation of local power brokers and experts.[1]

Global and local competitions overlapped as Spanish officials worked to acquire and apply imperially beneficial knowledge in the Gulf South. On the one hand, Spanish officials needed the cooperation of locals to obtain information about lands and resources that, officials believed, would help Spain in its global contest with other empires. On the other hand, locals tried to use Spanish officials' dependence on them to improve their own positions in local and regional competitions for status and survival. Span-

1. On the late eighteenth-century Spanish Gulf South as part of global imperial competitions, see David J. Weber, *Bárbaros: Spaniards and Their Savages in the Age of Enlightenment* (New Haven, Conn., 2005), 183, 204, 214–216; J. H. Elliott, *Empires of the Atlantic World: Britain and Spain in America, 1492–1830* (New Haven, Conn., 2006), 295, 353, 373; and Sylvia L. Hilton, "Spanish Louisiana in Atlantic Contexts: Nexus of Imperial Transactions and International Relations," in Cécile Vidal, ed., *Louisiana: Crossroads of the Atlantic World* (Philadelphia, 2014), 68–86.

ish officials' efforts to acquire and circulate reliable accounts about nature in the Gulf South hinged on, and sometimes failed because of, interactions with the region's inhabitants.[2]

Case studies of Spanish, West African, Swiss, and Creek men suggest that individuals and nations in the Gulf South approached knowledge production from a position of weakness that severely limited their potential to benefit from investigations of the natural world. This weakness was evident in the inability of Spanish officials to know and rule the region without local cooperation, the vulnerability of enslaved blacks to violence and exploitation, and the fragile place that natives held in the contested borderlands. Spanish officials like Antonio de Ulloa and Martín Navarro could sometimes amass valuable observations and, by circulating them to other bureaucrats, influence imperial policies. But matters were rarely so simple. Spanish governance depended on the domination of enslaved blacks and the cooperation of local whites and natives, and learned individuals from these groups were far more interested in improving their positions within colonial society than in promoting officials' agendas. Enslaved West African herbalists Carlos and Cipion hoped to avoid imperial officials entirely after getting caught up in competitions for status on their plantation. The Swiss planter Francis Philip Fatio and the Creek guide Yaolaychi, however, relied on engaging with Spanish officials to profit from their expertise. Still, the officials who depended on these informants struggled to extract useful facts from their written and oral narratives: Fatio's accounts exaggerated and obfuscated what he knew while Yaolaychi's stories interwove reports about minerals with mythology. Spanish imperialism did animate new investigations and interpretations of Gulf South nature, but this by no means guaranteed that any group in the borderlands, including the Spanish Empire, had the capacity to benefit from these insights. Efforts to acquire and

2. Scholars have revealed how British and French challenges to Spain's Atlantic empire inspired Spanish officials to launch new scientific initiatives as part of the so-called Bourbon reforms of the 1700s. They have also argued that these reforms, which stressed centralization, exacerbated competitions between creoles and peninsular Spaniards that encouraged Spanish American creoles to develop alternative ways of understanding nature and new political identities. Yet, in the Spanish Gulf South, the relationships that most influenced political and intellectual developments were not between colony and metropole but among the region's multiethnic residents. See Antonio Lafuente, "Enlightenment in an Imperial Context: Local Science in the Late-Eighteenth-Century Hispanic World," *Osiris*, 2d Ser., XV (2000), 155–173; and Jorge Cañizares-Esguerra, *How to Write the History of the New World: Histories, Epistemologies, and Identities in the Eighteenth-Century Atlantic World* (Stanford, Calif., 2001).

apply natural knowledge often revealed the limits of an individual's or empire's power.

Since the peoples of the Gulf South relied on each other for ways to acquire and benefit from natural knowledge, their efforts to translate information across cultural boundaries could infuse their written and oral accounts with material that was, at best, only partially understood. Many of these accounts were inherently ambiguous. Recipes for African herbal compounds that might heal or harm, reports that simultaneously shared and concealed data about natural resources, and monster stories rooted in both mythology and the history of colonialism confounded their audiences. Blacks, whites, and natives all attempted to derive actionable intelligence from narratives about the natural world that were, to a lesser or greater degree, anchored in intercultural encounters and misunderstandings, and they also retold or altered these stories in hopes of advancement or survival. Such narratives reflected and encouraged the crossing of cultural boundaries and, more importantly, informed the decisions of Spanish officials and borderland inhabitants alike.[3]

FROM *LA LOUISIANE* TO *LA LUISIANA*

Spanish ministers faced significant challenges as they prepared to take control of Louisiana after 1763. Indians dominated the province's vast interior, and power over the roughly fifty-five hundred whites (mostly creole French) and fifty-nine hundred blacks (largely of Senegambian descent) living in and around New Orleans was in the hands of French creole merchants and planters. Spanish policymakers, moreover, knew little about Louisiana's geography, native groups, or natural resources. Yet they were optimistic that two tried-and-true bases of Spanish imperialism—the zeal of scientifically minded officials and the tradition of negotiated authority—could overcome these obstacles and make Spanish Louisiana viable and valuable.[4]

The crown's appointment of Antonio de Ulloa as first governor of *la Luisiana* gave Spanish policymakers, cartographers, and naturalists cause to

3. On the narration and circulation of knowledge in colonial contexts, see especially Neil Safier, *Measuring the New World: Enlightenment Science and South America* (Chicago, 2008).

4. Gilbert C. Din, "Spanish Control over a Multiethnic Society: Louisiana, 1763–1803," in Jesús F. de la Teja and Ross Frank, eds., *Choice, Persuasion, and Coercion: Social Control on Spain's North American Frontiers* (Albuquerque, N.Mex., 2005), 49; Gwendolyn Midlo Hall, *Africans in Colonial Louisiana: The Development of Afro-Creole Culture in the Eighteenth Century* (Baton Rouge, La., 1995).

expect a flood of useful and curious knowledge about the region. Ulloa was one of several scientifically educated and reform-minded officials stationed throughout Spanish America's borderlands during the 1700s to make those posts more secure and profitable. These men initiated local projects, such as collecting potentially valuable specimens for Madrid's Royal Botanical Garden, that promised to advance their own careers while making colonists and the empire as a whole happier and better able to compete with Britain and France. But Ulloa's reputation far exceeded those of most, and perhaps all, of Spain's scientifically minded military and civil officials. In 1734, the crown had appointed Ulloa, then nineteen and fresh out of Cadiz's naval academy, to join Charles Marie de la Condamine's French geodesic mission to South America. Ulloa's 1748 publication of his astronomical, ethnographic, and natural historical observations from the expedition, including his discovery of platinum, earned him international recognition and a series of prominent naval and political positions. His professional credentials, fluency in French, experience working across imperial boundaries, and capacity to generate a broad range of scientific knowledge all made Ulloa seem like the ideal figure to order Louisiana and make its nature and peoples known to Spain.[5]

Effecting Spanish imperium had long depended on the cooperation of people in the colonies and adjusting rules to local realities, and Spain formalized this approach into policy for Louisiana. Carlos III's instructions to Ulloa were clear that Spanish rule would initially bring "no change in the system of [Louisiana's] government." Ulloa was left with the task of securing the collaboration of French soldiers, French officials, and Indian leaders. He had little choice: the force of ninety soldiers that accompanied him on his arrival in New Orleans on March 5, 1766, offered little coercive power. Ulloa tried to woo the province's remaining French soldiers into joining the Spanish military, but they refused and remained an independent source of power. He also refrained from officially claiming all his

5. Francisco de Solano Pérez-Lila, *La pasión de reformar: Antonio de Ulloa, marino y científico, 1716–1795* ([Cádiz], 1999), 13–14, 105–138; Weber, *Bárbaros*, 2; Daniela Bleichmar, *Visible Empire: Botanical Expeditions and Visual Culture in the Hispanic Enlightenment* (Chicago, 2012), 142–143; Juan José Saldaña, "Science and Public Happiness during the Latin American Enlightenment," in Saldaña, ed., *Science in Latin America: A History*, trans. Bernabé Madrigal (Austin, Tex., 2006), 51–92; John Preston Moore, *Revolt in Louisiana: The Spanish Occupation, 1766–1770* (Baton Rouge, La., 1976), 2–9; Charles Gayarré, *History of Louisiana: The French Domination: Two Volumes in One* (New York, 1866), II, 142–151.

powers as governor and kept the day-to-day supervision of the colony in the hands of the acting French governor, Charles-Philippe Aubry, in an effort to appease the inhabitants. Since winning the allegiance of Louisiana's Indians was one of Spain's highest priorities, Ulloa incorporated the French officials stationed in native towns into the Spanish bureaucracy in hopes of ensuring that the commercial and strategic relationships that had bolstered French Louisiana would endure. He also showered native leaders with gifts because Indian allies were essential to defending New Spain's frontiers, and the Choctaws, among other groups, "could destroy various settlements of the colony" if Spain failed to satisfy their demands.[6]

Although Ulloa did make strides toward gaining the attachment of Louisiana's Indians, its white leaders—a creole oligarchy of merchants, planters, and members of New Orleans' Superior Council—would not cooperate in their own colonization. United by a shared interest in amassing slaves and maintaining free (if illicit) trade, these men had held power since France effectively abandoned the colony in the 1730s. They resented Ulloa for implementing mercantile restrictions and excluding them from most of his political negotiations. In their minds, the governor's erudition in no way made up for his unwillingness to accommodate their demands. "Notwithstanding [Ulloa's] vast intellect, his talents, his learning, [and] his great reputation in all the academies of Europe," they believed, as Aubry put it, that, "instead of endeavoring to gain the hearts [of the people] (which is absolutely necessary in a change of government,) he has done all that could tend to alienate them." On October 29, 1768, the oligarchs' discontent erupted into open revolt when they riled up a crowd of about one thousand Louisianans that forced Ulloa to flee the colony. Although New Orleans' leaders briefly achieved independence, Spain crushed this precocious experiment in colonial revolution nine months later by temporarily foregoing negotiation in favor of force. The second governor of Louisiana, Alejandro O'Reilly, arrived with two thousand men—more than the white population of New Orleans—and executed the rebellion's ringleaders. Yet O'Reilly was eager to move past governance by coercion and soon pardoned most of the men involved in the revolt, loosened commercial policy, and ushered in a

6. "Royal Decree Commissioning Don Antonio de Ulloa Governor of Louisiana," May 21, 1765, AGI, SD, 86-5-21, in Lawrence Kinnaird, ed., *Spain in the Mississippi Valley, 1765–1794*, 3 vols. (Washington, D.C., 1946–1949), II, 1 ("no change"), "Governmental Expenses," 1767, AGI, SD, 86-6-6, II, 18 ("could destroy"); Weber, *Bárbaros*, 8–9; Moore, *Revolt in Louisiana*, 15, 42–58.

thirty-year period in which, despite a few hiccups, Spanish governors and Louisiana's merchants and planters cooperated in running the colony.[7]

Although O'Reilly ensured Spanish governance in Louisiana, it was Ulloa who integrated the colony's environments and peoples into intellectual conceptions of Spanish America. After returning to Spain, Ulloa pursued scientific projects that took advantage of his broad experiences throughout the Americas: he drafted a questionnaire for compiling a "complete knowledge of the Geography, Physics, Antiquities, Mineralogy, and Metallurgy of the Kingdom of New Spain" (Louisiana was part of the viceroyalty of New Spain) and published his *Noticias americanas* in 1772. The *Noticias* incorporated the knowledge Ulloa had accrued in Louisiana through a census, his discussions with French officials, experiments in Louisiana's crown-owned botanical garden, and his extensive travels to elaborate sweeping arguments about the geography, natural history, and ethnography of Spanish America and the planet as a whole. For example, he challenged grand geographic visions about the "rules of the climates" by comparing weather in Louisiana, Cuba, and Peru and finding that New Orleans was both hotter in summer and cooler in winter than Havana, a circumstance that only made sense in light of topography and undermined simplistic notions that temperature was directly proportional to distance from the equator.[8]

7. [Charles-Philippe] Aubry to [César Gabriel de Choiseul], duke of Praslin, Nov. 25, [1768], in Gayarré, *History of Louisiana,* II, 224 (quotes). A British official in West Florida who exchanged natural knowledge about the Mississippi Valley with Ulloa believed that the French inhabitants' hostility to the governor reinforced their reputation for ignorance and laziness, writing that Ulloa was "what the French call *une Esprit Parfait* . . . whose activity has already reflected much disgrace on the Indolence of the French." See George Johnstone to Hutchinson Muso, July 19, 1766, 7, Kislak Collection, MS 100, LOC. On the 1768 revolt, see Moore, *Revolt in Louisiana,* esp. 143–216; and Shannon Lee Dawdy, *Building the Devil's Empire: French Colonial New Orleans* (Chicago, 2008), 219–232.

8. "Compendio de las noticias que S.M. por su real ordén de 20 de Octubre proximo pasado ordena que se puntualisen para el completo conocimiento de la geografía, física, antiguedades, mineralogía y metalurgía de este reyno de Nueva España," Jan. 23, 1777, AGI/S Indiferente, L. 1549. fol. 1, quoted in Paula S. De Vos, "Research, Development, and Empire: State Support of Science in the Later Spanish Empire," *Colonial Latin American Review,* XV (2006), 66 ("complete"); Antonio de Ulloa, *Noticias americanas: Entretenimientos phisicos-historicos sobre la América meridional, y la septentrianal oriental* . . . (Madrid, 1772), 57–85, 67 ("rules"); Solano Pérez-Lila, *La pasión de reformar,* 205–206; Moore, *Revolt in Louisiana,* 66; Gayarré, *History of Louisiana,* II, 133–134, 164, 175–177; Alejandro O'Reilly to Julián de Arriaga, Oct. 17, 1769, AGI, SD, 80-1-7, in Kinnaird, ed., *Spain in the Mississippi Valley, 1765–1794,* II, 99–100.

But Ulloa was not just concerned with theory; he applied these observations toward developing Louisiana and making it valuable to the Spanish Empire. Lower Louisiana's widely fluctuating temperatures led him to the reluctant conclusion that it was not fit for sugar plantations, and, while indigo and tobacco would fare better, they were nevertheless inferior to those grown in Central America and Cuba, respectively, so would not bolster imperial commerce on the whole. Yet Louisiana was an ideal location for discovering and experimenting with other potentially valuable plants because, as in Peru, "One finds some things that are common around the *Equator,* and others ... equal to those of Europe, even though they are [all] in the same Country." Ulloa integrated local residents' knowledge into many of his botanical descriptions, such as noting that creoles used the "herb called *Stinky*" to kill intestinal worms and Indian women took snakeroot to improve the flow of breast milk. He added that "a catalog of Plants and their virtues and method of applying them [should be] made for each Country, so that they could be known and used." Not one to rely exclusively on local informants, Ulloa also argued that "the fields of Louisiana are very fertile and abound with plants: a circumstance that ought to lead to experimenting." His experimental agenda included finding a better way to extract wax from wax trees and testing samples of ginseng. These and other Louisianan plants would make the people of the Spanish Empire wealthier and healthier and, as an added bonus, less reliant on Dutch, English, and French imports.[9]

Ulloa's study of Spanish America's native peoples challenged facile theories about human difference and—by analyzing violence and mortuary practices—incorporated the Indians of Louisiana as subjects of Spanish rule and ethnography. He remarked that while there had been "no shortage of those intent on solving the variety of human colors," no theory had adequately accounted for the diversity among *"Whites, Blacks, and Reds."* By comparing the skin tone of Indians from Louisiana and Peru, Ulloa found that climate failed to explain the variety he observed. But he was nevertheless eager to assert that all natives shared common traits and to apply his ethnographic opinions about Gulf South Indians to all Amerindian groups. He considered Louisianan burial mounds proof that natives throughout the Americas honored their dead, adding to a long intellectual tradition of

9. Ulloa, *Noticias americanas,* 85, 98 ("One finds"), 112 ("herb"), 114 ("fields"), 119 ("a catalog"), emphasis in original. On botany's role in imperial commerce and competition, see Londa Schiebinger and Claudia Swan, eds., *Colonial Botany: Science, Commerce, and Politics in the Early Modern World* (Philadelphia, 2005).

studying the bones and mortuary practices of North America's natives in the context of Peruvian remains and tombs. He also looked to Louisiana Indians' practice of scalping not only to suggest that "in general this Nation is inhumane" but that Peruvian Indians would behave just as barbarously as "those of *Louisiana* [and] *Florida*" if Spanish laws did not force them to maintain a pretense of civility. Louisiana Indians, according to Ulloa, proved that Spanish rule in Peru was necessary and benevolent.[10]

As a governor, Ulloa's administration was too weak to impose Spanish rule. As a man of science, his observations and experiments helped redefine Louisiana's nature and natives as parts of, and justifications for, Spain's American empire. But Ulloa had the luxury of no longer being governor by the time he published these scientific writings, leaving other bureaucrats with the ongoing challenge of acquiring and applying knowledge that might turn this tenuously held border region into a prized part of the Spanish Empire.

THINKING GLOBALLY, ACTING LOCALLY

Since the sixteenth century, the production of imperially useful knowledge in the Spanish Empire had been organized largely through networks of bureaucrats. Some of these officials worked on their own initiative while others responded to questionnaires and instructions drafted by their superiors. Civil and military officials in the eighteenth-century Gulf South penned reports on the region's resources and inhabitants, conducted large-scale agricultural experiments, and organized natural history collections that shaped imperial policy and contributed to European science. As in other Spanish colonies, scientific projects in Louisiana emerged in overlapping provincial and international contexts as bureaucrats drew on their local experiences to develop their colonies and improve the empire's economic and strategic position globally.[11]

10. Ulloa, *Noticias americanas,* 305 *("Whites"),* 306 ("no shortage"), 312 ("in general"), 313 ("those of"), emphasis in original. On how Ulloa's earlier writings disparaging Quito's natives (such as his *Relación histórica del viage a la América meridional* [1748]) sparked debate among Europeans about the nature of Amerindians and ethnography alike, see Safier, *Measuring the New World,* 166–199. On Peruvian and North American remains, see Christopher Heaney, "A Peru of Their Own: English Grave-Opening and Indian Sovereignty in Early America," *WMQ,* 3d Ser., LXXIII (2016), 609–646.

11. De Vos, "Research, Development, and Empire," *Colonial Latin American Review,* XV (2006), 55–79; María M. Portuondo, *Secret Science: Spanish Cosmography and the New World* (Chicago, 2009); Daniela Bleichmar, "Atlantic Competitions: Botany in the

Martín Navarro's "Reflexiones políticas sobre el estado actual de la provincia de la Luisiana" (1782) was one of many reports penned by Spanish bureaucrats in the Americas to reform Spanish policy. Navarro was among the handful of officials who stepped off the boat with Antonio de Ulloa in 1766, and, unlike the governor, he enjoyed a long and successful career in the province. His trilingualism and competence attracted official attention, and Navarro eventually rose to intendant, the second highest colonial official, and made a fortune in land and slaves. Navarro's overriding concern throughout his twenty-two years in Louisiana was to improve agriculture and commerce and, therefore, attract a loyal population that would defend New Spain against Anglo ambitions. His "Reflections," which were based on his experiences during more than "fifteen years [of] serving His Majesty in this province," argued that Louisiana, unlike the rest of Spanish America, ought to have free trade and be open to immigration from people of all nations, including Protestants. Commerce and immigration depended on agriculture, and Navarro looked to both the Atlantic and the West Indies for settlers whose expertise could properly exploit this simultaneously Caribbean and North American province: migrants from Europe and the Canaries could cultivate wheat and linen while sugar planters from Santo Domingo and Martinique could leave their insect-ridden fields and, with their slaves in tow, establish plantations in Louisiana at the crown's expense. Although it would ultimately take the Haitian Revolution to attract a significant number of Caribbean planters to Louisiana, Navarro's "Reflections" did lead to sweeping reforms in Spanish policy. The crown allowed Protestant foreigners to settle in the colony and implemented the loosest trade restrictions in Spanish America. Navarro claimed that "there is no other province ... in America that should occupy the Ministry's attention as much as Louisiana" and that his inspiration for this empire-altering report derived "only [from] the zeal of a recognized honored patriot." For Navarro and many other colonial officials, the most patriotic projects promoted both local and imperial agendas.[12]

Eighteenth-Century Spanish Empire," in James Delbourgo and Nicholas Dew, eds., *Science and Empire in the Atlantic World* (New York, 2008), 237–245.

12. Martin Navarro, "Reflexiones políticas sobre el estado actual de la provincia de la Luisiana," 1782 (Biblioteca Nacional,—Mss. de Ultramar, no. 13), in M[anuel] S[errano] y S[anz], [ed.], *Documentos históricos de la Florida y la Luisiana, siglos XVI al XVIII* (Madrid, 1912), 361, 367 ("there is," "only [from]"), 368–369, 373, 374 ("fifteen years"), 378, 379. On Navarro, see Brian E. Coutts, "Martín Navarro: Treasurer, Contador, Intendant, 1766–1788: Politics and Trade in Spanish Louisiana" (Ph.D. Diss., Louisiana

Navarro cultivated and collected plants that, he hoped, would make the colony more prosperous and beneficial to Spain. From 1781 to 1786, he oversaw large-scale agricultural experiments that emerged, and ultimately failed, amid tensions between local experience and imperial visions. Spanish ministers sent flax experts from Granada to Louisiana and encouraged the cultivation of tobacco; but the flax experts proved unwilling to adapt their methods to the wet climate, and the viceroy of New Spain stunted tobacco exports by enforcing impractical packaging standards. Navarro fought against the viceroy's commercial restrictions and drew on his own observations and those of his regional network of commandants and planters to identify the best lands for growing these commodities. These initiatives worked, and flax and tobacco were well on their way to becoming significant staples when metropolitan ministers cut these projects short. Intent on balancing agriculture and commerce on an imperial scale, policymakers discouraged Louisianans from cultivating and exporting products that duplicated, or were inferior to, those from other parts of the empire.[13]

It was this same political economy that inspired Navarro's dedication to botany. Between 1783 and 1787, Navarro collected specimens and wrote detailed descriptions of "the species, qualities, and virtues of [Louisianan] plants." He carefully packaged them for transport to officials and scientific institutions in Spain in a 1786 shipment of nine boxes—eight of plants, one of seeds—and a 1787 delivery that included four boxes of plants, one of seeds, and twenty-one stuffed birds destined for Madrid's Royal Natural History Cabinet. At one level, Navarro was fulfilling instructions from Casimiro Gómez-Ortega, director of the Royal Botanical Garden, to ship

---

State University, 1981), esp. 5–6, 19, 65, 157, 177, 244, 324. For other such reports and how they shaped Spanish policy, see Francisco Bouligny, *Louisiana in 1776: A Memoria of Francisco Bouligny*, trans. and ed. Gilbert C. Din (New Orleans, 1977), esp. 29–38; Juan Filhiol, "Description of the Ouachita in 1786," trans. H. Wynn Ricky (1935), Juan Filhiol Papers: Reports, Correspondence (1783–1786, 1834–1835), J. Fair Hardin Collection, LLMVC; Carl A. Brasseaux, H. Dickson Hoese, and Thomas C. Michot, "Pioneer Amateur Naturalist Louis Judice: Observations on the Fauna, Flora, Geography, and Agriculture of the Bayou LaFourche Region, Louisiana, 1772–1786," *Louisiana History*, XLV (2004), 77–103; Esteban Miró, "Descripción de la Luisiana, 1792," in Jack D. L. Holmes, ed., *Documentos inéditos para la historia de la Luisiana, 1792–1810* (Madrid, 1963), 1–64, Joseph Piernas, "Proyecto de una nueva población en el Rio de Calcasieu en la provincia de la Luisiana, 1795," 131–169; William S. Coker, ed., *John Forbes' Description of the Spanish Floridas, 1804*, trans. Vicki D. Butt [et al.] (Pensacola, Fla., 1979); Helen Hornbeck Tanner, *Zéspedes in East Florida, 1784–1790* (Jacksonville, Fla., 1989), 137–148; and Hilton, "Spanish Louisiana in Atlantic Contexts," in Vidal, ed., *Louisiana*, 72–75.

13. On these agricultural experiments, see Coutts, "Martín Navarro," 448–486.

specimens of "the most special trees" and "medicinal or particularly beautiful plants" to Spain. Naturalists in Madrid were increasingly interested in obtaining physical specimens of American plants, and Spanish leaders saw botany as a way to decrease Spain's imports, strengthen its position in inter-imperial competitions, and legitimize their self-image as participants in the Enlightenment. Navarro was eager to forward these imperial agendas, but, as with other botanical collectors in the eighteenth-century Spanish Atlantic, his diligent attention to Gómez-Ortega's instructions also reflected local and personal goals: he hoped the plants he collected would become a source of wealth, attract settlers, and advance his own career. In the process, he helped make New Orleans into one of Spanish America's many scientific centers, cities from which officials organized research about their colony's resources (often through written instructions), compiled and analyzed local information, and used this knowledge to develop the province and empire alike. Although far smaller than Lima or Mexico City, New Orleans was nevertheless a node that blurred local / global and center / periphery distinctions in Spanish imperial science.[14]

Navarro drew on, organized, and integrated regional networks of North American informants and Atlantic networks of mariners to execute this botanical project. He traveled across much of the Gulf South in 1784 to negotiate alliances with the Creeks, Chickasaws, and Choctaws, and these encounters led to exchanges in which Navarro gifted measuring instruments to native leaders and received a wealth of botanical information and specimens. Navarro's descriptions of the plants he sent to the minister of the Indies, José de Gálvez, noted how Indians used the ashes of prickly ash trees to "destroy funguses in wounds" and made a fragrant resin from wax tree seeds that, according to his informants, only germinated in a martinet's stomach. Navarro's many creole friends and subordinates also provided him with living and preserved plants, including the "herb that coagulates water." Furthermore, he seems to have had some access to Anglo-American knowledge networks. Navarro described how "the doctors of Virginia produced

14. Martín Navarro to José de Gálvez, Feb. 22, 1786, AGI, SD, leg. 2610, no. 362, reel 74, 390 ("species"), Navarro to Gálvez, June 2, 1783, AGI, SD, leg. 2609, no. 171, reel 72, 404 ("most special"), Navarro to Gálvez, Mar. 26, 1787, AGI, SD, leg. 2611, no. 480, reel 74, 142; Ramon Rivera to Antonio Porlier, Aug. 17, 1787, Expedición botánica a los reinos del Perú y Chile, 1777–1787, reel 6, no. 126, AMNCN. On nodes of science, see Bleichmar, "Atlantic Competitions," in Delbourgo and Dew, eds., *Science and Empire*, 225–252; De Vos, "Research, Development, and Empire," *Colonial Latin American Review*, XV (2006), 55–79; and Antonio Lafuente and Nuria Valverde, "Linnaean Botany and Spanish Imperial Biopolitics," in Schiebinger and Swan, eds., *Colonial Botany*, 134–147.

very terrible effects" by improperly dosing patients with Indian pinkroot, and he offered Madrid naturalists his own assessment about the best way to administer this potent vermifuge.[15]

Navarro also organized the circulation of Louisianan specimens from New Orleans across the Atlantic. The instructions he provided for ship captains carrying "the boxes of plants and trees for [the King]" went into exacting detail about how to protect the living specimens from excessive heat, saltwater, and the cats, rats, and sailors that might try to dig them up. Since "each one of the trees that arrives green, will bring honor to the conductors and be of utility to the nation," the instructions made it clear that "all good patriots" ought to treat these plants as precious cargo.[16]

Imperial competition inspired official research and reports that, in turn, provided specimens and information that influenced imperial policies. Bureaucrats like Navarro, who worked at the intersection of North American and Atlantic networks of knowledge and goods, were pivotal to this method of ordering empire. Indeed, the correspondence of officials like Navarro could make it seem like information always traveled easily both within and beyond the Gulf South. But other colonial records, including court cases that captured multiple voices and the writings of white settlers, make it clear that on-the-ground realities—including Spain's regional weakness—could hamper intellectual exchanges and limit officials' ability to translate local expertise into policy. In the Gulf South, competitions among officials, slaves, planters, and natives did at least as much to shape the production and circulation of natural knowledge as the global contest among empires.

15. Martín Navarro, "Clase, y calidades de los árboles, arbustos, plantas, y semillas que comprehende esta noticia," enclosed with Navarro to Gálvez, Feb. 22, 1786, 393 ("destroy"), 395 ("herb"); Coutts, "Martín Navarro," 365–376.

16. "Ynstruccion que deberan observar los capitanes de los buques que ban encargados de la condución de los caxones de plantas, y arboles para S.M.," enclosed with Navarro to Gálvez, Feb. 22, 1786, 391 (quotes). On similar instructions meant to regulate the transoceanic shipment of natural history specimens in the French and British Empires, see Christopher M. Parsons and Kathleen S. Murphy, "Ecosystems under Sail: Specimen Transport in the Eighteenth-Century French and British Atlantics," *Early American Studies,* X (2012), 503–529. Navarro would remain a transatlantic broker of natural history after his 1788 retirement to Barcelona, negotiating an exchange of crystals and ores from Britain and the East Indies for "the gold and silver ores of Spanish America" with an Oxford geologist. See W[illiam] Thomson to M[artín] Navarro, Feb. 3, 1790, Real Gabinete, 1787–1815, reel 53, no. 117, AMNCN.

POISONING RELATIONSHIPS

In 1773, five black men enslaved in Spanish Louisiana found themselves on trial in New Orleans for conspiring to poison an overseer. The lawyers and officials assigned to the case probably thought this would be a pretty straightforward affair: as whites throughout the Americas knew and feared, enslaved blacks had both the botanical know-how to formulate poisons and plenty of motive to use them against their tormenters. But Spanish officials found few easy answers in 1773. According to courtroom testimonies and chemical experiments, the substance the accused men had concocted hardly seemed like a conventional poison. And instead of using the compound to kill white masters, the plotters claimed that they planned to administer it to a higher-ranking slave. Although the complexities of the case baffled Spanish officials, records from the trial offer a rare glimpse into the intellectual world of a Louisiana plantation. Plantations were among the many zones of interaction created by imperialism in the Gulf South, and the 1773 trial suggests that rivalries among slaves, African natural and spiritual expertise, and blacks' access to Louisiana nature all combined to influence how enslaved blacks pursued and applied knowledge.[17]

Plantation slavery had been central to Louisiana society since the French era. Blacks outnumbered whites in the province since the 1730s, and whites depended on blacks' labor and expertise to pursue their vision of turning lower Louisiana into a Caribbean-style plantation society. For example, French Louisianans imported black indigo specialists from the Caribbean and, it seems, specifically sought Senegambian slaves proficient in the environmental engineering skills needed to turn wetlands into rice plantations.

17. Throughout the Atlantic world, black healers, herbalists, and other experts challenged the conditions of slavery and the European visions of modernity that sought to naturalize their oppression. Poisoning was the most direct and, for whites, terrifying way that enslaved blacks used what they knew about nature to contest colonial social hierarchies. See Stephan Palmié, *Wizards and Scientists: Explorations in Afro-Cuban Modernity and Tradition* (Durham, N.C., 2002); Laura de Mello e Souza, *The Devil and the Land of the Holy Cross: Witchcraft, Slavery, and Popular Religion in Colonial Brazil* (Austin, Tex., 2003); Jason R. Young, *Rituals of Resistance: African Atlantic Religion in Kongo and the Lowcountry South in the Era of Slavery* (Baton Rouge, La., 2007); James H. Sweet, *Domingos Álvares, African Healing, and the Intellectual History of the Atlantic World* (Chapel Hill, N.C., 2011); Diana Paton, "Witchcraft, Poison, Law, and Atlantic Slavery," *WMQ*, 3d Ser., LXIX (2012), 235–264; and Pablo F. Gómez, *The Experiential Caribbean: Creating Knowledge and Healing in the Early Modern Atlantic* (Chapel Hill, N.C., 2017).

The dream of emulating Caribbean plantation society became even more pronounced during the Spanish period. Louisiana's population was about 55 percent black throughout the Spanish era, and, thanks to the most successful of Spain's efforts to populate Louisiana and develop its economy, 60 to 75 percent of the enslaved were newly arrived from Africa.[18]

The 1773 poisoning case reveals that a constellation of place-specific power relationships among Africans, creole blacks, planters, overseers, and officials could affect how enslaved blacks used and transmitted knowledge. The events that would land the five enslaved men in court began when Francisco, a black "creole of Illinois" and second overseer on Francisco Bellile's indigo plantation, decided he wanted to kill Augustin, another black creole and the plantation's first overseer. Francisco's goal was to eliminate Augustin in order to take his place in the plantation's chain of command. But Francisco's plot to climb the social ladder would only work if he avoided detection, so he sought the help of Carlos and Cipion, two Mande-speaking slaves who had recently been imported from Upper Guinea, and promised them new clothes if they concocted a substance to eliminate Augustin on the sly. They came to agree that an alligator, a convenient Louisianan substitute for an African crocodile, was the key ingredient. The enslaved hunter they recruited from a neighboring plantation, Bernardo, reported that he was unable to get them a live alligator (a dubious claim that suggests he did not want to collaborate with the conspirators). So Carlos, Cipion, and Francisco found a dead alligator, collected its heart and gall, and combined these with herbs to form a paste that, once dried, would be a powerful *grisgris,* a term that could have meant a poison or a power-endowed object. But Big Luis, the plantation's enslaved blacksmith, ratted the three conspirators out to Bellile before their compound had finished curing, and the planter had Francisco, Cipion, Carlos, and Bernardo arrested and imprisoned in New

18. On black indigo experts, see Tivas de Gourville to [Jérôme Phélypeaux], comte de Pontchartrain, June 1712, in Dunbar Rowland and Albert Godfrey Sanders, eds., *Mississippi Provincial Archives,* II, *French Dominion: 1701–1729* (Jackson, Miss., 1929), 69–70; Jean-Françoise-Benjamin Dumont de Montigny, *The Memoir of Lieutenant Dumont, 1715–1747: A Sojourner in the French Atlantic,* ed. Gordon M. Sayre and Carla Zecher, trans. Sayre (Chapel Hill, N.C., 2012), 380–382; and Pierre F. X. de Charlevoix, *Charlevoix's Louisiana: Selections from the History and the Journal,* ed. Charles E. O'Neill (Baton Rouge, La., 1977), 33. On Senegambian rice experts, see Hall, *Africans in Colonial Louisiana,* 29, 34, 121–127. On Spanish Louisiana's black population, see ibid., 277, 286; and Daniel H. Usner, Jr., *Indians, Settlers, and Slaves in a Frontier Exchange Economy: The Lower Mississippi Valley Before 1783* (Chapel Hill, N.C., 1992), 112–113, 116.

Orleans where, in accordance with Spanish law, they were tried and provided with legal defense.[19]

Louisiana governor Luis de Unzaga took slave poisons seriously, and he sought to uncover all possible details about the grisgris and the men who created it as part of a larger effort to curb the threat that the black majority posed to whites' fragile supremacy in Louisiana. The governor's fears were merited. Slaves from Upper Guinea had been poisoning whites in Louisiana since the 1720s, and officials and physicians in French Louisiana, like those during the Spanish period, had tried to make sense of these compounds through chemical tests, experiments, and autopsies. The 1773 case was one of at least four other poisoning trials in the lower Mississippi Valley during Spanish dominion, and, as in many other slave societies, these cases raised the specter of mass uprisings and thus inspired brutal punishment. The court threatened Big Luis with torture for merely hesitating to reveal the plot because, according to officials, "the knowledge of the conspiracy had reached all the negroes and became an outcry among them" by the time he told Bellile. The actual conspirators had even more cause to despair, particularly after the prosecuting attorney pressed to make their "punishment ... serve as a public example." Spanish officials enforced spectacular sentences against blacks who threatened their still-young authority in Louisiana, including entombing the body of one slave who killed his master in a leather bag with the remains of a monkey, dog, snake, and rooster in a cruel mockery of a Mandinga pouch, a charm that many Mande-speakers viewed as a tool of resistance.[20]

19. Don Francisco Bellile Legal Case, 1773, transcribed and translated from Judicial Records of the Spanish Cabildo, LSM, no. 77306121, by Laura Porteous, M625, 36–55 (quote 44), LRC. A narrative summary of this source has been published as Laura L. Porteous, "The Gri-Gri Case," *LHQ*, XVII (1934), 48–63. Nearly all plantation overseers in Louisiana were enslaved black creoles, and slaves throughout the Americas tended to use poisons against each other more often than against their owners. See Ira Berlin, *Many Thousands Gone the First Two Centuries of Slavery in North America* (Cambridge, Mass., 1998), 200; Susan Scott Parrish, *American Curiosity: Cultures of Natural History in the Colonial British Atlantic World* (Chapel Hill, N.C., 2006), 278; and Philip D. Morgan, *Slave Counterpoint: Black Culture in the Eighteenth-Century Chesapeake and Lowcountry* (Chapel Hill, N.C., 1998), 614. On the legal rights of slaves, see Gilbert C. Din, *Spaniards, Planters, and Slaves: The Spanish Regulation of Slavery in Louisiana, 1763–1803* (College Station, Tex., 1999), 53.

20. Don Francisco Bellile Legal Case, 51 ("knowledge"), 52 ("punishment"); Hall, *Africans in Colonial Louisiana*, 163; Derek Noel Kerr, "Petty Felony, Slave Defiance, and Frontier Villainy: Crime and Criminal Justice in Spanish Louisiana, 1770–1803" (Ph.D.

Although Spanish sources described the 1773 grisgris as a kind of biological poison, West Africans and their descendants viewed grisgris more broadly as a category of force-endowed objects—including amulets, powders, and pouches containing herbs, animal parts, or Koranic or biblical scripture—with the power to protect or cause harm. Most black creoles in colonial Louisiana were descended from Senegambian Mande-speaking people who shared much of their culture and cosmology with the Upper Guineans who contributed to the "re-Africanization" of Louisiana during the Spanish era.[21] According to Antoine-Simon Le Page Du Pratz, who observed the colony's enslaved blacks during the 1720s and 1730s, West Africans in French Louisiana were "very superstitious and attached to their prejudices and to the charms they call *gris-gris*." Anthropologists suggest that Mande charms like grisgris were powerful because they harnessed *nyama*, the life force that all beings possess. Some specialists, including blacksmiths, were endowed with an extraordinarily strong nyama that facilitated esoteric expertise while others, like hunters and herbalists, could harness the nyama of animals through their remains. These animal remains had their own active power that blurred the lines between a passive nature and one with agency. Crocodiles and, presumably, alligators were particularly rich with nyama and presented Mande experts in Upper Guinea and Louisiana with ready sources of power that, like other plant or spirit-based knowledge, could be used to heal or harm. *Boro*, for example, meant both "medicine" and "poison" in the Mandinka language, and such uncertainties perplexed blacks and whites alike. When testifying in their own defense,

---

Diss., Tulane University, 1983), 152–156; Andrew McMichael, *Atlantic Loyalties: Americans in Spanish West Florida, 1785–1810* (Athens, Ga., 2008), 102–126; Din, *Spaniards, Planters, and Slaves*, 52–54, 60; Usner, *Indians, Settlers, and Slaves*, 107–138; Souza, *Devil and the Land of the Holy Cross*, 129–141; Young, *Rituals of Resistance*, 118–126; Matt Schaffer, "Bound to Africa: The Mandinka Legacy in the New World," *History in Africa*, XXXII (2005), 349–350.

21. Hall, *Africans in Colonial Louisiana*, 277–302 ("re-Africanization," 302). Gwendolyn Midlo Hall's suggestion that the Mande-speakers imported to French Louisiana largely shared a "Bambara" culture and ethnic identity that previously existed in Africa and traveled to Louisiana has been criticized. "Bambara" might have been a new ethnic identity developed among diverse West Africans during the Middle Passage and enslavement in Louisiana. See Stephan Palmié, review of *Africans in Colonial Louisiana: The Development of Afro-Creole Culture in the Eighteenth Century*, by Gwendolyn Midlo Hall, *Africa: Journal of the International African Institute*, LXIV (1994), 168–171. See also Schaffer, "Bound to Africa," *History in Africa*, XXXII (2005), 322–323, 328; and John Thornton, *Africa and Africans in the Making of the Atlantic World, 1400–1800*, 2d ed. (Cambridge, 1999), 187–189.

Bernardo and Cipion claimed that Francisco had tricked them into thinking that the alligator remains were for a healing ointment, not "a 'gri gri' to remove the overseer." European officials in many slave societies recognized the close relationship between healing and hurting in African pharmacopeia and, lacking the power to differentiate these practices by European standards, banned slaves from practicing any kind of medicine on pain of death. For whites, the ambiguities inherent in accounts of African compounds justified oppressive policies.[22]

The enslaved men involved in the 1773 plot agreed that they were making a grisgris but had no shared understanding of what that meant or why it would work. Lorenzo, one of the enslaved creoles who revealed the conspiracy, described Guinea-born Carlos as the expert who created the grisgris "to kill the overseer" and believed that once the mash of alligator parts and herbs had dried into a powder "all of those who drink it would die." Lorenzo equated the grisgris with a poison, and, more specifically, his account suggests some familiarity with Mande *kòròtiw*, harmful powders placed in drinks or sent by winds. Cipion, one of the two conspirators from Upper Guinea, claimed in court that he knew nothing about making grisgris, but Francisco's testimony hinted that Cipion did have specialized knowledge because, after collecting the alligator, Cipion supposedly said "now I have the 'utility' give me a knife to take out the heart." Cipion—or,

22. [Antoine-Simon] Le Page Du Pratz, *Histoire de la Louisiane . . .*, 3 vols. (Paris, 1758), I, 334 ("very superstitious"); Don Francisco Bellile Legal Case, 41–43, 47 ("a 'gri gri,'" 42). On grisgris, see Hall, *Africans in Colonial Louisiana*, 45; Patrick R. McNaughton, *The Mande Blacksmiths: Knowledge, Power, and Art in West Africa* (Bloomington, Ind., 1993), 58, 62; Jane G. Landers, *Atlantic Creoles in the Age of Revolutions* (Cambridge, Mass., 2010), 62; Young, *Rituals of Resistance*, 126. On nyama, see McNaughton, *Mande Blacksmiths*, 15–17, 126; Hall, *Africans in Colonial Louisiana*, 51–52; Young, *Rituals of Resistance*, 114; and Pascal James Imperato, *African Folk Medicine: Practices and Beliefs of the Bambara and Other Peoples* (Baltimore, 1977), 31–34. On boro, see Schaffer, "Bound to Africa," *History in Africa*, XXXII (2005), 342. On the power of African medicinal experts and conjurers to both heal and harm, see McNaughton, *Mande Blacksmiths*, 12; and Parrish, *American Curiosity*, 295. For a description of some African formulas that incorporated animal parts to heal or harm, see João Curvo Semmedo, "Memorial de varios simplices," in *Polyanthea medicinal: Noticias galenicas e chymicas repartidas em tres tratados* (Lisbon, 1716), 11 (my thanks to Benjamin Breen for this source). On bans against African medicine, see Hall, *Africans in Colonial Louisiana*, 164–165; James E. McClellan, III, *Colonialism and Science: Saint Domingue in the Old Regime* (Baltimore, 1992), 136; and Richard B. Sheridan, *Doctors and Slaves: A Medical and Demographic History of Slavery in the British West Indies, 1680–1834* (Cambridge, 1985), 96.

maybe, Francisco—might have understood this "utility" to be similar to a *daliluw,* what one twentieth-century Mande blacksmith described as "the thing that can make something work." Such things included herbs, animal matter, or words that packaged nyama and enabled specialists to use it toward particular goals. Carlos, whose account was filtered through a court-appointed "Mandringa" interpreter and two European ones, held that it was Francisco and Cipion who "made the 'gri gri'" and that he knew "it was the gall of the crocodile with the heart [but] that he did not know how it was made to kill people." These diverse understandings of the grisgris, which ranged from Mande esoteric knowledge to hybridization and ignorance, defy any easy characterization about the persistence of specific aspects of African natural knowledge in the New World.[23]

Yet the grisgris case does indicate that standards of intellectual authority among Louisiana blacks both derived from Africa and emerged amid encounters engendered by imperialism. Place, where one could go and where one was from, structured colonial blacks' knowledge networks just as it did those that connected colonial whites with Europeans. In both cases, Old World specialists valued individuals with access to local specimens who, in turn, looked to those specialists to interpret and use the materials they provided. Such exchanges could improve the social standing of individuals in both groups. The enslaved Louisianans with the most access to nature beyond the plantation were those with the most freedom, including hunters like Bernardo, cypress cutters, runaways, and creole overseers like Francisco. Thus Francisco could procure the necessary ingredients but depended on African experts to make what he called "the drugs of their country." He told the court that "it was agreed ... that with a crocodile's heart they would prepare [the grisgris] but that he could not be instructed in the preparation because they would not tell him." "As a creole," Francisco explained, "he did not know this drug which the lately imported negroes had knowledge of." After several weeks in prison, Francisco altered his story and blamed the creole blacksmith Big Luis for instigating the plot, which now included poisoning their white master as well as Augustin, and testified that he had "treated many times with Big Luis" over how to make and administer "the drug" but relied on Carlos and Cipion to produce it. Big Luis insisted that Francisco implicated him out of revenge for revealing the plot, but it is possible that Francisco would have considered Big Luis qualified to

23. Don Francisco Bellile Legal Case, 38 ("kill," "all of those"), 42 ("gall"), 48 ("now I have"), 49 ("Mandringa"), 50 ("made"); McNaughton, *Mande Blacksmiths,* 42–46 ("the thing," 43). On *kòròtiw,* see ibid., 44.

help develop the grisgris because he was a blacksmith and thus, according to many Mande-speakers, had exceptional powers to manipulate nature, social relationships, and (of course) metal. If Big Luis did inherit the esoteric knowledge of Mande blacksmiths, or if Francisco knew enough about this tradition to seek out his skills alongside those of men born in Guinea, it would suggest that black creole Louisianans maintained a significant appreciation of West African expertise.[24]

One of the central roles of African experts in the Americas was as healers who both treated ailments and worked to build new communities that restored at least some of the social harmony that enslavement destroyed. Since the 1773 trial centered on accusations of poisoning, the vast majority of the testimony concerned the potentially harmful applications of the substance and not its potential role in healing. But Bernardo and Cipion both claimed that they had sought the alligator to procure its fat for use as a balm to ease a pain in Francisco's leg. It is possible that Francisco (a creole) and Cipion (an African) had misunderstood each other: perhaps both agreed to make a grisgris, but Francisco understood this to mean a deadly tool to employ for selfish ends while Cipion believed he was creating a medicine, one that he might have even considered a small step toward coalescing a new community on Bellile's plantation. Cipion's inability to speak French or Spanish, Francisco's loquacity during the trial, and Louisiana officials' predisposition to see any slave-made compound as dangerous, however, combined to all but silence the possibility that Cipion saw himself as a personal and communal healer instead of a self-interested poisoner.[25]

Spanish coercive power, blacks' investments in colonial society, and the very language needed to narrate knowledge all limited learned blacks' capacity—and even desire—to establish themselves as experts who could heal communities or challenge European authority. Considering the savage punishments Louisianan courts inflicted against other accused slaves, Cipion and Carlos no doubt feared torture and execution and, thus, accused each

24. Don Francisco Bellile Legal Case, 41 ("the drugs"), 42, 48 ("it was agreed"), 49 ("treated"); McNaughton, *Mande Blacksmiths*, 3, 41–42, 47. On exchanges among white creoles in America and European natural philosophers, see Parrish, *American Curiosity*, esp. 103–134. On enslaved individuals benefitting from access to forests, see Ras Michael Brown, "'Walk in the Feenda': West-Central Africans and the Forest in the South Carolina-Georgia Lowcountry," in Linda M. Heywood, ed., *Central Africans and Cultural Transformations in the American Diaspora* (Cambridge, 2002), 289–317; and Usner, *Indians, Settlers, and Slaves*, 106–108.

25. Don Francisco Bellile Legal Case, 41–42. On how African healers built new communities among diverse diasporic Africans, see Sweet, *Domingos Álvares*.

other of being the real expert capable of making a grisgris. Their attorney, Francisco Broutin, would later declare that Cipion should not be punished because he had only collected the alligator parts and herbs that Big Luis had known were "necessary to make the poison" and had only consented to participate in the conspiracy out of "the fear that he had of Big Luis and Francisco." Big Luis, for his part, was born in New Orleans and had a Catholic kinship network of godchildren. Although he might have been able to capitalize on his standing as a blacksmith to rally Louisiana blacks against white power—much as the Virginia blacksmith Gabriel would try to do in 1800—he made the safer choice of integrating himself into colonial hierarchies instead of risking his neck and soul by committing murder. Lastly, many Mande believed that spoken words were themselves rich with nyama and had an agency of their own. This was particularly true for words about magic and other means of manipulating nature, topics that, as one anthropologist has put it, "should never be talked about unguardedly" because "bits of knowledge about sorcery vocalized might well lead to chaos and destruction." Enslaved individuals' fears of the repercussions of speaking truth to power, their differing visions about the best means of enduring life in a slave society, and, ironically, their awareness that words mattered all limited the circulation and influence of West African understandings of nature in the Americas.[26]

White doctors were also hesitant to speak authoritatively about the grisgris and its properties. Governor Unzaga's first move during the 1773 case was to order French doctor François LeBeau and Spanish surgeon Juan Rubio to examine "the quality of the poison and its activity," and these medical men drew on experimentation and the enslaved men's testimonies to determine if the mass was, by their standards, poisonous. LeBeau was particularly qualified for this task: he was in charge of examining pharmacists to determine their capacity to derive active compounds from natural substances, and this was, in essence, his duty during the grisgris trial. LeBeau and Rubio found the partially dried grisgris "stinking" and "concluded that . . . the mass was composed of some parts animal and some vegetable," but they could not "affirm if it is poison as they do not know the species . . . of which it is composed." They conducted two experiments in which they

26. Don Francisco Bellile Legal Case, 46, 50, 53 ("necessary"); McNaughton, *Mande Blacksmiths*, 12 ("should never"), 16, 43; Brown, "'Walk in the Feenda,'" in Heywood, ed., *Central Africans and Cultural Transformations*, 294; James Sidbury, *Ploughshares into Swords: Race, Rebellion, and Identity in Gabriel's Virginia, 1730–1810* (Cambridge, 1997), 83–86.

fed the wormy goop to a dog and locked it in a room, and, in both cases, the dog emerged unharmed. Yet this did not convince LeBeau and Rubio that the substance was not poisonous because "without knowing or understanding its compound they cannot say if it would cause greater injuries to human nature," and, based on the slaves' statements, they understood that the mass had not been dried "to the point necessary so that the poison could cause its effect." Their methods of inquiry were sound by European standards but could neither confirm nor deny that the compound was potentially poisonous.[27]

The trial climaxed in November as the prosecution and defense made their closing arguments. Both sides admitted that there was a conspiracy. The prosecution asked the court to punish Big Luis "severely" and sentence Carlos, Cipion, Francisco, and even the hunter Bernardo (who had only the most tangential role in the affair) to death. Big Luis chose his own lawyer, and the attorney for the four other men, Broutin, fingered him as "the Motor of the Conspiracy." Broutin argued that Carlos was in fact the expert capable of making the grisgris for Big Luis but added that Carlos was "crazy" and a "brute, ignorant, untrained negro" who had not produced anything verified as poison nor would he have been "capable of using it" if he had. In other words, Broutin pled that Carlos should be declared innocent because he was too ignorant—too African—to compose a specialized African poison.

27. Don Francisco Bellile Legal Case, 36 ("quality"), 37 ("stinking"), 43, 44 ("without"). Like many physicians throughout the Americas, French, Spanish, and Anglo medical men in colonial Louisiana valued the botanical knowledge of blacks and their descendants. Spanish Louisiana's Charity Hospital, for example, employed a black horticulturalist to oversee its botanical garden and a "Mulatto named Domingo ... intelligent in surgery" as bloodletter. But, while white officials and men of science were curious about what blacks knew about the properties of plants and animals, this curiosity did not supersede the mandate to oppress blacks: the survival of the colony was contingent on exploiting their labor, trade, and expertise. See *Constituciones para el nuevo Hospital de Caridad, construído en la ciudad de Nueva Orleans* (Madrid, 1793), xiv, xxvi ("Mulatto"). On whites' interest in blacks' natural knowledge, see Hall, *Africans in Colonial Louisiana,* 126; Rudolph Matas, *The Rudolph Matas History of Medicine in Louisiana,* ed. John Duffy, 2 vols. ([Baton Rouge, La.,] 1958–1962), I, 323–325; and Londa Schiebinger, *Plants and Empire: Colonial Bioprospecting in the Atlantic World* (Cambridge, Mass., 2004), 75–80. LeBeau had been physician and botanist to the king during the French era and would continue to serve Louisiana's government into the U.S. era. See Matas, *Rudolph Matas History of Medicine,* ed. Duffy, I, 66, 123, 165, 168, 186. On how a similar "tone of uncertainty" pervaded court records of poisoning cases in Martinique, see John Savage, "'Black Magic' and White Terror: Slave Poisoning and Colonial Society in Early 19th Century Martinique," *Journal of Social History,* XL (2007), 636.

Carlos, however, would die in prison later that month, and Unzaga would not sentence the surviving slaves until January 1774. The sources do not reveal the court's final verdicts, but the brutality of the punishments handed down to slaves in similar cases, even those in which the accused were found innocent, suggest that the conspirators were most likely killed.[28]

The grisgris affair probably contributed to a larger shift in Europeans' attitudes toward African natural and spiritual knowledge in Louisiana. Despite their geopolitical fears of the expansion of Anglo influence, New Orleans officials altered Spanish slave laws in 1778 so that they became less concerned with Protestantism, a prominent theme in the French *Code Noir*, and more devoted to suppressing African spiritual practices. Colonial officials recognized that they did not understand most aspects of how Louisiana's increasingly African black population approached nature, but they suspected that African-derived beliefs might somehow empower the slaves whose labor undergirded colonial rule.[29]

The social relationships that imperialism created in early America strongly shaped the ways enslaved individuals understood and utilized nature. The grisgris case makes it clear that some important aspects of West African intellectual life, particularly Mande standards of expertise, persisted in Louisiana. But it also suggests that the Afro-Louisianan experts involved in the conspiracy were not able or willing to apply their knowledge toward forming or leading communities. Many learned blacks throughout the Atlantic world were instrumental to the cohesion of new collective identities—including new ethnic groups—that helped enslaved peoples endure or challenge slavery, but the men involved in the grisgris conspiracy seemed more interested in personal advancement. This selfish application of knowledge did have parallels among Mande-speakers in Africa, particularly in that using esoteric knowledge for individualistic ends was both a potent means of promoting one's interests and the source of social discord and collective disasters like enslavement. But such practices were not simply traditional or African. They were modern and American: like Euro-American science, African Americans' knowledge and networks emerged amid, and were divided by, personal, regional, and imperial contests for power.[30]

28. Don Francisco Bellile Legal Case, 52 ("severely"), 54 ("crazy"), 55; Kerr, "Petty Felony, Slave Defiance, and Frontier Villainy," 153–156.

29. Thomas N. Ingersoll, "Slave Codes and Judicial Practice in New Orleans, 1718–1807," *Law and History Review*, XIII (1995), 47–48.

30. The literature on the relative influences of African traditions and New World circumstances on the formation of African American culture is vast. See, for example, Sidney W. Mintz and Richard Price, *The Birth of African-American Culture: An Anthro-*

The failure of the conspiracy and the broader unwillingness of many enslaved experts in Louisiana to challenge the institution of slavery (instead of just specific hierarchies within it) reflect the weakness of enslaved blacks. Weakness limited the ability of all groups in the region to pursue and use knowledge in the ways they would have wished, but, as the most vulnerable part of the colony's population, enslaved experts like Carlos and Cipion faced the greatest obstacles and risks. Access to specialized knowledge did not lead these learned Africans to greater influence or status. It led to incarceration and, probably, death.

"LET THE LORDS BE LUMBER CUTTERS!"

Spanish officials had an even weaker hold over East Florida (which Spain repossessed from 1784 to 1821) than they did over Louisiana. Spain depended on the cooperation of local experts and power brokers to know and govern the province, and the most influential non-Indian inhabitants were European planters. Imperial officials did attempt to co-opt planters' authority and information, but officials also had to deal with the frustrating fact that planters tried to find ways of increasing their own wealth and power by exploiting Spain's reliance on their expertise. By working to control and manipulate accounts of Gulf South resources, planters limited officials' ability to develop the region and integrate it into Spain's political economy.[31]

Planters' intellectual authority and power over lands and people could persist across changes of sovereignty, and plantations established in West and East Florida during the British era (1763–1783) were a crucial source

---

pological Perspective (Boston, 1992); Morgan, *Slave Counterpoint*, 441–658; and Thornton, *Africa and Africans in the Making of the Atlantic World.* On community formation and how the selfish application of esoteric knowledge could splinter social groups, see Joseph C. Miller, "Retention, Reinvention, and Remembering: Restoring Identities through Enslavement in Africa and under Slavery in Brazil," in José C. Curto and Paul E. Lovejoy, eds., *Enslaving Connections: Changing Cultures of Africa and Brazil during the Era of Slavery* (Amherst, N.Y., 2004), 81–121; Sweet, *Domingos Álvares;* and McNaughton, *Mande Blacksmiths*, 11–14, 42. Works that argue African American intellectual life was inextricable from modernity and Atlantic slavery include Palmié, *Wizards and Scientists;* Young, *Rituals of Resistance;* and Diana Paton and Maarit Forde, eds., *Obeah and Other Powers: The Politics of Caribbean Religion and Healing* (Durham, N.C., 2012).

31. On how the control of information networks was both vital to imperialism and contested, see Joshua Piker, *The Four Deaths of Acorn Whistler: Telling Stories in Colonial America* (Cambridge, Mass., 2013), esp. 65–66; Katherine Grandjean, *American Passage: The Communications Frontier in Early New England* (Cambridge, Mass., 2015); and Alejandra Dubcovsky, *Informed Power: Communication in the Early American South* (Cambridge, Mass., 2016).

of political and intellectual continuity as these provinces came back under Spanish rule. Enterprising planters—including the notorious Dr. Andrew Turnbull, who forced some twelve hundred Greeks and Minorcans to make indigo under the whips of black overseers—flocked to the British Floridas to experiment with cash crops and labor regimes. Many of these planters remained loyal to Britain and left the Floridas with the change of flags, but some stayed behind and presented Spanish officials with ready-made hubs of racial domination, staple production, and European science. Jacob Nash, who owned electrical apparatus and a library of scientific books, and William Dunbar, who experimented with Caribbean plants and would establish an astronomical observatory on his plantation, were among the British planters who remained in West Florida after it became Spanish. Only five substantial landholders chose to remain in East Florida as it transitioned back to Spanish rule, the most wealthy and influential of whom was Francis Philip Fatio.[32]

Fatio's wealth, influence, and knowledge were grounded in his plantations, especially his ten thousand-acre New Switzerland on the Saint Johns River. Born in Switzerland in 1724, Fatio fought for the Swiss Guard, lived as a viscount in Sardinia, became a Londoner in 1761, and, in 1771, caught Florida fever and moved to British East Florida with his family. His diverse economic pursuits—growing staples, ranching, lumbering, trading, smuggling—made him the richest man in East Florida by the mid-1780s. The persistence of New Switzerland—which, in 1787, had eighty-two slaves, about one-third of all the slaves along the Saint Johns River—helped ensure that white power carried over into the Spanish era in at least part of the

32. On planters in British East Florida, see Patricia C. Griffin, "Blue Gold: Andrew Turnbull's New Smyrna Plantation," in Jane G. Landers, ed., *Colonial Plantations and Economy in Florida* (Gainesville, Fla., 2000), 43, 52; and Christopher P. Iannini, *Fatal Revolutions: Natural History, West Indian Slavery, and the Routes of American Literature* (Chapel Hill, N.C., 2012), 179–188. On Nash, see Archives of the Spanish Government of West Florida, I, Spanish Records for the Years 1782–1789, 180–194, and IX, Spanish Records for the Year 1805, 179–180, trans. Marie Claire Berthelot, Works Progress Administration Survey of Federal Archives in Louisiana (1937), LLMVC. On Dunbar, see William Dunbar to Messrs Thompsons of Kingston, Jamaica, June 18, 1775, "Extracts," 9; and Dunbar to Andrew Ellicott, Mar. 1, 1801, AEP, V, Correspondence, 1026. On Fatio and the four other landholders to remain, see Jane Landers, *Black Society in Spanish Florida* (Urbana, Ill., 1999), 73; and Paul E. Hoffman, *Florida's Frontiers* (Bloomington, Ind., 2002), 234. Similarly, almost no Spaniards had chosen to remain in Florida when it transitioned to British rule in 1763. See Robert Olwell, "Seeds of Empire: Florida, Kew, and the British Imperial Meridian in the 1760s," in Elizabeth Mancke and Carole Shammas, eds., *The Creation of the British Atlantic World* (Baltimore, 2005), 264.

province. Fatio's scientific pursuits also depended on exploiting slaves. He used slave-harvested wealth to buy books and instruments, study astronomy and astrology, and build a hospital that treated enslaved blacks and Spanish soldiers. Fatio was well aware of how valuable his local knowledge was to officials trying to implement imperial rule. As he told a Spanish official in 1787, "Florida was very little known" during the first decade of the British era because "no industrious Spanish families [had] remained who could teach, instruct, or direct the new possessors." The message was not subtle: spanning imperial eras qualified him to manage the development of Florida after it once again became a Spanish colony.[33]

Fatio quickly established himself as a power broker who profited from Spanish governance. When Governor Vicente Manuel de Zéspedes claimed East Florida on July 12, 1784, Spain gained a province that was mostly controlled by Indians and had neither enough white settlers to fulfill its strategic purposes—defending New Spain and sea-lanes in the Gulf of Mexico—nor known commodities that would attract settlers and strengthen the imperial economy. Zéspedes's reliance on Fatio as an established authority became clear only two days after the change of flags, when the governor appointed him and the Indian trader John Leslie as the judges who would decide property disputes among departing British subjects. Fatio used this position to buy up Briton-owned land and to perform power. British subjects were soon complaining that Fatio "assumes a jurisdiction of a very extensive nature" and, instead of following due process, "prejudges Causes, and decides by whim and caprice." The departing British governor even put the question to Zéspedes, that, "if Mr. Fatio exercises this extensive jurisdiction by the direction and approbation of your Excellency, . . . where [is] the line . . . to stop?" Zéspedes probably could not have answered this; he also relied on Fatio to feed Spanish soldiers, defend the Saint Johns River fron-

33. Francis P[hilip] Fatio to Joseph del Río, June 26, 1787 (Spanish trans. of French original), EFP, sect. 23, bundle 98G6, no. 44, reel 37 (quotes); William Bartram, "Travels in Georgia, and Florida, 1773–74; A Report to Dr. John Fothergill," [ed.] Francis Harper, American Philosophical Society, *Transactions*, New Ser., XXXIII, part 2 (1943), 145; Susan R. Parker, "I Am Neither Your Subject nor Your Subordinate," in Jacqueline K. Fretwell and Parker, eds., *Clash Between Cultures: Spanish East Florida, 1784–1821* (Saint Augustine, Fla., 1988), 43, 46, 56–57; Parker, "Success through Diversification: Francis Philip Fatio's New Switzerland Plantation," in Landers, ed., *Colonial Plantations and Economy in Florida*, 69–70, 73, 75–76; Parker, "Men without God or King: Rural Settlers of East Florida, 1784–1790," *FHQ*, LXIX (1990), 144, 148–149; Landers, *Black Society in Spanish Florida*, 80, 162–163; William Scott Willis, ed., "A Swiss Settler in East Florida: A Letter of Francis Philip Fatio," *FHQ*, LXIV (1985), 174–177, 187.

tier, and lend money to the perpetually cash-poor treasury. In exchange, Zéspedes turned a blind eye to his illicit commerce and stationed troops at New Switzerland to help prevent Seminoles from stealing Fatio's slaves.[34]

Fatio's written descriptions of Florida's resources were balancing acts in which he provided just enough information to pique officials' interests while withholding or lying about details that might affect his profits. He was especially interested in getting rich from, and determining what officials knew about, forest products such as lumber, tar and pitch, and the turpentine from his own still. As a planter and timber baron, Fatio exemplified British surveyor Bernard Romans's call to "let the lords be lumber cutters!" in East Florida. Fatio penned reports to the British and, later, Spanish governments about integrating Florida's trees into their Atlantic economies. His 1785 "Description of East Florida" drew on his experience commercializing timber products during the British era to argue that Florida tar and turpentine would free Spain from Baltic imports and hinder U.S. expansion by attracting "loyal inhabitants capable of withstanding the first encroachments and of delaying their [the United States'] progress." Fatio sent this report to Zéspedes who forwarded it to the minister of the Indies in Madrid, where it was placed "at the foot of the throne" and inspired a royal order to explore the quality and profitability of Florida's timber products, a mission that fell to navy lieutenant José del Río Cosa in 1787.[35]

Fatio's efforts to control the narrative about Florida's natural resources

34. Patrick Tonyn to [Thomas Townshend, first Viscount Sydney], Dec. 6, 1784, in Joseph Byrne Lockey, [ed.], *East Florida, 1783–1785: A File of Documents Assembled, and Many of Them Translated* (Berkeley, Calif., 1949), 322 ("assumes"), Tonyn to Vicente Manuel de Zéspedes, Oct. 11, 1784, 373 ("if Mr. Fatio"), Zéspedes to Bernardo de Gálvez, Feb. 28, 1785, 461; Tanner, *Zéspedes in East Florida,* 48–50; Parker, "I Am Neither Your Subject nor Your Subordinate," in Fretwell and Parker, eds., *Clash Between Cultures,* 51–52; David J. Weber, *The Spanish Frontier in North America* (New Haven, Conn., 1992), 275–279.

35. Bernard Romans, *A Concise Natural History of East and West-Florida . . .* (New York, 1776), 117 ("let the lords"); Francis Philip Fatio, "Description of East Florida," Mar. 18, 1785, in Lockey, ed., *East Florida,* 479–482 ("loyal," 480); Fatio to Vicente Manuel de Zéspedes, June 28, 1787, EFP, bundle 195M15, reel 82, 501 ("at the foot"); Jack D. L. Holmes, "Jose Del Río Cosa," *Tequesta,* XXVI (1966), 40–41. For Fatio's 1782 report to British officials, see "Considerations on the Importance of the Province of East Florida to the British Empire (on the Supposition That It Will Be Deprived of Its Southern Colonies) by Its Situation, Its Produce in Naval Stores, Ship Lumber, and the Asylum It May Afford to the Wretched and Distressed Loyalists. 1782. PRO CO 5:560," in Lawrence Feldman, ed., *Colonization and Conquest: British Florida in the Eighteenth Century* (Baltimore, 2007), 12–19.

that circulated beyond the province clashed with the goals of imperial officials. Del Río's instructions were to "treat" with Fatio to produce a report about the tree-derived products valuable "to the royal service and the development of this Country," and del Río sent Fatio a questionnaire about "this very interesting branch of knowledge" and spent ten days at New Switzerland in hopes of collecting specimens. But Fatio never showed up to meet del Río at New Switzerland, withheld samples of woods and resins, and, in his terse replies to the questionnaire's commercial queries, listed the inflated costs that naval stores had commanded during the American Revolution. Fatio chastised del Río for acting as if he were the expert despite having only "lightly crossed a small part of the province" and needing to rely on Fatio's "brilliance" to complete his commission. He was indignant that del Río took him for a mere "guide" and told the lieutenant that he did "not consider himself in any manner subject to your orders, nor your *inferior*." Fatio had penned his 1785 "Description" in hopes of getting richer, but, when it inspired an expedition that threatened to produce an alternate account about Florida nature that could undermine his credibility and potential profits, he switched gears from sharing knowledge to concealing it. The potential profits of transoceanic commerce not only promoted the spread of information, as historians have argued. They also distorted it and hampered its circulation.[36]

Fatio's control over narratives about Florida nature declined as more voices became involved and texts capturing these voices traveled further from his plantation. Thus, while he could limit del Río's access to specimens at New Switzerland and blast del Río's credibility in his response to the questionnaire, del Río simply edited these rants out of the copies he sent to his superiors. The lieutenant also wrote his own report, and, while he did confirm that Florida pines could "destroy the pitch trade of the Baltic," he added that the few trees suitable for lumber were too expensive because "only one planter [Fatio] on the S. John's River is producing them, and

36. Vicente Manuel de Zéspedes to F[rancis] P[hilip] Fatio, July 14, 1787, EFP, bundle 195M15, reel 82, 451 ("treat"), José Cosa del Río to Zéspedes, June 26, 1787, no. 42, bundle 98G6, reel 37 ("this very"), Fatio to del Río, June 26, 1787, bundle 98G6, reel 37 ("lightly"), emphasis in original, Fatio to Zéspedes, July 12, 1787, bundle 195M15, reel 82, 435; Parker, "I Am Neither Your Subject nor Your Subordinate," in Fretwell and Parker, eds., *Clash Between Cultures*, 50. On the role of commerce in early modern science, see Pamela H. Smith and Paula Findlen, eds., *Merchants and Marvels: Commerce, Science, and Art in Early Modern Europe* (New York, N.Y., 2002); and Harold J. Cook, *Matters of Exchange: Commerce, Medicine, and Science in the Dutch Golden Age* (New Haven, Conn., 2007).

supply and demand roughly determine the value, for without an alternative or competitor, there can be no just price." Seizing a rare opportunity to curb Fatio's influence, Governor Zéspedes accused Fatio of having insulted and obstructed del Río simply because the "arrival of an official commissioned to verify the certainty of what you expressed in your Memorial" had made Fatio's "importance vulnerable." Zéspedes assured Fatio that "I will always esteem, and make correct use of, whatever information that it pleases you to present to me that is conducive to the better service of the King, and good of Florida," but berated him for "imagining things that do not exist, nor ever existed," instead of sharing his knowledge fully and honestly. Zéspedes might have hoped these condemnations (part of a fifteen-page letter devoted to faulting Fatio's conduct) would strengthen his hand in future negotiations with the planter, and the governor sought further leverage by making Fatio's behavior a matter of state. He told Minister of the Indies José de Gálvez that Fatio's "interests are no less powerful than his vanity" and blamed Fatio for trying to undermine del Río's mission after the "baseless ideas" in Fatio's report inspired an expedition rather than "amazed assent." This was, the governor explained, evidence of Fatio's "deranged mind."[37]

But officials in Madrid did not take del Río's or Zéspedes's writings as the final word on the potential usefulness of East Florida's trees and, in 1792, once more looked to planters for local knowledge. The new governor, Juan Nepomuceno de Quesada, circulated another questionnaire to the province's planters to satisfy another royal order to identify "species of trees known in this Province," their size, the quality of their lumber, their use as dyes, and "the known medicinal virtues of their Gums, Resins, leaves, fruits, or barks." Quesada "very particularly" asked eight planters, including Fatio, because they were "the few able to provide intelligence about this matter." Francisco Xavier Sanchez, a cattle rancher whose residence dated back to the British era, offered a detailed response that described oaks and ash trees good for construction and explained how locals made medicine from pine resin through "the same operation that [produces] Turlington's English drops." Sanchez added that even though cypress lumber was quite valuable, it was so difficult to harvest that "it was used for nothing more than Taxa-

37. José del Río, "Observations Corresponding to the Measures for Re-establishing East Florida in a Flourishing State . . . ," in Holmes, "Jose Del Río Cosa," *Tequesta*, XXVI (1966), 41–52 ("destroy," 50, "only one," 48); Zéspedes to Fatio, July 14, 1787, EFP, bundle 195M15, reel 82, 443, 446 ("I will always"), 447 ("imagining"), 448 ("arrival," "importance"), 451, Vicente Manuel de Zéspedes to José de Galvez, July 16, 1787, bundle 43D1, no. 200, reel 17, 154, 158 ("interests," "baseless," "amazed," "deranged"). Del Río's edited copy of Fatio's reports follows after Fatio to del Río, June 26, 1787, bundle 98G6, reel 37.

namy *[sic]*." Only two of the seven other planters responded to Quesada's instructions, and one of them, Fatio, did not describe a single tree. Although Fatio promised to tell Quesada "the little I have learned in the course of a very long, thoughtful, and laborious life" when he next visited Saint Augustine, his written reply—the artifact that actually circulated among Spanish officials—noted only that he was "a decrepit old man little learned in Botany and less in Medicine" without any useful information to contribute. After reviewing the responses to his questionnaire, Quesada grumbled that, "as there is not one among [the respondents] whose curiosity passes beyond the limits of their own benefit, the present [descriptions] will poorly satisfy the patriotic aims of the Court." Florida's planters had once more proven a disappointing source for imperially beneficial knowledge.[38]

Planters and other power brokers sought to control the flow of knowledge beyond their spheres of influence, and, by 1792, Fatio's experience with Spanish officials had led him to believe that the key to personal enrichment was keeping what he knew about resources and their value to himself. Other planters, such as William Dunbar in West Florida, would reach the opposite conclusion that sharing an abundance of scientific information among competing empires bolstered wealth and prestige. Either way, planters had great potential to shape and undermine what officials knew about Gulf South nature because imperial rule and economic development relied heavily on planters' power, knowledge, and productions. Just as Fatio's political and intellectual authority stemmed largely from having persisted across a change of sovereigns, so, too, would the planter power brokers who established themselves during the Spanish domination prove pivotal to U.S. imperialism.

Fatio's New Switzerland, one of Florida's most enduring hubs of European power, succumbed to Seminole raiders in 1812. The plantation's fall was a stark reminder that whites' supremacy within East Florida was limited almost entirely to spaces near the colonial capital of Saint Augus-

38. Juan Nepomuceno de Quesada to [Juan MacQueen, Francisco Felipe Fatio, Andres Atkinson, Thomas Raling, Thomas MacIntosh, Ricardo Lang, Francisco Sanchez, and Guillermo Pengree], Sept. 24, 1792, EFP, bundle 23J2, no. 300, reel 9, 222 ("species," "the known," "very particularly," "the few"), Francisco Xavier Sanchez to Quesada, Apr. 1, 1792, bundle 23J2, no. 300, reel 9, 230, 231 ("Taxanamy"), 232 ("same operation"), ("Taxanamy" is in English in this otherwise Spanish report), F[rancis] P[hilip] Fatio to Quesada, Mar. 31, 1792, bundle 23J2, no. 300, reel 9, 227 ("little," "decrepit"), Quesada to Luis de las Casas, Sept. 24, 1792, bundle 23J2, no. 300, reel 9, 221 ("as there is"). The other response was Andrew Atkinson to Quesada, May 12, 1792, EFP, bundle 23J2, no. 300, reel 9, 237–240.

tine. Native groups—Creeks on the northern frontier and Seminoles in the peninsula's interior—held sway throughout most of the province.[39]

MINERALOGY, MYTHS, AND MONSTERS

In 1790, a Creek Indian named Yaolaychi told stories that motivated a Spanish mineralogical expedition, guided how its members pursued and circulated knowledge, and influenced the ways Spanish officials and experts in Saint Augustine analyzed the information it generated. Historians have argued that European men of science consistently stripped indigenous knowledge of its narrative and cosmological contexts to better fit desirable elements of it into scientific and commercial frameworks. Yet native stories were coherent, memorable, and—in certain circumstances—compelling packages of information, and the collections and observations that the expedition members generated in the field remained associated with these narratives even after the expedition's return to Saint Augustine. Europeans' difficulties extracting valuable knowledge from Yaolaychi's tales not only reflected the power of Indian storytelling but also the feebleness of European rule. Spanish weakness in Florida was evident in officials' lack of control over the province's interior, their reliance on Indian informants to access and circulate natural knowledge, and their unwillingness to take actions that might upset the native allies who shielded Florida from Anglo rivals. But just because Spain was weak in the Florida borderlands did not mean Indians were strong: native informants ran enormous risks when they tried to profit from presenting knowledge to officials who could opt to punish instead of pay them. And, according to the expedition members, Spaniards and Indians proved equally powerless in the face of a monstrous being that blocked their access to knowledge and resources.[40]

39. On the destruction of New Switzerland, see Parker, "Men without God or King," *FHQ*, LXIX (1990), 144.

40. Anthropologists and historians of science have explored some of the ways that native stories shaped and moved the local knowledge that developed during encounters in colonial spaces. See Julie Cruikshank, *Do Glaciers Listen?: Local Knowledge, Colonial Encounters, and Social Imagination* (Vancouver, B.C., 2005); and Neil Safier, "Global Knowledge on the Move: Itineraries, Amerindian Narratives, and Deep Histories of Science," *Isis*, CI (2010), 136. On how Euro-American men of science worked to extract data from indigenous contexts, see Schiebinger, *Plants and Empire;* and Cook, *Matters of Exchange.* Neither indigenous nor European approaches to nature were fully bounded or coherent systems; as David Turnbull has noted, "All knowledge traditions remain messy and complex, replete with unbridged gaps and overlapping spatialities." See Turnbull, "Travelling Knowledge: Narrative, Assemblage, and Encounters," in Marie-Noëlle Bour-

The mineralogical expedition began to take shape on February 15, 1790, when an Indian-language interpreter named Alonso Gil visited the Saint Augustine office of Bartolomé Benítez y Gálvez, the royal treasurer of East Florida, to share some information he had heard from Yaolaychi, a Lower Creek Indian, about a deposit of shiny rocks in the northwestern part of the province. Benítez, who had experience with mineralogy and chemistry, believed these metals would prove to be silver or mercury, resources that would attract much-needed settlers and funds to East Florida. Benítez had arrived in Florida only a few months earlier after a brief stint as intendant of Ilocos in the Philippines, and, as he told governor Vicente Manuel de Zéspedes, he decided to organize an expedition to the mineral deposit because it "could result in the good of the State and these Provinces, and give credit to my patriotic spirit."[41]

Although Bourbon Spain funded many massive scientific expeditions, Benítez was obliged to finance this venture out of his own pocket and at his own risk. Zéspedes gave him permission to explore the mine and provided passports for the expedition members, yet the governor was unwilling to support the endeavor financially until the Indian's claims had been "proven to the satisfaction of expert men." Benítez's exploratory party did not consist of experts. It included only Yaolaychi, Gil, and a young official named Bartolomé de Castro. The treasurer lamented that Gil and Castro were "ignorant men" unqualified "to carry out a charge that certainly merits talents, enlightenment, and ideas in metallurgy and mineralogy."[42]

Benítez claimed that his poor health prevented him from leading the expedition, so he put together a detailed set of instructions that spelled out where the commissioners should explore, what they should collect, and how to preserve what they found. These instructions gave Benítez hope that the

---

guet, Christian Licoppe, and H. Otto Sibum, eds., *Instruments, Travel, and Science: Itineraries of Precision from the Seventeenth to the Twentieth Century* (London, 2002), 287.

41. Bartolomé Benítez y Gálvez to Vicente Manuel de Zéspedes, Apr. 7, 1790, EFP, sect. 89, bundle 353, no. 18, reel 173, 442–443 (quote, 443), Benítez to the king of Spain, Dec. 30, 1791, sect. 16, reel 30. In July 1789, the crown had sent royal orders to bureaucrats throughout the empire to collect and write reports about potentially valuable natural productions. See Bleichmar, "Atlantic Competitions," in Delbourgo and Dew, eds., *Science and Empire*, 235–236.

42. Bartolomé Benítez y Gálvez to Vicente Manuel de Zéspedes, Feb. 17, 1790, EFP, sect. 16, reel 30 ("proven"), Benítez to Zéspedes, Apr. 7, 1790, 444 ("ignorant"), Benítez to Zéspedes, Feb. 18, 1790. For an overview of Bourbon Spain's scientific expeditions, see Iris H. W. Engstrand, *Spanish Scientists in the New World: The Eighteenth-Century Expeditions* (Seattle, Wash., 1981).

expedition members would concentrate on procuring the information he considered most valuable instead of relying on their own judgment about which aspects of Floridian nature merited attention. The expedition members were to travel to the spring in Yaolaychi's account, gather shiny rocks, collect water and mud in sealed bottles, perform "excavations," preserve herbs in paper, and note the region's trees. Benítez would analyze these collections and observations on their return.[43]

Castro, Yaolaychi, and Gil were not quite as unqualified as Benítez portrayed them to be: a promising young official, an Indian guide, and an experienced translator made a suitable team for accomplishing the expedition's goals. Castro was a dependent in Benítez's household at the time of the expedition and would go on to become a public attorney, major supervisor of royal works at Saint Augustine, and a successful slave-owning planter. Gil had worked as an interpreter in Florida for at least three years and had several more years of experience interacting with nearby Indian groups. Both Castro and Gil were described as Spaniards throughout the expedition's sources, though Gil might have had mixed Indian and European parentage. Yaolaychi was from the town of Hitchiti on the Chattahoochee River in present-day southwestern Georgia, and he described himself as "a poor man." Yaolaychi and his family of seven, like most Hitchitis, had left the town to hunt for the winter, and Lower Creeks often hunted deer and bears in Florida well into February. He understood a little Spanish, but Hitchiti was his primary language.[44]

43. This list of "Ynstrucciones" was included in Benítez's copy of the expedition's diary. See "Diario que hicieron los comisionados Alonzo Hill, ynterprete de yndios, y Don Bartulome de Castro y Ferrer, para la expedición de averiguar la verdad de quanto expusó el yndio llamado *Yaolaychi* de la nación *Yuchises* natural del pueblo de *Gechiti*," Apr. 6, 1790, EFP, sect. 89, bundle 353, no. 17, reel 173, 433.

44. On Castro, see Bartolomé Benitez y Gálvez to Vicente Manuel de Zéspedes, May 17, 1790, EFP, sect. 16, reel 30, José María Coppinger to Juan Ruiz de Apodaca, Feb. 22, 1816, sect. 13, reel 19; and James G. Cusick, *The Other War of 1812: The Patriot War and the American Invasion of Spanish East Florida* (Gainesville, Fla., 2003), 174, 208. On Gil, see Interrogation of Alonso Gil, Apr. 9, 1790, EFP, sect. 89, bundle 353, no. 19, reel 173, 467, Zéspedes to Gonzalo Zamorano, Apr. 22, 1787, sect. 15, reel 24, and Alonso Gil to Enrique White, Jan. 19, 1797, sect. 44, reel 79. On Yaolaychi, see "Diario," Apr. 6, 1790, 433, 437 ("poor man"); and Benítez to Zéspedes, Apr. 7, 1790, 443. I have found no evidence regarding Alonso Gil's parentage. Although some official interpreters in Florida were of mixed Indian and European heritage, there were also several ethnically European translators. See Tanner, *Zéspedes in East Florida*, 92. On how individuals of mixed race, natives, and free blacks could sometimes achieve social status as "españoles," see

The sources are not clear about Yaolaychi's motives for informing Gil about the mine and participating in the expedition, but, by placing what is known about him in the context of Hitchiti society, it is possible to hazard some speculations. According to U.S. Indian Agent Benjamin Hawkins, Hitchiti was a place of "poverty and indolence" in the 1790s, and its lack of fences indicated that the townspeople had not begun ranching, one of Creek Country's latest sources of wealth. Poor Hitchitis were often economically dependent on wealthy local traders, and the skins Yaolaychi collected during a winter in Florida probably would not have been sufficient to get him out of their debt. By offering to reveal a mine to the Spanish, Yaolaychi was perhaps pursuing an alternate source of income: like other Creeks who acted as guides or traded medicinal plants for cash, Yaolaychi might have sought to profit from selling local knowledge.[45]

The Seminoles were the most formidable group in East Florida, but Creeks maintained a significant presence in the northern part of the province and claimed to possess large swaths of it as their exclusive hunting grounds. Spanish officials recognized that Florida's interior was Indian country and beyond their control, but they hoped to make the best of their position by turning Indian neighbors into Spanish-aligned buffers against U.S. expansion. Florida officials thus formalized many of the Seminoles' and Creeks' territorial claims through treaties and worried about impinging on their boundaries for fear of losing these crucial alliances. Zéspedes was particularly eager not to cause any offense in early 1790 when Alexander McGillivray, the self-proclaimed head of the Creek Nation, was showing signs of abandoning his political and commercial ties with Spain in favor of a closer attachment to the United States. The governor therefore

---

Weber, *Spanish Frontier in North America,* 326–328. On hunting, see Robbie Ethridge, *Creek Country: The Creek Indians and Their World* (Chapel Hill, N.C., 2003), 135; and Angela Pulley Hudson, *Creek Paths and Federal Roads: Indians, Settlers, and Slaves and the Making of the American South* (Chapel Hill, N.C., 2010), 20.

45. Benjamin Hawkins, "A Sketch of the Creek Country in the Years 1798 and 1799," in Thomas Foster, ed., *The Collected Works of Benjamin Hawkins, 1796–1810* (Tuscaloosa, Ala., 2003), 64s (quote). On fences, see Claudio Saunt, *A New Order of Things: Property, Power, and the Transformation of the Creek Indians, 1733–1816* (Cambridge, 1999), 57, 172–175. Although Spanish officials described Yaolaychi as part of the Uchise (Lower Creek) "nation," he would have identified himself primarily as a Hitchiti, and any political or economic reasons underlying Yaolaychi's actions should be put in the context of that town. See "Diario," Apr. 6, 1790, 433 ("nation"); Joshua Piker, *Okfuskee: A Creek Indian Town in Colonial America* (Cambridge, Mass., 2004), esp. 105, 154–156; and Ethridge, *Creek Country,* 93, 175, 179.

only granted passports to the expedition members after Benítez promised him that the mine was not situated within recognized native boundaries.[46]

The commissioners began to stray from Benítez's instructions about which types of information they should attend and record almost as soon as they left Saint Augustine. After settling into camp on February 20, the very first evening of their journey, Yaolaychi told the two Spaniards a story that would have a profound effect on the rest of the expedition.

The Story, Version One: Told February 20, 1790

> The Indian said, that the Animal about which he had spoken to [Gil] at this place [Saint Augustine], (who told me [Benítez] nothing about this matter; and he [Gil] had believed it to be a caiman, crocodile, or lizard); was as big as a one year old calf: that it was ferocious, and that it had killed many Indians; eight men from the town of Afasqui had joined together with several dogs to hunt it, they found it and faced it, and shot at it five or six times, but the balls would not enter its body . . . , [Yaolaychi] told that only one [of the Indians] escaped by the road; the others perished because the Animal was very quick, and that with its breath it attracts, and the Spirit cuts the people.

After recounting this story, Yaolaychi told Gil and Castro that he would guide them to the area near the deposit of shiny metals, but that he would not go near the beast's cave himself.[47]

46. Andrew K. Frank, "Taking the State Out: Seminoles and Creeks in Late Eighteenth-Century Florida," *FHQ,* LXXXIV (2005), 10–27; Ethridge, *Creek Country,* 30, 120; Hudson, *Creek Paths and Federal Roads,* 1, 19; Saunt, *New Order of Things,* 128, 152; Weber, *Bárbaros,* 75, 204, 214. McGillivray would officially ally with the United States as a result of the Treaty of New York on August 7, 1790. See John Walton Caughey, *McGillivray of the Creeks* (Norman, Okla., 1938), 37–44; and Tanner, *Zéspedes in East Florida,* 210–215. For passports, see Benítez to Zéspedes, Feb. 17, 1790, EFP, sect. 16, reel 30.

47. "Diario," Apr. 6, 1790, 434 (quote). Yaolaychi's tellings of the stories cannot be separated from the voices of the individuals who translated and recorded them. Gil translated the two versions told during the expedition, and Castro wrote them down. Benítez made the extant copy of the diary, and it is sometimes written from his perspective. Regarding the "eight men," the number four and its multiples are common in Creek stories and sacred in Creek cosmology. See Greg Urban and Jason Baird Jackson, "Mythology and Folklore," in Raymond D. Fogelson, ed., *Handbook of North American Indians,* XIV, *Southeast* (Washington, D.C., 2004), 717. Monsters with reptilian features and the power to attract prey were part of a shared mythological tradition in the Southeast. See George E. Lankford, "The Great Serpent in Eastern North America," in F. Kent Reilly,

The three men arrived near the cave on February 28, and, despite Yaolaychi's protests, the two Spaniards compelled him to visit the cave alone to collect the mineralogical samples that the treasurer had requested. The entry in the expedition's diary following Yaolaychi's excursion to the cave interweaves the story of the hunters and the beast with Yaolaychi's account of his own experiences.

The Story, Version Two: Told February 28, 1790

[Yaolaychi spoke] about the veracity of [the area near the spring] having minerals ... and he swore by Feosé, which is God, about the proofs he gives: By He that is there above, He that sees all, that I do not lie; I do not deceive you; and that the wind was wrong for beginning anything on this day, because it goes towards the spot where the beast lives, which would perceive them with its sense of smell, and the danger was great, because of the powers of attraction that it has.

At seven in the morning they [Gil and Castro] determined to send the Indian alone to the spot, to look for earth and stones ... giving him a pick, the best horse and a shot gun, the two [Spaniards] staying at camp: He left, and returned frightened at 12:00 ..., they asked for his news, he said: that arriving at the spot he had mentioned, where he should have taken the stones, mud, water, and the rest that they would prepare [for analysis in Saint Augustine], believing the wind favorable, he tied the horse to a pine tree, and went forth on foot, he heard the Beast roaring a short distance away, which surprised him, dreading that it would smell him, or see him, and take away his life, that returning to the site where he left his horse, he found in the path the bones of a deer, dead, because of the Beast, which devoured it, and plants that had been uprooted by the same, with which the two travelers formed an idea of [Yaolaychi's] prediction being true.

Asking him [Yaolaychi] about the Beast, its shape, and name he said that its name was Achuguilipalascó, which means Big Ferocious Animal; and he described its shape by its separate parts, being unknown to [the Spaniards], who then asked [Yaolaychi] for its color, he said that its head is similar to that of a Bear, but without ears, or that it has them very close to its head, and the head, and neck, is the color of gold, that its tongue is split in the middle in the shape of scissors, or like that of a serpent, that

---

III, and James F. Garber, eds., *Ancient Objects and Sacred Realms: Interpretations of Mississippian Iconography* (Austin, Tex., 2007), 114.

> the body is similar to that of a bear, with a very short tail, that its skin is very hard, having been verified by shooting it with bullets, and it reflects the balls ... ; the Indian affirming, that an expedition had been formed in the town of Afasqui to kill it, composed of eight men of his [Yaolaychi's] same nation, and there was among them a mighty war captain and nine dogs, they attacked the Beast, which facing up to them, killed them, first the captain, and six of his companions, only one escaping .... This same Indian [that escaped the beast] returned to his town, recounted the misfortunes, with which he induced the utmost terror in those people, that they no longer go near that place [the cave of the beast], ... the Indian [Yaolaychi] added one more thing to his relation, that its sense of smell is extraordinary; it is able to perceive from long distances, and its breath is so active, that it discovers birds, deer, and any other animal, and even with man, it conquers, demoralizes, and amazes until not leaving the least action to him.

Yaolaychi said that approaching the cave at this time was very dangerous because the only way to drive the beast away was by setting fire to the undergrowth, and a recent burning in the area had made this impossible. Castro and Gil then asked if Achuguilipalascó was "the only beast of its species known in these Provinces," and Yaolaychi responded that "he only knew of four: One near the Cherokee Indians, another near the town of Nocosuque, another far ... in the Northern part, and that which is here nearby."[48]

Twentieth-century ethnographers have recorded Hitchiti, Muskogee, Yuchi, and Seminole tales—collectively known as Monster Lizard stories—in which the scenario, action, and beast correspond with Yaolaychi's narrative. In these versions, a group of hunters and their dogs used fire to flush out the monster, which was often described as glittering and impervious to bullets. The Monster Lizard would kill most of the dogs and hunters before being defeated by esoteric magic, a clever trick, or—as Yaolaychi would relate during his interrogation—the timely intervention of a tiger. It is also likely that stories of a monster like Achuguilipalascó had older roots: a Mississippian pipe bowl unearthed in Alabama portrays a bear-shaped beast that resembles the one Yaolaychi described. Creeks inherited and perpetuated several elements of Mississippian culture, and Yaolaychi might have been the latest in a long line of southeastern natives to narrate esoteric knowledge of monsters to demonstrate his own expertise and bolster his influence.[49]

48. "Diario," Apr. 6, 1790, 436–437 ("[Yaolaychi spoke] about"), 437 ("the only beast").

49. On the Monster Lizard Story in comparative perspective, see John R. Swanton, *Myths and Tales of the Southeastern Indians* (Washington, D.C., 1929), 272. The

FIGURE 8. Mississippian Piasa Effigy Pipe. Circa 1300–1500 C.E. National Museum of the American Indian. Smithsonian Institution. 17/893 Clarence B. Moore Collection

Generally speaking, southeastern Indians did not draw sharp distinctions between a particular fact and its social, historical, and extranatural contexts, and they often expressed the complex interplay of these factors in the form of stories. Based on fieldwork among the Yuchis, a nation that shares many storytelling traditions with Creeks and other southeastern Indians, folklorist and anthropologist Jason Baird Jackson has argued that stories, including the Monster Lizard story, connect the mythical past to ritual life and the present circumstances of the community. Stories are thus, not dead or closed narratives of an inaccessible past, but bridges through which present and past, quotidian and extranatural can be brought together in adaptive and creative ways.[50]

Yaolaychi's story might have expressed southeastern Indians' interpretations of the relationships among precious metals, dangerous powers, and European colonialism. Indian groups throughout the Spanish American borderlands told stories in which deposits of minerals or buried treasure were closely tied with monsters or spirits. Such tales, it seems, reflected the world-changing effects of Europeans' violent search for American gold and silver and might have been meant to warn natives against revealing mineral

---

above summary is based on "Hitchiti Story 15: The Monster Lizard," ibid., 96–97, "Creek Story 20: The Hunter, the Monster Lizard, and the Panther," 27–28, "Creek Story 21: The Hunter, the Monster Lizard, and the Panther (Second Version)," 28–29; "The Red-Mouthed Lizard and the Hunters [told by Maxey Simms]," in Günter Wagner, *Yuchi Tales,* American Ethnological Society, *Publications,* XIII (New York, 1931), 106–112, "How the Yuchi Kill the Red-Mouthed Lizard [told by Andy Johnson]," 238–241; Hedvig Tetens Evans, ed., "Seminole Folktales," *FHQ,* LVI (1978), 474; and "Tiger Helps Man Defeat a Giant Lizard," in Earnest Gouge, *Totkv Mocvse, New Fire: Creek Folktales,* ed. Jack B. Martin, Margaret McKane Mauldin, and Juanita McGirt (Norman, Okla., 2004). On monsters in the Mississippian era, see Vernon James Knight, Jr., "Some Speculations on Mississippian Monsters," in Patricia Kay Galloway, ed., *The Southeastern Ceremonial Complex, Artifacts and Analysis: The Cottonlandia Conference* (Lincoln, Neb., 1989), 205–210.

50. On the interconnectedness of knowledge, its contexts, and stories, see John Reed Swanton, *Religious Beliefs and Medicinal Practices of the Creek Indians* ([Washington, D.C., 1928]), 489; Julie Cruikshank, *The Social Life of Stories: Narrative and Knowledge in the Yukon Territory* (Lincoln, Neb., 1998); Bill Grantham, *Creation Myths and Legends of the Creek Indians* (Gainesville, Fla., 2002), 10; and Clara Sue Kidwell, "Native American Systems of Knowledge," in Philip J. Deloria and Neal Salisbury, eds., *A Companion to American Indian History* (Malden, Mass., 2002), 87, 96–99. For Yuchi storytelling, see Jason Baird Jackson, *Yuchi Ceremonial Life: Performance, Meaning, and Tradition in a Contemporary American Indian Community* (Lincoln, Neb., 2003), 15–19, 100, 236–238.

treasures. These narratives were prevalent in northern Florida, often centering on the ruins of the Spanish mission of San Luis de Apalachee, which Creek slave raiders allied with South Carolina had destroyed in 1704. As late as 1824, one Floridian writer described how "the Indians have preserved a superstitious story, which keeps them at an awful distance from San Louis *[sic]*." "They say," he reported, "that the Spaniards, on quitting San Louis, buried their church ornaments and with them some bottles of medicine, (magic,) which would prove fatal to them if they touched [it]." Accounts such as this expressed enduring associations between disease, death, Spaniards, colonialism, and precious metals.[51]

Yaolaychi's skill as a performer, the settings in which he told his stories, and the congruencies between the three expedition members' understandings of nature all enhanced the capacity of the Monster Lizard tale to influence how the commissioners pursued and related knowledge. Like other native storytellers, Yaolaychi modified his presentations in response to his audience and their reactions. The second telling was thus far more detailed than the first because Gil and Castro now feared the nearby beast and pressed Yaolaychi for more details about it. Although many of the nuances of Yaolaychi's narrative were probably lost in translation, the way that he performed the story might have made an encounter with the Monster Lizard seem imminent. Furthermore, as historian Steven Hahn has noted, "The stage upon which the storyteller recites the tale ... might be viewed as an arena of power, where the storyteller may establish or reinforce his or her authority as a keeper of wisdom." By speaking up near the spot where the action of both the Monster Lizard legend and the expedition were situated, Yaolaychi emphasized his access to Florida nature while also strengthening his influence as a narrator.[52]

51. "Old Settlements in Florida," *East Florida Herald* (Saint Augustine, Fla.), Sep. 25, 1824 (quotes). For other native stories from the borderlands that associated Spaniards' search for minerals with dangerous powers, see John Sibley to William Dunbar, Apr. 2, 1805, in Eron Rowland, [ed.], *Life, Letters, and Papers of William Dunbar ...* (Jackson, Miss., 1930), 171; and Dunbar to Thomas Jefferson, Aug. 22, 1801, TJP. For associations between mines, conquest, and supernatural beings in South America, see Michael T. Taussig, *The Devil and Commodity Fetishism in South America* (Chapel Hill, N.C., 2010), 143–228.

52. Steven C. Hahn, "The Cussita Migration Legend: History, Ideology, and the Politics of Mythmaking," in Thomas J. Pluckhahn and Robbie Ethridge, eds., *Light on the Path: The Anthropology and History of the Southeastern Indians* (Tuscaloosa, Ala., 2006), 65 (quote). On the performative aspects of Creek storytelling and politics, see Saunt, *New Order of Things*, 31–33. For Gil and Castro pressing Yaolaychi for details,

Although Yaolaychi most likely did have a different understanding of nature than Gil and Castro, they all probably shared a belief that encounters with otherworldly powers were possible and particularly likely to occur at the specific spaces of the expedition: forest paths and bodies of water. Many Creek adventure stories—including most tales about the Monster Lizard—began with men travelling on forest trails and ended with deadly encounters with monsters, and bodies of water were thresholds to the Lower World and, thus, the abodes of dangerous beings. Most early modern Europeans would have agreed that forest-dwelling beasts like the Monster Lizard could exist. Like Yaolaychi, Gil and Castro would have experienced forests and water sources within multiple mythological and historical contexts, and spiritual associations could mark their scientific practices within these settings. Gil, for example, would claim to have smelted some metallic rocks near the spring, a basic mineralogical operation that he infused with Christian elements by "putting sticks in the form of a cross" in the fire below the minerals. Moreover, Gil and Yaolaychi had extensive contact with both European and native peoples in the Southeast, and, at least in Gil's case, his experience with Hitchiti speakers and their stories might have made him more open to believing Yaolaychi's tale. As Gil later told the governor, he "had heard talk for more than seven years that there was a wicked beast infesting the territory of the Shawnee Indians ... but that the first notice he had of a wicked beast in this province, was from [Yaolaychi]."[53]

---

see "Diario," Apr. 6, 1790, 436–437. As ethnographer Rodney Frey has suggested, "The creative power of language, coupled with the various techniques used by storytellers, all coalesce to help encourage the listeners of the stories to become participants in them, traveling the same trails alongside Salmon, Coyote, or Burnt Face." See Frey, "Oral Traditions," in Thomas Biolsi, ed., *A Companion to the Anthropology of American Indians* (Malden, Mass., 2004), 164.

53. "Diario," Apr. 6, 1790, 440 ("putting"); Interrogation of Alonso Gil, Apr. 8, 1790, 461–462 ("had heard"). Gil made it clear that he situated this cross independently of the logs burnt for heat. For eighteenth-century Creeks, "A spirit path and a path through the woods could be one and the same and a journey that appears mundane at our historical remove could take on universal importance in the right circumstances." See Hudson, *Creek Paths and Federal Roads*, 6 (quote), 14–21. Creek cosmology included three worlds: Upper, Lower, and Middle. Humans inhabited the Middle World but were in constant contact with the other two. See Grantham, *Creation Myths*, 21. For congruencies between native and European experiences traveling through woods, see James H. Merrell, *Into the American Woods: Negotiators on the Pennsylvania Frontier* (New York, 1999), 22–27, 139. For early modern Europeans' views on monstrosities, see Lorraine Daston and Katharine Park, *Wonders and the Order of Nature, 1150–1750* (New York, 1998), esp.

Like accounts of the 1790 expedition itself, Yaolaychi's stories expressed aspects of how power and knowledge were interconnected during and after journeys into the woods. The heroes in some versions of the Monster Lizard story, especially Shawnee and Yuchi variations, were medicine men or hunters whose triumph over the beast depended on, or led to the development of, esoteric knowledge about potent words or plants. The revelation of new sources of knowledge and power was less pronounced in Creek stories. Yaolaychi's telling and subsequent accounts of the 1790 expedition seem to be the only recorded variations of the story in which the Monster Lizard resided near mineral treasures. This detail might have reflected historical associations between dangerous forces, Spanish adventurers, and precious metals, or perhaps it was intended to resemble European stories about treasure-guarding beasts. Yet one point of Yaolaychi's story was clear: the Monster Lizard stood between the commissioners and the acquisition of mineralogical knowledge that, Benítez hoped, would promote Spanish power.[54]

Yaolaychi, Castro, and Gil made only a few observations and collections during the eight-day journey to the spring, but their activities were mostly in keeping with the treasurer's instructions. Castro and Gil gathered samples of useful plants, including China oranges and sarsaparilla, and described the topography and trees. After listening to Yaolaychi's story about his excursion to the cave on February 28, however, Castro's and Gil's inclination to follow scientific conventions began to wane. As Castro wrote in the diary, the Spaniards "were full of amazement and confusion with the reasoning of the Indian, and lost the effective desires of the commission for which they came." Based on Castro's notes and the itinerary in the diary, it seems they had arrived at or near Ichetucknee Springs, now a state park between the Suwannee and Santa Fe Rivers. According to recent mineralogical analyses, there is no evidence of silver, gold, or mercury at this site, but there is an abundance of phosphorescent algae. On March 1, Yaolaychi led the two Spaniards to the spring where he claimed they would find the shiny metal, and they proceeded to collect some stones, mud, and water that they put in sealed containers and "presented for tests" on their return to Saint Augus-

173–214. For Europeans' understandings of forests and water, see Simon Schama, *Landscape and Memory* (New York, 1996), 207–239, 257–267. On how symbolic and religious associations could influence scientific practices, see James Delbourgo and Nicholas Dew, "Introduction: The Far Side of the Ocean," in Delbourgo and Dew, eds., *Science and Empire in the Atlantic World*, 12–13.

54. Jackson, *Yuchi Ceremonial Life*, 213–214, 223–228, 236–238.

tine. Yet the commissioners failed to "do any excavations" near the spring and could not find any of the shiny rocks that Yaolaychi had promised.[55]

Benítez's instructions faded further from the commissioners' minds as Castro and Gil grew upset about the lack of precious metals. The Spaniards "attacked [Yaolaychi] anew, saying that he was a liar," and, as on other occasions, Yaolaychi became furious when the Spaniards questioned his credibility. He told them that he would risk death to prove his veracity and "that if they had any honor they would join him in searching for the place of the beast, but if not then he would go alone, and if he did not return tonight ... it would mean that he has been killed." Yaolaychi then marched off towards the cave, and, reluctantly, Castro and Gil followed.[56]

When they arrived near the beast's cave, Yaolaychi stripped off his clothes, put on a loincloth, and covered his face and neck with musk oil. He advised Gil and Castro to follow suit, and, "in effect they imitated him," removing their clothes and donning loincloths fashioned from their scarves. This shift in physical appearance was so appropriate that it could pass for a literary device: they had become the hunters in their own version of the Monster Lizard story. Yaolaychi then led the two Spaniards towards the cave, showing them "many footprints and traces of the Beast, tracks that were the same as a bear's, but much larger." Darting from one pine tree to the next for cover, the three men approached "the cave of the mine." When they were about one hundred yards away, Yaolaychi froze in his tracks, and, "speaking low to the interpreter, and signaling with his finger that the Animal was there, he withdrew immediately, fleeing with the greatest haste, full of fear." The Spaniards soon realized why. Castro recorded that "they saw the beast" prowling around its den; its body was "very brilliant, silver-plated and smooth, rays of the sun reflect [off it,] just as in the glass of a mirror." These were the only observations that the Spaniards were able to make, for they soon followed Yaolaychi's example and ran away. Lacking the courage to confront the lizard, the three men retreated to Saint Augustine.[57]

55. "Diario," Apr. 6, 1790, 434–435, 437 (quotes). Castro noted that "the Indians call the spring *Oycayagua,* spring of Water." See ibid., 438 ("the Indians"); Thomas M. Scott et al., *Springs of Florida,* Florida Geological Survey, Bulletin no. 66 (Tallahassee, Fla., 2004), 30–36, 71–76; and Brenda J. Herring and Walter S. Judd, "A Floristic Study of Ichetucknee Springs State Park, Suwannee and Columbia Counties, Florida," *Castanea,* LX (1995), 347.

56. "Diario," Apr. 6, 1790, 437 ("attacked"), 438 ("that if").

57. Ibid., 438 (quotes). Gil later recalled that Yaolaychi "seemed very worried about taking off his boots and shirt, and putting on a loincloth," and it is possible that Yaolaychi considered this specialized clothing as essential to surviving the encounter with the

In their diary, reports, and eventual interrogations, all three expedition members insisted they had seen Achuguilipalascó. Royal Treasurer Benítez had expected the expedition to produce natural knowledge that would bolster Spain's geopolitical power. According to the commissioners, however, it was the alleged power of a beast defined through Hitchiti stories that determined the scope and success of their mineralogical activities. Native storytelling, particularly in Indian-dominated spaces, could undermine imperial officials' presumptions that efforts to acquire knowledge would lead to greater power.[58]

Benítez's goal for the expedition had been to accrue enough evidence of valuable natural productions to convince Governor Zéspedes to approve of a second expedition to the spring, and, in the days immediately following the commissioners' return, Benítez feared that they had failed. He complained that the botanical samples Gil and Castro had gathered were improperly preserved and useless. The mineral samples were equally disappointing, and Benítez's chemical analyses of the rocks, mud, and spring water found little evidence of precious metals. Twelve days after the expedition's return, however, Benítez's luck apparently changed. He was "meditating over the observations that I should practice to ascertain something good from the waters, mud, rocks, and plants that they had presented to me" when Gil brought him "three little pieces of a metal that seemed like gold due to its specific gravity and fine color." Gil claimed that he had discovered and smelted these nuggets near the spring while his two companions were asleep and had since forgotten about them. After conducting some experiments, Benítez reported that this metal was "very fine copper . . . , gold of low quality, or platinum." Ten days later, Gil told Benítez that all three expedition members had observed "shiny, silvery little globes" beneath the spring water but had not collected any of this substance because they were scared of entering the cold water. Benítez thought that these globules "may well have been Mercury, [which] can be more important to us than any other metal." He rued their failure to bring any of this material back to Saint

---

beast: at least one Creek version of the Monster Lizard story included a similar change of dress before a hunter speaking powerful words was able to outmatch the beast in a footrace. The musk oil was probably meant to mask Yaolaychi's smell. See Interrogation of Alonso Gil, Apr. 9, 1790, 463; and "Creek Story 19: The Monster Lizard," in Swanton, *Myths and Tales*, 26.

58. Interrogation of Yaolaychi, Apr. 4, 1790, EFP, sect. 89, bundle 353, no. 19, reel 173, 451, Interrogation of Bartolomé de Castro, Apr. 8, 1790, sect. 89, bundle 353, no. 19, reel 173, 456, Interrogation of Alonso Gil, Apr. 9, 1790, sect. 89, bundle 353, no. 19, reel 173, 466.

Augustine as another unfortunate "result of commissioning two ignorant men for an undertaking of the most considerable importance."[59]

Despite the legitimizing power of his experiments and mineralogical expertise, Benítez worried that Gil's belated and shady presentation of the new materials might not appear credible enough to the governor to merit a second expedition. He therefore looked to the expedition's other major discovery, the Big Ferocious Animal, to add validity to the commissioners' mineralogical collections and observations. Instead of isolating the mineralogical aspects of the expedition from the zoological ones, Benítez made the beast a central focus of his report to the governor and used his analysis of it as a way to verify the truth of the expedition's mineralogical claims. As he informed the governor, "although it could be doubted if this metal is truly extracted as [Gil] declares," he would "make for Your Lordship the same reflections [about it] that [were] in order for the beast." Even in Saint Augustine, the process of confirming mineralogical facts was intertwined with the natural history of a beast in a Hitchiti story.[60]

Benítez looked to famed European naturalists, Indian experts, and Saint Augustine's learned elite to add weight to his claim that the expedition members had been honest in describing the beast and, therefore, their mineralogical observations and collections could be valid as well. Spanish and other European men of science shared an increased interest in monstrosities during the late eighteenth century and, by contemporary standards, Benítez was right to argue that the beast might exist. Although, as he told the governor, some of his peers in Saint Augustine had denied the beast's reality because "the famous Buffon, and other authors, do not describe an animal that is in all respects similar," Benítez retorted that nature had not yet presented "all of its wonders" and that there were many hitherto unverified kinds of "monstrosities, that mixed-species will produce." He even chided his critics for "admiring the same Buffon, who says he has spent sixty years observing rare prodigies of nature, and finds himself more ignorant about it than on the first day of his studies." The treasurer drew on Indian experts to further support his case and, on March 22, interviewed "Ocaspa of the Seminole Nation, because he was one of those that had information

59. "Diario," Apr. 6, 1790, 439–440 ("meditating"), 440 ("three," "shiny"), 441 ("may well," "result"); Benítez to Zéspedes, Apr. 7, 1790, 443–444, 446 ("very fine"). There was a state-sponsored search for mercury throughout Spanish America in the 1780s and 1790s; some linguists in New Spain used the knowledge they believed was embedded in Indian place-names to search for mercury deposits. See Cañizares-Esguerra, *How to Write the History of the New World,* 308.

60. Benítez to Zéspedes, Apr. 7, 1790, 446.

on the beast of which the travelers had spoken." Speaking through an interpreter, Ocaspa said that "he had not seen it, but that his countrymen have informed him, that there certainly is [a beast]" and that "its body shines like a mirror." From Benítez's perspective, Ocaspa's testimony might have been more authoritative than Yaolaychi's: Ocaspa presumably lived in Florida, and his account of the monster came from a prominent Creek, Philatouche, whose social status lent credibility to the tale. Benítez also hosted a kind of scientific conference in which some of Saint Augustine's most learned men—a group that included the captain of engineers, a physician at the royal hospital, a Cuban pharmacist, an Irish "Professor of Chemistry," and the "very knowledgeable" planter Francis Philip Fatio—witnessed "experiments with the metal" and discussed the beast's characteristics. As Benítez assured the governor, "the presence of persons that I considered learned in the material" made his experiments and zoological deliberations "authoritative." These men concluded that even though the beast "is lacking some of the signs of the true figure of a bear, it is not surprising that it would be this animal, and some mix with another; much information agreeing that it is ferocious and that there are four in these provinces." These educated Europeans in the colonial capital based all of their knowledge of the beast on narratives derived from the Monster Lizard story.[61]

61. Ibid., 445 ("famous," "all," "monstrosities," "admiring," "Professor," "very," "experiments," "presence," "authoritative," "is lacking"); "Diario," Apr. 6, 1790, 440 ("Ocaspa," "he had not," "its body"). On natural history and monstrosities, see Paula De Vos, "The Rare, the Singular, and the Extraordinary: Natural History and the Collection of Curiosities in the Spanish Empire," in Daniela Bleichmar et al., eds., *Science in the Spanish and Portuguese Empires, 1500–1800* (Stanford, Calif., 2009), 271–289; Stephen T. Asma, *On Monsters: An Unnatural History of Our Worst Fears* (Oxford, 2009), 123–140. "The famous Buffon" was French naturalist Georges-Louis Leclerc, comte de Buffon (1707–1788). On Philatouche, a Lower Creek leader of mixed African and Indian descent, see Christina Snyder, *Slavery in Indian Country: The Changing Face of Captivity in Early America* (Cambridge, Mass., 2010), 236–237. The captain of engineers, physician, pharmacist, and professor were, respectively, Mariano de LaRocque, Bernardo de Madrid, Ramón de Fuentes, and Thomas Travers. On the demography of Saint Augustine, see Sherry Johnson, "The Spanish St. Augustine Community, 1784–1795: A Reevaluation," *FHQ*, LXVIII (1989), 38–39. On the ways that experiments and observations relied on credible witnesses to establish matters of fact, see Steven Shapin and Simon Schaffer, *Leviathan and the Air-Pump: Hobbes, Boyle, and the Experimental Life* (Princeton, 1985), esp. 55–60. For how some of North America's most learned naturalists saw scientific value in Indian monster stories, see Benjamin Smith Barton, "Indian Account of a Remarkably Strong and Ferocious Beast, Which (They Say) Existed in the Northern Parts of the State of New-York . . . ," in *The Philadelphia Medical and Physical Journal*, I, part II (1805), 164;

Benítez argued that the expedition had made significant discoveries "in the three kingdoms, animal, vegetable, and mineral," including a bizarre beast unknown to natural history, the "useful commodity" sarsaparilla, and precious metals. Yet he recognized that all of his conclusions about the expedition's findings relied on whether or not the three commissioners were telling the truth about what they saw and did near the springs. Benítez therefore proposed interrogating the expedition members, particularly Yaolaychi, to verify their credibility. He told the governor that "it will be necessary to detain the Indian, and his family, to further justify the reality of all of this; so as not to deprive the State, this Province, and the Public, of the benefits that could result [from] not dismissing all that has been said as fabulous, and the metal [as] a fiction." Zéspedes agreed. From April 4 to 9, Zéspedes interrogated Yaolaychi, Castro, and Gil about the minerals, territories, and monster they allegedly encountered during the expedition. Like Benítez's laboratory and the spring, the interrogation room was a site where narratives and power relations shaped the circulation and validation of natural knowledge.[62]

The interrogations made it clear that information about valuable minerals was potentially critical to regional geopolitics. The governor's overriding concern in early 1790 was maintaining good relations with Creek leaders, and, like before the expedition, he tried to ensure that the travelers were not violating Indian territorial boundaries. He asked all three commissioners if the spring was "in the district that the Indians believe to be theirs." Although Gil and Castro were unsure if that land belonged "to the Indians or us," Yaolaychi responded that "the spring proper is inside the limits of the Indian hunting grounds." Natives remained more populous and powerful than the Spaniards in Florida, and Benítez voiced anxiety about "the grave damage that would result" if Indians controlled the mineral wealth at the spring. Regional competitions with Indian nations could influence Spanish officials' pursuit of natural resources as much as Atlantic competitions with European empires.[63]

Powerful Creeks such as Alexander McGillivray would have taken the presence of mines in their territory, and a betrayal of their existence to

---

and Paul Semonin, *American Monster: How the Nation's First Prehistoric Creature Became a Symbol of National Identity* (New York, 2000), 101, 182–183, 308–310.

62. Benítez to Zéspedes, Apr. 7, 1790, 447–448 (quotes, 447).

63. Interrogation of Alonso Gil, Apr. 9, 1790, 467 ("in the district"); Interrogation of Bartolomé de Castro, Apr. 8, 1790, 458 ("to the Indians"); Interrogation of Yaolaychi, Apr. 4, 1790, 449 ("spring"); Benítez to Zéspedes, Apr. 7, 1790, 446 ("grave").

Spanish or U.S. officials, very seriously. During this period, some wealthy Creeks were working to turn the loose confederation of Creek towns into a Creek nation, a political entity against which individual Creeks could commit crimes, and Yaolaychi was well aware that some Creeks would construe his activities as criminal. He reportedly claimed that the Creeks "have some knowledge of the value of the mine" and feared that the nation's leaders would execute him for revealing it to the Spanish. Zéspedes asked Gil if Yaolaychi "would not want his compatriots to be aware of the motives and ends of the expedition." Gil replied "that it is certain, that the Indian [had told him this] because the rest of the Indians, even though they knew about the brilliant rocks, have never wanted to give any part of them to the English, nor the Spaniards; and if they knew that [Yaolaychi] had made it known, they would kill him." Creek leaders, like European ones, hoped to control the flow of information and resources across national boundaries.[64]

It is possible that Creek leaders would have killed Yaolaychi for revealing the location of a mineral deposit, a threat evident in later Creek stories about mines within their territory. These stories were meant to keep individual Creeks from exploring and revealing sites of mineral wealth, but they also expressed how Creek leaders understood natural resources in the context of regional geopolitics. Writing from Creek country in 1801, adventurer John Devereux DeLacy told Thomas Jefferson that "it is said and asserted as a fact that there is a gold mine in this Country" and that "an Indian took a small quantity of the ore to the Spaniards at Augustine who having essayed it begged of the Indians to bring him a larger quantity and that he would pay him well for it." However, when news of this came "to the ears of the Head of the Nation they assembled in general Council and decreed the death of any person that should ever attempt to shew discover or open the mine or take carry or give any of the ores to any white persons whatsoever but especially to the Spaniards whose inordinate thirst for it they considered as the sources of all the evils inflicted on them and the many oppressions exercised by the Spaniards in this Country heretofore." According to DeLacy, the Creeks feared that Spanish knowledge of the mine could lead to war and "the loss of their Country and independence." He concluded that

64. Benítez to Zéspedes, Apr. 7, 1790, 446 ("have some knowledge"); Interrogation of Alonso Gil, Apr. 9, 1790, 467 ("would not"). On changes in Creek society, see Saunt, *New Order of Things*. For a different way that storytelling and conceptions of the land overlapped "to promote compliance with standards for acceptable social behavior," see Keith H. Basso, *Wisdom Sits in Places: Landscape and Language among the Western Apache* (Albuquerque, N.Mex., 1996), 41.

the Creeks have therefore "wisely resolved to shut [the mine] up for ever if such thing there be." DeLacy's story might have been about Yaolaychi, and the detail that the Spaniards offered to pay him for the metal suggests that Yaolaychi's motives were, in fact, financial. More significantly, DeLacy found that Creek leaders' fears about their nation's weakness, combined with enduring associations between precious metals and Spanish violence, guided their decision to keep mineralogical knowledge hidden.[65]

The stakes of discovering precious metals were high for Spanish leaders as well, so just as Benítez had done in his report, Governor Zéspedes incorporated information from stories of the beast to better determine the veracity of the expedition's questionable mineralogical findings. Yaolaychi was the first of the expedition members that Zéspedes interrogated, and, when the governor asked him to describe the beast, Yaolaychi told his story for a third time.

The Story, Version Three: Told April 4, 1790

Questioned [Zéspedes]: What did you do when you saw the beast, and give a description of it.

Responded [Yaolaychi]: That he saw the beast at the distance of some 200 rods, . . . it has a round head, not so big as the rest of it, like that of a bull, without ears, and without horns, that it is about three feet high, with four legs, similar to those of a caiman, but longer, with some very long and sharp claws: the tail is also similar to a caiman, but shorter, and that its movement is so swift, that not even deer can escape it: the thickness of its body is about six feet, and its length about two rods, he cannot figure out how to express its color; the best that he can do is compare it to the rays that a metallic mirror gives off in the sun, and he has heard it said, but he does not know it from experience, that bullets do not penetrate it.

Questioned: If there are other beasts of the same species in this province.

Responded: That there could be many, and that he has heard that there is one fully similar near the stream of Rollestown . . . .

Questioned: If he knows if the beast has killed or eaten men.

Responded: That he does not have any proof of what the beast has killed or eaten other than what he has heard said, and in particular he remem-

65. John D[evereux] DeLacy to Thomas Jefferson, Dec. 18, 1801, TJP, [8], [9] (quotes), [15].

> bered having it recounted that two Indians having on one occasion left to hunt bears, with many dogs, they encountered the beast, or another of its species, it climbed into a tree, and the dogs rushed to attack it, it grabbed hold of them, and killed them with its tongue, and once the dogs were dead, it went down from the tree, and overcoming one of the men, took him and dragged him a good distance until by chance it met with a tiger, and while [the tiger] attacked the beast [the man] was able to escape, alive but very badly hurt: [Yaolaychi] adds that the beast has a tongue divided in two, and that it wields [its tongue] so effectively that it makes flying eagles come to its mouth.

Yaolaychi included the story of the hunters on all three occasions that he described the monster, yet he altered how he employed evidence about the story and the beast during his interrogation. In the first two tellings, he had sworn to the "proofs" and "veracity" of his description of the monster and included the story of the hunters as evidentiary context that buttressed his claims. In his interrogation, however, he drew distinctions between what he knew "from experience" and what he had "heard said" and made it clear that he had "no proof" regarding some aspects of the Monster Lizard story. Yaolaychi's authority and impact as a storyteller, and the epistemologies on which he based his authority, depended on his audience and his stage. In the Indian-controlled spaces of Florida's interior, his storytelling shaped the expedition's knowledge-gathering practices and the Spanish commissioners' accounts of their activities and findings. In Spanish-dominated Saint Augustine and, particularly, during an official interrogation, Yaolaychi modified his relation to better conform to the expectations of the powerful men questioning him. Neither natives nor Spaniards had a single set of methods for understanding or describing nature in the Florida borderlands; power and place influenced how Indians and Europeans alike approached the acquisition and relation of knowledge.[66]

Yet, even in a Saint Augustine interrogation room, Yaolaychi's story con-

66. Interrogation of Yaolaychi, Apr. 4, 1790, 451–452 ("Questioned [Zéspedes]: What did you do," 451, "with its tongue," 452); "Diario," Apr. 6, 1790, 436 ("proofs," "veracity"). John Hambly translated Yaolaychi's responses during the interrogation into English and another interpreter converted these words into Spanish. Rollestown was about twenty-seven miles southwest of Saint Augustine on the east bank of the Saint Johns River. The intervention of a tiger or panther was common to several versions of the Monster Lizard stories recorded in the twentieth century, and Yaolaychi's account during his interrogation was particularly similar to a Hitchiti version recorded by John Swanton. See "Hitchiti Story 15: The Monster Lizard," in Swanton, *Myths and Tales*, 96–97.

tinued to affect European narratives about the expedition and the knowledge it produced. Both Gil and Castro repeated for the governor how they had stripped down, stalked the beast, and seen a mirror-colored monstrosity. They both insisted that the beast looked just as Yaolaychi had described it, and Gil supplemented his testimony with information from Yaolaychi's story about the hunters and the monster, an account that he related as if it were fact.[67]

Indeed, the Monster Lizard story might have determined how Gil and Castro carried out and recounted all of their knowledge-gathering activities near the spring. Yaolaychi told the governor that "the two white men" had not actually visited the spring, observed the beast themselves, nor conducted mineralogical work at the correct spot because they were too afraid of the monster to go anywhere near it. He claimed that he had informed Gil about the beast before the start of the expedition, and "Gil responded that he was not a fearful woman, and that the beast would not frighten him: yet experience verified the contrary." According to Yaolaychi's testimony, his stories had such power over Castro and Gil that all their accounts of the expedition amounted merely to elaborations on his own tales. Yaolaychi's Monster Lizard story, as mediated by translators and situational power relations, was more than a vehicle through which native natural knowledge circulated to Europeans; it constituted and circumscribed the knowledge that commissioners in the field and analysts in Saint Augustine produced.[68]

On April 10, Governor Zéspedes sent Benítez his final decisions regarding the validity of the mineralogical, zoological, and geopolitical informa-

67. Interrogation of Bartolomé de Castro, Apr. 8, 1790, 456; Interrogation of Alonso Gil, Apr. 9, 1790, 466. Gil and Castro were not the first individuals of European descent in Florida to narrate themselves as the protagonists in stories centered around monstrous lizards that terrorized Indians. A French officer stationed in Florida in the 1560s described several "monsters," including "a lizard ... with the head and neck of a serpent; it also had wings and it flew wherever it perceived some man, or some woman, or a child, and devoured them ... everyone in the country told us that the lizard was not causing so much trouble at that time, and that they had several times tried to kill it, which they were unable to do." The officer was pleased to report, though, that the Frenchmen killed the monster and its young with arquebus bullets, and "the inhabitants were very happy over being delivered from that evil beast, and thanks was given to God." See Giles de Pysière, "Discourse on the Enterprise and Pillage That the Criminals in the Isle of Florida Did to Their Captains and Governors.... with a Description of the Wild Beasts ...," in Charles E. Bennett, ed., *Laudonniere and Fort Caroline: History and Documents* (Gainesville, Fla., 1964), 74–75 ("monsters," 74, "inhabitants," 75).

68. Interrogation of Yaolaychi, Apr. 4, 1790, 450 ("two white men"), 451, 452 ("Gil responded").

tion embedded in the commissioners' narratives and whether he would grant the treasurer's request to fund a second expedition to the spring. Zéspedes accepted that the mineralogical collections and observations that Gil presented in Saint Augustine were true and potentially valuable. Yet the governor told Benítez that, although "I agree in the reality of the mine, my natural sincerity obliges me to express to you that I in no way give credit to the existence of a quadruped animal, whose skin repels bullets and reflects light like mirrors in the sun." Zéspedes did not concur with Benítez and other learned Saint Augustinians that the Monster Lizard could be real. He argued instead that Gil and Castro believed they were telling the truth about seeing and stalking the monster but that this belief had developed because Yaolaychi had worked his wiles on "the imagination of the explorers." "The cunning Indian," the governor wrote, "used a trick to turn the kind of animal nature regularly produces into a Monster that never existed except in the stories that, as you know, the old Indians, just the same as us whites, tell to amuse their grandchildren." Zéspedes recognized that Yaolaychi's "cunning" was to blur distinctions between what "nature regularly produces" and "stories" so effectively that his tales could become the basis for natural knowledge among whites and Indians alike.[69]

The governor thought that the expedition's mineralogical findings were valid and potentially valuable, but he understood that any collection of natural resources had to be approached within a larger geopolitical context. His primary concerns as governor were maintaining peace with the region's natives and forestalling Anglo-American incursions, and he decided that, "even if the mine really were of mercury, or gold, far from being in the interest of the royal service, it would actually be exceedingly harmful to it." Based in part on Yaolaychi's testimony, Zéspedes determined that the spring was located "in a district which the Indians consider to be theirs," and he denied Benítez's request for a second expedition because he feared that violating the boundaries of the Creeks' hunting grounds would alien-

69. Vicente Manuel de Zéspedes to Bartolomé Benítez y Gálvez, Apr. 10, 1790, EFP, sect. 16, bundle 78, no. 17, reel 30 (quotes). On Europeans' perceptions of Indians' natural knowledge as cunning, see Parrish, *American Curiosity,* 215–258. The theory that supposed encounters with mythical powers could be explained as byproducts of fear, and the skill of certain cunning individuals to exploit that fear, was well established among educated Europeans in the eighteenth century. See Frank E. Manuel, *The Eighteenth Century Confronts the Gods* (Cambridge, Mass., 1959); and James Delbourgo, "Fugitive Colours: Shamans' Knowledge, Chemical Empire, and Atlantic Revolutions," in Simon Schaffer et al., eds., *The Brokered World: Go-Betweens and Global Intelligence, 1770–1820* (Sagamore Beach, Mass., 2009), 284, 317.

ate Spain's Creek allies and could even lead to a war among natives, Spaniards, and Anglo-Americans. So, just like the "Head of the Nation" in the Creek story that DeLacy recorded in 1801, Zéspedes proclaimed the mine off-limits. The expedition had revealed congruencies between native and European approaches to nature, and the way Spanish and Creek leaders reacted to the mineral knowledge it generated suggests key similarities in their approaches to regional geopolitics: long-term political concerns and fears of exposing their own vulnerability trumped the potential benefits of exploiting sites of mineral wealth. Colonizers were not necessarily interested in or capable of producing power-promoting knowledge, a pursuit that depended on an empire's strength on the ground, diplomatic clout, and geopolitical goals. Instead of making Spain's presence in Florida stronger, the 1790 expedition highlighted the empire's regional weakness.[70]

Power relationships—Zéspedes's authority as governor and the exigencies of regional geopolitics—determined what naturalists in Madrid and elsewhere would learn about Florida's nature: I have found no evidence that Spanish accounts of the expedition traveled to Europe. Yet, while Spain's recorded history of the expedition was finished, stories about it, and the story within it, persisted after Spain gave up its claim to East Florida. In 1824, a Seminole chief told Anglo-American naturalist James Pierce about "a monster, with a large serpent's body shining like silver, whose breath is destructive to all that approach" that "occupies a large sink or cave in East Florida, guarding a mine . . . ." "The Spanish authorities," the author concluded, "made a fruitless search for this treasure a few years hence." Although Pierce dismissed this story as "an instance of Indian credulity," native stories about the 1790 expedition—such as those recorded by DeLacy in 1801 and Pierce in 1824—demonstrated that its events, historical and mythical associations, and political significance remained known among Florida Indians and, through their narratives, Euro-Americans.[71]

70. Zéspedes to Benítez, Apr. 10, 1790 ("even if the mine," "in a district"); DeLacy to Jefferson, Dec. 18, 1801, TJP, [9] ("Head of the Nation"). On Zéspedes's goals, see Tanner, *Zéspedes in East Florida*, 220.

71. James Pierce, "Notices of the Agriculture, Scenery, Geology, and Animal, Vegetable and Mineral Productions of the Floridas, and of the Indian Tribes," *American Journal of Science*, IX (1825), 135 (quotes). The Monster Lizard story has proven even more resilient and remains a significant part of the folklore and ritual life of southeastern Indian groups, many of whom now reside in western North America. Although the mineralogical knowledge that native storytelling generated in 1790 has been forgotten, the story itself has continued to inspire knowledge about native identity. See Jackson, *Yuchi Ceremonial Life*, 111–116, 213–214, 234–240.

The case of the 1790 expedition, therefore, not only suggests that knowledge production and power relations were interconnected in borderland encounters but also that power (and weakness) could determine the itineraries through which knowledge moved beyond borderlands and across time. Written accounts that fit information into scientific frameworks were the primary means through which natural knowledge traveled from the Americas to Europe, but perhaps this had less to do with any inherent advantage to European methods of relating information than it did with the fact that the Atlantic ocean was a space dominated by European empires and mediated by their technologies. Within the Americas, particularly the vast parts of the hemisphere in which European empires were weak, Indian storytelling might have circulated knowledge—and the narrative, historical, and spiritual contexts in which it was embedded—more effectively than European texts.

Indian-European power relations at the individual, local, and regional levels did have an impact on how European colonists and natives incorporated and related each other's knowledge. These potential influences were, however, contingent on several factors, including the settings in which knowledge was presented, the congruencies between native and European approaches to nature, and the narrative forms in which knowledge was embedded. Power relations, place, and narrative—and not any set difference between Indian and European epistemologies—shaped the pursuit, circulation, and validation of natural knowledge in the Gulf South.[72]

• • •

For diverse inhabitants and Spanish officials alike, access to difficult-to-acquire knowledge remained as tantalizing a path to prestige and power in the late eighteenth-century Gulf South as it had been in the early colonial era. But attempts to benefit from natural knowledge often proved frustrating or even dangerous. This is because no group, including the Spanish Empire, was powerful enough to access, share, verify, and apply power-promoting knowledge in isolation. Weakness drove European officials to rely on self-interested power brokers and informants, inspired enslaved blacks to seek better positions within a slave society (but also exposed them

72. On the role of storytelling in political relations, see Piker, *Four Deaths of Acorn Whistler.* On native power in Florida, see Weber, *Bárbaros,* 75, 204, 214; and Frank, "Taking the State Out," *FHQ,* LXXXIV (2005), 10–27. On how political and military power enabled Indian conventions to hold sway in intercultural relations with Euro-Americans in borderland regions, see Juliana Barr, *Peace Came in the Form of a Woman: Indians and Spaniards in the Texas Borderlands* (Chapel Hill, N.C., 2007); and Pekka Hämäläinen, *The Comanche Empire* (New Haven, Conn., 2008).

to brutal punishments), and motivated natives to circulate knowledge despite the risk of angering both European and Indian authorities. All of these individuals packaged knowledge in narratives that reflected and perpetuated the crossing of cultural boundaries and, moreover, they all attempted to draw on these often ambiguous accounts as they made decisions that shaped their lives and, ultimately, the region's history. Encounters inspired by imperialism—violence, exchanges, competitions, and engagements with nature and history—conditioned the content and trajectories of these narratives. These same forms of encounter would continue to affect knowledge and narratives in the Gulf South during the early years of U.S. expansion, when the United States was yet another relatively weak contestant for the region, and into the antebellum era, when Anglo-American power was firmly established.

{ 3 }

ASTRONOMY AND U.S. EXPANSION IN THE LOWER MISSISSIPPI VALLEY

• • •

Less than twenty years after achieving independence from Britain, the United States began expanding into Spanish America. In 1795, U.S. diplomats obtained part of West Florida from Spain, and, by 1810, the United States had incorporated the rest of the lower Mississippi Valley.

U.S. expansion into the lower Mississippi Valley, and the encounters that resulted from it, evinced and inspired many of the ways that officials and experts in the United States used natural knowledge to promote territorial growth. This was particularly apparent in the case of astronomy, an observational science critical to claiming and organizing the earth's surface. As in European empires, officials and experts in the early republic measured the heavens to survey international boundary lines, establish the precise location of important spots, navigate between ports, and produce accurate maps. And although it is fairly obvious that these practical applications of astronomy bolstered imperialism, even the basic science of observational astronomy had imperial foundations and consequences: much as in other powers, polycentric networks of technology and information in the United States supported astronomy beyond urban observatories and enabled the circulations and calculations needed to turn observational data into useful geographic knowledge.[1]

Yet Anglo-Americans did not simply export scientific practices to the United States' new territories. U.S. expansion encouraged peaceful and violent encounters among Spaniards and Anglos, masters and slaves, inhabitants and administrators, and whites and Indians that shaped the practice

1. On astronomy and imperialism in European powers, see William J. H. Andrewes, ed., *The Quest for Longitude* (Cambridge, Mass., 1996); Alex Soojung-Kim Pang, *Empire and the Sun: Victorian Solar Eclipse Expeditions* (Stanford, Calif., 2002); María M. Portuondo, *Secret Science: Spanish Cosmography and the New World* (Chicago, 2009), 210–256; and Simon Schaffer, "'When the Stars Threw Down Their Spears': Histories of Astronomy and Empire," Tarner Lectures (2010), Trinity College, University of Cambridge, MP3, http://sms.cam.ac.uk/collection/741056.

of astronomy in the Gulf South and, moreover, influenced how astronomy and imperialism overlapped in the United States on the whole. The earliest such encounters occurred during the Florida boundary survey of 1798 to 1800, an astronomical expedition that marked the first of many times the United States would extend its own boundaries at Spanish America's expense. Astronomers serving Spain and the United States approached this survey in part as a competition to prove their respective nations' scientific excellence. Nevertheless, both sides depended on local slaves to run the boundary, and both sides proved too weak to perform their region-ordering astronomical work in the face of native resistance. In the years after the survey, eastern officials and men of science corresponded with experts in the Mississippi Valley to develop astronomical methods that encouraged the acquisition and exploration of the continent, and some local astronomers, especially the planter William Dunbar, proved adept at leveraging their power over enslaved blacks into national prominence. More broadly, U.S. expansion into the borderlands meant that the United States and its scientific community came to include Spanish American peoples and territories. Far from looking down on these new additions, Anglo-American officials and astronomers were eager to collaborate with former Spanish subjects and even looked to the Spanish Empire as a model for how the federal government should develop its own scientific institutions and practices.[2]

2. Historians of science have focused far more attention on the United States' postcolonial attachments to Europe, especially Britain, than on the United States' own continental expansion. Postcoloniality does help explain some aspects of transatlantic scientific exchanges: men of science in the United States did covet European recognition, rely on Europe for many instruments and theories, and worry that their nation was intellectually backward compared to France and Britain. But overemphasizing the postcolonial aspects of the early republic has led to an incomplete and misleading picture of natural knowledge in the United States, a rising imperial power whose own expansion affected knowledge production among Anglo-Americans and the peoples they encountered. For postcolonial readings of natural knowledge in the early United States, see George Basalla, "The Spread of Western Science," *Science*, [New Ser.], CLVI (1967), 614, 617; Norriss S. Hetherington, "Mid-Nineteenth-Century American Astronomy: Science in a Developing Nation," *Annals of Science*, XL (1983), 61–80; and Kariann Akemi Yokota, *Unbecoming British: How Revolutionary America Became a Postcolonial Nation* (Oxford, 2011), esp. 153–191. The diverse community of astronomers that developed amid expansion long anteceded the corps of professional astronomers that historians have identified as heralding the rise of American astronomy in the late nineteenth century. See John Lankford, *American Astronomy: Community, Careers, and Power, 1859–1940* (Chicago, 1997), 1, 6.

Nevertheless, historians have obscured, ignored, and forgotten the ways that expansion into Spain's northern territories affected the pursuit and application of natural knowledge in the United States. This historiographical neglect has several roots: Anglo-Americans became increasingly dismissive of the intellectual abilities of borderland peoples in later decades of the 1800s; historians have approached the continent's borderlands as spaces without a meaningful history of knowledge; and, most fundamentally, U.S. historians have largely taken Spanish backwardness and irrelevance for granted. The incorporation of Spanish places and peoples from 1798 to 1810, however, suggests that practitioners and practices in the borderlands did matter to the history of knowledge at the national level and, even more basically, that astronomy in the early United States was more important and interesting than historians have realized. In the United States, imperialism and astronomy developed together.[3]

THE FLORIDA BOUNDARY

From 1798 to 1800, astronomers representing Spain and the United States surveyed the boundary between the Spanish Floridas and the United States that had been settled through the 1795 Treaty of San Lorenzo. Although several histories have explored the context and consequences of this international boundary, they have viewed meridian astronomy—the practice of observing the movement of celestial objects to determine geographic positions or draw lines across the spherical earth—as simply the means to a geopolitical end. Yet astronomy itself was a meaningful field of competition between Spain and the United States, even as both sides cooperated to survey the borderline along the 31st parallel. The astronomical practices and rhetoric of both Spanish and U.S. experts were expressions of national pride and assertions of geopolitical power. Furthermore, the survey itself revealed the messiness of the Gulf South's international and interethnic relationships, a complexity that belied the boundary's presumption of order. Anglos served

3. The few historians who have focused on early American astronomy have tended to conclude that it was underdeveloped and, therefore, more or less irrelevant to either U.S. history or the larger history of science. See Deborah Jean Warner, "Astronomy in Antebellum America," in Nathan Reingold, ed., *The Sciences in the American Context: New Perspectives* (Washington, D.C., 1979), 59; Stephen G. Brush, "Looking Up: The Rise of Astronomy in America," *American Studies,* XX, no. 2 (Fall 1979), 41–67; John C. Greene, *American Science in the Age of Jefferson* (Ames, Iowa, 1984), 129–157; and Marc Rothenberg and John Lankford, "United States of America, Astronomy," in Lankford, ed., *History of Astronomy: An Encyclopedia* (New York, 1997), 538.

as astronomers for Spain and the United States, both commissions relied on slaves, and Indians both worked for and resisted the boundary survey.[4]

For Spain, astronomy had been a source of prestige and power since the earliest decades of overseas expansion, and maritime competition from Britain further spurred Spanish astronomy during the eighteenth century. Military-trained astronomers in the Bourbon era worked to improve navigation and secure control over the empire's many contested borderlands. Spanish astronomers thus built royal observatories, tried to solve the problem of finding longitude at sea, and conducted expeditions throughout the Americas, including several surveys of the Gulf coast.[5]

Yet Manuel Gayoso de Lemos, governor of Spanish Louisiana and West Florida, employed local Anglos, not scientifically trained Spanish officers, to fill the leading positions of the Florida boundary survey. He appointed West Florida planter William Dunbar as Spain's co-commissioner and official astronomer. Stephen Minor, a Virginia-born captain in the Spanish army living near Natchez, was Spain's other commissioner, and Minor took over as lead astronomer after Dunbar quit the survey in August 1798. Despite lacking any previous astronomical experience, Minor picked up the basics quickly—Gayoso praised his "progress in Astronomy"—and, by the

4. On the Florida boundary, see John Craig Hammond, *Slavery, Freedom, and Expansion in the Early American West* (Charlottesville, Va., 2007), 13–29; J. C. A. Stagg, *Borderlines in Borderlands: James Madison and the Spanish-American Frontier, 1776–1821* (New Haven, Conn., 2009), 13–51; and David Narrett, *Adventurism and Empire: The Struggle for Mastery in the Louisiana-Florida Borderlands, 1762–1803* (Chapel Hill, N.C., 2015), 233–256. On science and national pride, see Carol E. Harrison and Ann Johnson, "Introduction: Science and National Identity," *Osiris*, XXIV (2009), 8. On how Europeans used astronomical accomplishments as a means of comparing the relative advancement of civilizations, see Simon Schaffer, "The Asiatic Enlightenments of British Astronomy," in Schaffer et al., eds., *The Brokered World: Go-Betweens and Global Intelligence, 1770–1820* (Sagamore Beach, Mass., 2009), 49–104. On the importance of international boundaries as sites where national identities were both defined and challenged, see Rachel St. John, *Line in the Sand: A History of the Western U.S.-Mexico Border* (Princeton, 2011), esp. 9.

5. Portuondo, *Secret Science*, 211–245; Jorge Cañizares-Esguerra, *Nature, Empire, and Nation: Explorations of the History of Science in the Iberian World* (Stanford, Calif., 2006), 57; Horacio Capel [Sáez], *Geografía y matemáticas en la España del siglo XVIII* (Barcelona, 1982), 195–205, 295; Francisco José González González, *Astronomía y navegación en España: Siglos XVI–XVIII* (Madrid, 1992), 129–201; Jack D. L. Holmes, ed., *José de Evia y sus reconocimientos del Golfo de Mexico, 1783–1796* (Madrid, 1968); Robert S. Weddle, *Changing Tides: Twilight and Dawn in the Spanish Sea, 1763–1803* (College Station, Tex., 1995).

last stages of the survey in 1800, Minor could observe and measure lunar distances, zenith transits, and occultations of Jupiter's moons. Gayoso also hired Thomas Power, an ethnically Irish Canary Islander, as the commission's surveyor and secretary. The preponderance of Anglos in Spanish West Florida might have obliged Gayoso to rely on them for regional scientific work, yet the English-educated governor also seems to have preferred hiring Anglos to lead Spanish astronomical expeditions throughout North America. He proposed in 1798 that Spain employ "the two famous travelers" James Mackay and John Evans, British subjects who had previously explored the upper Missouri River for Spain, to lead a boundary survey from Nootka Sound to the Missouri River with the aim of securing Louisiana's borders.[6]

Gayoso believed that Spain's international prestige was related to perceptions of its astronomical competency. The governor therefore went to great lengths to hire Dunbar as Spain's astronomer because he was convinced that the Scottish-born planter would reflect honor on his own administration and the Spanish Empire on the whole. Dunbar, whom Gayoso described as an "excellent astronomer," had received a scientific education in Scotland, moved to British West Florida in 1773, and served Spain as surveyor of the Natchez District during the 1780s and 1790s. Gayoso agreed to pay Dunbar the handsome annual salary of thirty-five hundred pesos for his work on the boundary, a sum that equaled 31 percent of the total moneys paid out per year to all of the commission's surveyors, surgeons, interpreters, and laborers combined. Gayoso told Dunbar that he warranted this high pay for being "so well calculated to fulfill so important a charge for which is required science with every other quality worthy of public trust: you possess them all, in a degree to do honour to any country." Gayoso even thought that his own honor and public reputation were at stake in hiring a qualified astronomer like Dunbar. As he wrote to Minor during the planning stages of the expedition, "Dunbar is a man of principles and the only astronomer in the whole province, if he can not act I must send for an Astronomer for I'll not be laughed at, I know as much of Astronomy as that."

6. Manuel Gayoso de Lemos to Francisco Saavedra, Nov. 22, 1798, AHN, leg. 3900, apartado 4, no. 1, 1356 ("two famous"); Gayoso to Stephen Minor, Oct. 23, 1798, Gayoso de Lemos Letters, LLMVC ("progress"); Stephen Minor, Report on Observations, Cumberland Island, Apr. 9, 1800, MPA, SD, VII, January 1800 to August 1802, reel 61V, 46–68, PKY. For biographical details on Dunbar, see Arthur H. DeRosier, Jr., *William Dunbar: Scientific Pioneer of the Old Southwest* (Lexington, Ky., 2007). Minor was promoted to major during the boundary expedition, making him Major Minor of Minor Manor.

Gayoso's reliance on Dunbar for both expertise and pride would give the planter significant leverage in their ongoing negotiations.[7]

Gayoso was not the only Spanish official in the region to draw rhetorical connections between the astronomical expedition and the grandeur of the Spanish monarchy. Minor and Power repeatedly requested more funding and higher salaries because they believed it was not "compatible with the dignity of our sovereign and the national decorum and honor" for Spanish employees to be paid less than their U.S. counterparts. Juan Ventura Morales, the intendant of Louisiana, employed similarly nationalistic language to refuse them. Morales admitted that U.S. commissioner Andrew Ellicott did indeed receive a larger salary than Minor, yet he argued that such short-term rewards were inconsequential compared with the ongoing benefits of serving the Spanish crown. When Ellicott returned to Pennsylvania, Morales claimed, he would "retire to his house without a salary and without any more distinction than that which other scientists *[científicos]* of his class enjoy in Republican Governments." Morales's prediction proved true: Ellicott nearly went broke after the expedition.[8]

Spanish officials and men of science working on the boundary line were offended by the U.S. commissioners' presumptions of superiority and were eager to demonstrate their prowess. As Power told Gayoso: "It is a common practice with these people [U.S. commissioners] to boast of their activity, and inveigh against the lentitude of the Spaniards, turning us into ridicule by saying we always go *poco á poco* [slowly]. Thank God we turned the tables upon them in this Expedition." Despite his Irish parentage, Power was prickly about stereotypes of the lazy Spaniard and proud to outdo the U.S. commission in efficiency. Power especially hated Ellicott—who he described as a "designing, artful, cunning, tho' violent, impudent, and unforgiving man"—and insulted his scientific capabilities as well as his character. He wrote to Gayoso that "Yr Excelly may perhaps be a little surprised that I

7. Manuel Gayoso de Lemos to Manuel Godoy, June 6, 1798, AHN, leg. 3900, apartado 4, no. 21, 1235 ("excellent"); Gayoso to William Dunbar, June 26, 1798, PC-LLMVC, leg. 215-B, reel 75, Ser. 106, 152 ("so well calculated"); Gayoso to Stephen Minor, Feb. 5, 1798, Gayoso de Lemos Letters ("Dunbar"). On Dunbar's pay, see "Relación de los empleados nombrados por el señor gobernador gral. de esta provincia para la demarcación de límites con los Estados Unidos, y sueldos que disfrutan," Apr. 27, 1798, PC-LLMVC, leg. 576, reel 40, Ser. 87, 744–745.

8. Stephen Minor to Enrique White, Oct. 22, 1799, EFP, sect. 29, reel 43 ("compatible"); Juan Ventura Morales to Manuel Gayoso de Lemos, Jan. 30, 1799, AHN, leg. 3902, apartado 5, no. 14, 1206 ("retire"); Greene, *American Science in the Age of Jefferson*, 138.

should load [Ellicott] with so many hateful epithets; [but] I will even go so far as to deny his having any kind of merit except that of being a *Mechanical or practical Astronomer*—the attainment of which demands neither genius nor brilliant parts." U.S. men of science during this era tended to take pride in the practicality and usefulness of their work, but Power derided it as devoid of theoretical innovation. Still, Spanish officials' pride was often in tension with a lurking sense that their technologies and training were inferior to those of the U.S. commission. On visiting a temporary observatory on the boundary line, Governor Gayoso was disappointed to find that the U.S. team had more impressive apparatus and that Ellicott seemed more accomplished than any of Spain's commissioners.[9]

As in the Spanish Empire, experts in the United States took national pride in their astronomical work and used it to promote territorial expansion. Men of science in the United States had been thinking about how astronomy could help their nation develop into a continental power since independence. Writing from Massachusetts in 1783, the Reverend Phillips Payson claimed that "the extensive territories of the United States of *America,* are a foundation in nature for a vast empire.—The geography of its interior parts, though of great importance, is, at present, but little better than conjectural: To perfect which, and fix the interesting boundaries and lines, the best, and indeed the only proper method is, that of astronomical observations." Since the production and circulation of astronomical data "promote[d] the purposes of navagation *[sic]* and geography," astronomy ensured "special advantages" for the nation's commercial and political power. Astronomy—particularly the practical approach to observation and instrument making exemplified by the strident patriot David Rittenhouse—was both the prestige science of the early republic and a popular pursuit among educated individuals from all walks of life. Patriotism infused astronomy so thoroughly in the United States that it influenced the way Anglo-Americans saw the night sky. For example, the first star chart published in the United States traced the "Flying Squirrel" and the "Bust of Columbus" among the

9. Thomas Power to Manuel Gayoso de Lemos, Nov. 9, 1798, PC-LLMVC, leg. 215-B, reel 75, Ser. 106, 69 ("common practice") ("Poco á poco" is the only Spanish phrase in this otherwise English letter), Power to Gayoso, Sept. 29, 1798, leg. 215-B, reel 75, Ser. 106, 62 ("designing"), emphasis in original. "Practical" referred to applied, as opposed to theoretical, astronomy. On practical and theoretical astronomy in the United States, see Warner, "Astronomy in Antebellum America," in Reingold, ed., *Sciences in the American Context,* 57. For Gayoso's impressions, see Gayoso to Manuel Godoy, June 6, 1798, AHN, leg. 3900, apartado 4, no. 21, 1234–1242.

constellations. Patriotism was also evident throughout the Florida boundary survey, and it shaped how commissioner Ellicott approached his scientific work and interacted with local inhabitants and officials.[10]

Ellicott's astronomical expedition to run the Florida boundary was an imperial project. This was most obvious in that it established a new southern limit that added more territory, white settlers, plantations, slaves, and Indian groups to the United States. This 31° boundary would then serve as a baseline for cadastral surveys that, by establishing property boundaries, helped structure further Anglo-American settlement. The expedition also developed the United States' geographical knowledge of its new Gulf South territories, as Ellicott's meridian observations set the positions of several sites "with as much precision as any point within the United States."[11]

Ellicott used astronomy to exhibit the United States' political power and technological sophistication. A few days after arriving in the then-Spanish city of Natchez in February 1797, Ellicott performed an astronomical ceremony of possession that laid claim to the city and sent the message that he represented a scientifically advanced nation. Ellicott's men raised the U.S. flag to indicate that Natchez, which was north of the 31st parallel, was now part of the United States, and, "on the day the flag was hoisted, a course of astronomical observations was begun." These observations, in which Ellicott employed his six-foot zenith sector, large telescope, and state-of-the-art clock would have been the most elaborate scientific display that most of the city's residents had ever seen. This performance had its desired effect: one Natchez resident referred to the U.S. flag as "that emblem of peace and science which has recently been displayed to us." Ellicott even extended his astronomical observations across the boundary to sites within Spanish

10. Phillips Payson, "Some Select Astronomical Observations Made at Chelsea, Latitude 42° 25,' and 26" in Time East of the University at Cambridge," *Memoirs of the American Academy of Arts and Sciences: To the End of the Year, M,DCC, LXXXIII,* I (Boston, 1785), 124 ("extensive"), emphasis in original; W[illiam] Croswell, *Description and Explanation of the Mercator Map of the Starry Heavens* ... (Boston, 1810), [iv]–v ("Flying Squirrel," [iv], "Bust," v). On astronomy's prestige and Rittenhouse, see Brooke Hindle, *The Pursuit of Science in Revolutionary America, 1735–1789* (New York, 1974), 167, 229–230, 255, 333–339. For the impact of patriotism on natural history and engineering, respectively, see Andrew J. Lewis, *A Democracy of Facts: Natural History in the Early Republic* (Philadelphia, 2011); and Ann Johnson, "Material Experiments: Environment and Engineering Institutions in the Early American Republic," *Osiris,* XXIV (2009), 55.

11. Andrew Ellicott, *The Journal of Andrew Ellicott* . . . (Philadelphia, 1803), 137 (quote). For details on Ellicott's astronomical methods, calculations, and observations, see "Appendix: Astronomical, and Thermometrical Observations . . . ," ibid.

America, using his spectacular apparatus to perform the United States' scientific capabilities in the Spanish cities of New Orleans, Mobile, and Pensacola.[12]

The expedition presented Ellicott with an opportunity to research which astronomical techniques would be most effective in the United States' borderlands, and he conducted experiments to determine how well the astronomical practices of metropolitan observatories could be recreated in contested territories. As historian Simon Schaffer has argued, part of the prestige of meridian astronomy was its supposed detachment from social and environmental influences. Although this isolation was not in fact as complete as astronomers wished, the ordered spaces of observatories were nevertheless considered manifestations of astronomy's separation from earthly chaos, and imperial efforts to build observatories in distant colonies were part of wider efforts to order these regions. Ellicott claimed that one of his main goals during the survey was "to determine by experiment, what reliance might be placed in observatories made at temporary stations, without any of the conveniences annexed to permanent observatories." This was an expansion-oriented test of U.S. astronomers' capacity to translate the order of metropolitan observatories to the borderlands where, according to Ellicott, "notwithstanding the exertions of the French, British, and Spanish governments the country ... is yet in a state of infamy or rather decrepitude." The failure of his temporary observatories to protect him or his instruments from the weather, mosquitoes, and human sabotage under-

12. Ellicott, *Journal*, 44 ("on the day"), Address of Narsworthy Hunter, quoted ibid., 70 ("emblem"). On how astronomical performances acted as ceremonies of possession, see Patricia Seed, *Ceremonies of Possession in Europe's Conquest of the New World, 1492–1640* (New York, 1995), 100–148; and Neil Safier, *Measuring the New World: Enlightenment Science and South America* (Chicago, 2008), 1–4, 23–56. On Ellicott's observations and instruments, see Ellicott, *Journal*, "Appendix," 12, 37, 44–48. Ellicott's sophisticated instruments also meant his observations differed from those of Spain's commissioners because their apparatus, particularly Dunbar's astronomical circle, were better suited for observing the transits of different heavenly bodies. See William Dunbar to Manuel Gayoso de Lemos, June 11, 1798, PC-LLMVC, leg. 215-B, reel 75, Ser. 106, 153; and Robert J. Malone, "Everyday Science, Surveying, and Politics in the Old Southwest: William Dunbar and the Influence of Place on Natural Philosophy" (Ph.D. Diss., University of Florida, 1996), 179. For a detailed description of Dunbar's astronomical practices on the line, see Dunbar, "Descripción topográfico de la linea divisoria que yo D. Guillermo Dunbar, astronomo nombrado en favor de S.M. señale con concurrencia del comisario por los Estados Unidos de Norte América," Aug. 31, 1798, PC-LLMVC, leg. 215-B, reel 75, Ser. 106, 158–161. A slightly varying English version of this source is in Eron Rowland, ed., *Life, Letters, and Papers of William Dunbar ...* (Jackson, Miss., 1930), 78–82.

lined the truth that astronomers and their technologies were never wholly isolated from encounters with other people and terrestrial nature, especially not in such a politically unstable region.[13]

U.S. officials and experts in the lower Mississippi Valley developed an anti-Spanish rhetoric that combined elements of the Black Legend with specific references to Spain's astronomical incompetence. This was strikingly apparent in how Ellicott tried to deflect rumors of his own sexual deviance by claiming that Spaniards' ignorance of astronomy made them credulous. Thomas Freeman, one of the U.S. surveyors on the expedition, asserted that Ellicott engaged in ménages à trois with the commission's laundress, Betsy, and his own son, who had accompanied the elder Ellicott on the expedition as an assistant surveyor. Freeman grumbled: "My colleague E is full as contented and contemptible as ever; he has as usual given up all society for his *blubbermouthed* washer woman, himself, the *cub-brute* his son ... eat their *Bacon and cabbage* in their own way and sleep together in—every way! So much for Don. A.—I was going to say *Damnable!*" Ellicott tried to deflect Freeman's accusations—which also included misplacing the boundary and pilfering funds—by refocusing attention onto how Freeman and those who believed him were scientifically incompetent and, thus, unreliable. He told Secretary of State Timothy Pickering that Freeman "has not made a single observation since we came to this country tho' ... he is nevertheless by Genl. Matthews and Col. Pawnel declared not inferior to Newton!" "This is not strange," he concluded, as "they may possibly have less scientific knowledge than he has." Although Ellicott blamed the supposed gullibility of these officers on their ignorance of science, he categorically accused all Spaniards of inherently lacking the scientific—and, specifically, astronomical—capabilities needed to disregard such slanders. He wrote that "these insinuations I am confident would have but little weight with the people of the U.S. but the case is very different with the Spaniards, naturally jealous, and uninformed in science, particularly so far as it relates to astronomical operations." Ellicott was not just drawing a territorial

13. Ellicott, *Journal*, "Appendix," 76 ("to determine"); Andrew Ellicott to Timothy Pickering, June 1799, AEP, III, Letters Sent, 1797–1800, 759 ("notwithstanding"); Simon Schaffer, "A Pattern Science," in *"When the Stars Threw Down their Spears,"* Lecture 1, Tarner Lectures (2010); Schaffer, "Keeping the Books at Parramatta Observatory," in David Aubin, Charlotte Bigg, and H. Otto Sibum, eds., *The Heavens on Earth: Observatories and Astronomy in Nineteenth-Century Science and Culture* (Durham, N.C., 2010), 120–128. On how astronomical fieldwork in colonial spaces depended on an empire's capacity to exercise power, see Pang, *Empire and the Sun*, 121–144.

boundary: he was bounding Anglo-Americans as a nationality with distinct traits, and it served both his personal reputation and national pride to use contrasts with Spaniards to define U.S. citizens as scientifically competent.[14]

Dunbar's, Minor's, and Ellicott's observations fixed the line's position, but their astronomical project depended on their coercive power over enslaved blacks. During the planning stages of the expedition, Minor claimed that "it is my opinion that negroes employed as labourers and ax men are preferable to white men, and I should wish to employ 10 or more of my own and perhaps Dunbar as many." In 1798, Minor, Dunbar, and two other local planters rented out their slaves—twenty-four in total—for the expedition, and the Spanish and U.S. commissions split the costs of supporting the slaves and paying a daily wage to their owners. The commissioners on both sides relied on these slaves most heavily during the first months of the survey, when disease and midsummer toil in bayous near the Mississippi River left only twenty-five of the fifty total white laborers fit for work. These slaves were not, however, permanent employees, and the white commissioners all seem to have rued the moment when the slaves were "called home by their Masters" in early September to pick cotton, the lower Mississippi Valley's promising new staple crop.[15]

Dunbar employed the expedition's slaves more heavily than any of the other commissioners. He engaged all of the survey's twenty-four slaves for the dangerous task of "cutting thro' the swamp" during the westward extension of the line to the Mississippi River in 1798. As Minor noted, the ostensible "motive for employing the negroes in this venture [was] to not expose

14. Thomas Freeman to John McKee, Jan. 27, 1798, John McKee Papers, 1792 to 1825, box 1, folder 4, no. 297, LOC ("My colleague"), emphasis in original; Andrew Ellicott to Timothy Pickering, Nov. 8, 1798, AEP, III, Letters Sent, 1797–1800, 722 ("these insinuations"), 724 ("has not made," "this is not strange," "they may possibly"). Freeman made these same accusations more publicly as part of James Wilkinson's 1811 court martial. See Deposition of Thomas Freeman, Aug. 20, 1811, in James Wilkinson, *Memoirs of My Own Times,* II (Philadelphia, 1816), app. XXXII.

15. Stephen Minor to Manuel Gayoso de Lemos, Jan. 20, 1798, PC-LLMVC, leg. 215-B, reel 75, Ser. 106, 96 ("my opinion"), William Dunbar to Manuel Gayoso de Lemos, Aug. 4, 1798, leg. 215-B, reel 75, Ser. 106, 156 ("called"). On the enslaved blacks employed on the line, see Thomas Power, "Lista de revista de los negros empleados en los trabajos de la expedicción para la demarcación de límites en los Estados Unidos de América," July 5, 1798, PC-LLMVC, leg. 576, reel 40, Ser. 87, 647; Juan Ventura Morales to Francisco Saavedra, July 31, 1798, AHN, leg. 3902, apartado 4, no. 234, 1066; and Andrew Ellicott to Timothy Pickering, July 29, 1798, AEP, III, Letters Sent, 1797–1800, 688.

the white laborers to such a sickly locale," but Dunbar volunteered for this "arduous and disagreeable task" because it would bring his slaves nearer to his plantation and back into his fields before the cotton harvest. Dunbar was at the forefront of Natchez's incipient cotton boom, and it seems that he was so anxious to profit from his crop that he risked his own life and those of his slaves by running the line through the swamp and toward his plantation.[16]

Dunbar's exploitation of slave labor during the expedition transposed the order and discipline of his plantation onto the otherwise poorly controlled sites of the survey's fieldwork. Whereas Ellicott tried and failed to imitate metropolitan stability during the survey through his experimental observatory, Dunbar largely succeeded in recreating the racially ordered world of his plantation. For one, he insisted that supposedly docile blacks were superior to the expedition's white employees, writing to his wife that "the line goes on briskly since the arrival of the negroes; the white people are a mutinous set, not easily managed." Dunbar also acted simultaneously as astronomer and slave master during the expedition, taking self-consciously paternalistic responsibility as the slaves' provider and healer despite that Spain employed a commissary and a surgeon. He explained to Ellicott: "It would be a pity to discourage the ardour of the black workmen, by obliging them to work with bad tools. There was left a sick negro at your Camp, if he is able to ride here, I will endeavour to cure him; I have recovered all those who came invalids from your camp." Dunbar took his persona as master cum astronomer so seriously that, while the efficiency of his slaves inspired pride, he was embarrassed when their performance reflected poorly on him. He told his wife that "I am quite ashamed of York being here as a hand, he is a disgrace to see." Dunbar's transposition of plantation order onto the survey both regulated the expedition's labor force and structured his public identity.[17]

16. Stephen Minor to Manuel Gayoso de Lemos, Aug. 24, 1798, PC-LLMVC, leg. 215-B, reel 75, Ser. 106, 20 ("cutting"), Minor to Gayoso, July 29, 1798, leg. 50, reel 111, Ser. 143, 1137 ("motive").

17. William Dunbar to Dinah Dunbar, June 23, 1798, WDP, Letters, Accretion ("line," "I am quite"); William Dunbar to Andrew Ellicott, Aug. 4, 1798, AEP, II, Letters Received, 1798–1800, 289 ("It would be a pity"). Dunbar also provided medical care at his plantations and conducted autopsies to determine the cause of his slaves' deaths. See William Dunbar, Journal, 1776–1780, WDP, 38. For how other masters administered medical care to slaves as a way to perform paternalism, see Philip D. Morgan, *Slave Counterpoint: Black Culture in the Eighteenth-Century Chesapeake and Lowcountry* (Chapel Hill, N.C., 1998), 322.

Dunbar's experience with slave labor during this expedition would lead him to recommend that the U.S. government employ local slaves for its future boundary surveys. Soon after learning of the Louisiana Purchase, Dunbar told Thomas Jefferson that "should it be found necessary to run a line of demarcation I would suggest the impropriety of sending white laborers with the commission from the seat of Govt—as in the case of Mr. Ellicot—negros hired here will be more tractable and will execute much more labor—a proof of which we had on the line of 31°." Although Jefferson did not take Dunbar's advice on this occasion, this letter suggests one of the many ways that Dunbar worked to perpetuate local scientific practices—which, in this case, centered on violence against enslaved blacks—across the change of government.[18]

Like the Spanish commissioners, Ellicott relied on black slaves to overcome the hardships of astronomical fieldwork in the Gulf South. Yet this fact has not dampened his historical reputation as an abolitionist and patron of black scientific practitioners. Ellicott did indeed rail against slavery and promoted the talents of Benjamin Banneker, the African American surveyor with whom he ran the boundaries of the District of Columbia in 1791. The discrepancy between Ellicott's view of slaves during the Florida boundary survey and historians' praise for his attitude toward blacks is largely a product of Ellicott's own self-representation in his published *Journal.* He described himself as an unwilling accomplice to the use of slave labor who was "reduced to the necessity of either abandoning the business for some time, or employing slaves; the latter was adopted." This passively voiced apology conflicts with the records of the Spanish commission, in which Minor wrote that the prevalence of the "fever-ague" among the white assistants "caused Mr. Ellicott to propose to me to engage jointly 24 Negros for two or 3 months." Although Ellicott had protested the hiring of slaves during the planning stages of the expedition, he had not done so out of any altruistic impulse. Rather, as Governor Gayoso noted, Ellicott simply feared that "putting Negros at work with their white men" would create conflict among the laborers and upset the progress of the survey. In his unpublished correspondence from the expedition, Ellicott echoed racialized understandings of bodily difference that had long justified plantation slavery. He wrote that "man and almost all other animals ... lose their vigour" in the region's enervating heat and humidity but that "White men are much more affected than Black" by the environment. He added that enslaved blacks in the Span-

18. William Dunbar to Tho[mas] Jefferson, Sept. 30, 1803, WDP, Letter Book, 1802–1805, 14.

ish borderlands "fall far short of the Blacks in our country in manly athletic exercise," an oddly patriotic boast that suggests how closely race, nation, and climate were connected.[19]

Although enslaved blacks enabled the expedition to progress despite numerous environmental and physical challenges, the surveyors could not perform astronomical observations in the face of concerted Indian resistance. The Choctaws of the lower Mississippi Valley showed signs of hostility against the U.S. commissioners who, after all, had summarily claimed all of their territory above the 31st parallel. Ellicott wrote in 1797 that the Choctaws were "very insolent" and "walked about the camp with drawn knives," yet his preconceived suspicions of Spanish and Indian intrigue might have led him to misread this threat. The outgoing Spanish governor described this incident as simply a drunken confrontation between individual Choctaws and a few of Ellicott's "servants," but the paranoid Ellicott believed that Gayoso "had made [the Choctaws] inebriated, to cause them to put him and his people to the sword; . . . what littleness in a man, who boasts of the title of the representative of a nation." Spain's commissioners also feared Choctaw violence, and Minor penned a list of questions for Gayoso regarding how he should react to various hypothetical Choctaw interferences with the survey. Minor wanted to know how to respond if Choctaw delegates "demand to be made acquainted with the motives of our running the line through their territory," how to "act if they insist on suspending the progress of our expedition," and what to do "if they show a disposition amicable to us, but insulting and hostile to our friends the A[mericans]." Although the limited Choctaw threats were focused on the U.S. commission more than Spain's, that the commissioners for both sides were of British descent seems to have been a source of confusion for some Choctaws. One Choctaw who visited the Spanish surveyors' camp and was "eying every object with the most eager curiosity . . . asked [Thomas] Power if he was a Spaniard, to which he replied by pointing [and saying] we are all Spaniards here."[20]

19. Ellicott, *Journal,* 2, 180 ("reduced"); Stephen Minor to Manuel Gayoso de Lemos, June 11, 1798, PC-LLMVC, leg. 215-B, reel 75, Ser. 106, 2 ("fever-ague"); Gayoso to Minor, Feb. 5, 1798, Gayoso de Lemos Letters ("putting Negros"); Ellicott to Pickering, June 1799, AEP, III, Letters Sent, 1797–1800, 759 ("man and almost all"); Silvio A. Bedini, *The Life of Benjamin Banneker* (New York, 1972); Andro Linklater, *The Fabric of America: How Our Borders and Boundaries Shaped the Country and Forged Our National Identity* (New York, 2007), 95, 110, 135.

20. Ellicott, *Journal,* 45 ("very insolent"); Francisco Luis Hector, barón de Carondelet, [to Thomas Power], Apr. 23, 1797, in Daniel Clark, *Proofs of the Corruption of*

Although the Choctaw threat never materialized, the commissioners faced far greater opposition when the expedition moved into the Creek and Seminole dominated spaces along the East Florida boundary in 1799. Although various native factions supported or resisted the expedition, the Euro-American astronomers were often unclear about which Indian groups had sided with them and which groups were opposing them. Thus, while Ellicott thought that most Upper Creeks supported the survey and most Lower Creeks and Seminoles resented it, he also believed that a large group of Upper Creek warriors were threatening to attack the expedition. The Spanish and U.S. commissions, moreover, came to rely on a corps of Creek warriors for protection. As Power reported, the commissioners had met with six hundred Creeks who "consented to our running the Line thro' their Country and are to send to *[sic]* chiefs of the first rank and twenty warriors to protect us against the Seminoles, none of whom were present, and who declare they will not suffer us to tread on their land." Power added that the U.S. and Spanish commissioners jointly gave these Creeks four hundred dollars of credit at Panton, Leslie, and Company after the conference. The wealthiest Creeks, men who were then in the process of developing a Creek Nation with police, laws, and jurisdictions, tended to support the boundary survey because it promised a means of better securing their property.[21]

The Creeks and Seminoles who resisted the survey targeted the expedition's technologies, materials, and security. These natives were particularly keen on messing with Ellicott's timekeepers, including one incident in which "the minute hand of the clock was moved by an impertinent young Indian." Along with repeatedly stealing the expedition's horses, the Indi-

---

Gen. James Wilkinson ... (Philadelphia, 1809), Appendix, 104, note no. 44 ("servants"); Stephen Minor to Manuel Gayoso de Lemos, [July?] 19, 1798, PC-LLMVC, leg. 50, reel 111, Ser. 143, 1110 ("demand"), 1112 ("eying"). On southeastern Indians' astronomical practices, see George E. Lankford, *Reachable Stars: Patterns in the Ethnoastronomy of Eastern North America* (Tuscaloosa, Ala., 2007).

21. Thomas Power to Manuel Gayoso de Lemos, May 11, 1799, PC-HNOC, leg. 216-B, reel 78, Ser. 106, 103 (quote). For Ellicott's perceptions of the Creek and Seminole threat, see Ellicott, *Journal*, 215, 219, 220, 225; and Jack D. L. Holmes, "The Southern Boundary Commission, the Chattahoochee River, and the Florida Seminoles, 1799," *FHQ*, XLIV (1966), 312–341. For names and rations of the "individuals of the Indian guard," see Thomas Power, "Relación de las raciones que el contratista Dn. Roberto Cochran ha suministrado ... á los yndividuos de la escolta india de la expd. de límites," July 1, 1799, PC-LLMVC, leg. 576, reel 40, Ser. 87, 706. On wealth and property among the Creeks, see Claudio Saunt, *A New Order of Things: Property, Power, and the Transformation of the Creek Indians, 1733–1816* (Cambridge, 1999), 181–183.

ans who opposed the boundary also burgled Ellicott's artificial horizon and, more generally, took "every article in our camp they could lay their hands on." Yet the most direct form of opposition to the survey was armed resistance, and the threat of violence ultimately succeeded in forcing the expedition members to abandon their efforts to complete the 31st parallel boundary. Ellicott claimed that bands of Lower and Upper Creeks numbering in the hundreds stood in the expedition's path and, despite the intercession of the influential U.S. Indian agent Benjamin Hawkins, would not be assuaged. Ellicott wrote that "the hostile disposition of the Indians, and an attempt to plunder our camp, compelled us to relinquish our design, and leave the country." Despite all the assertions of national and scientific preeminence that Spanish and U.S. astronomers exchanged over the course of the expedition, Indian power on the ground ultimately determined its success.[22]

ASTRONOMICAL METHODS AND CONTINENTAL EXPANSION

The struggle of European astronomers and instrument makers to develop a method of finding longitude at sea is perhaps the most well-known example of how science and technology served the needs of European expansion. The Englishman John Harrison's invention of the marine chronometer in the late eighteenth century did much to solve this problem by enabling navigators to compare their local time to that of meridians in Greenwich, Paris, or Cadiz and, after observing celestial motions and consulting ephemerides, determine their longitude. U.S. astronomers contributed little to the scientific and technological innovations needed to find longitude at sea and have thus been excluded from this archetypal story of science and imperialism. Yet, when American astronomy is analyzed in the context of the United States' own territorial expansion, and not as a postcolonial shadow of Europe's, U.S. astronomers and officials appear deeply engaged in developing creative solutions to problems specific to U.S. imperialism. Whereas European empires were concerned with developing technologies and meth-

22. Ellicott, *Journal,* 214 ("every article"), ibid., "Appendix," 39, 104 ("the minute hand"), 115–116 ("hostile"). On how other indigenous peoples stole or destroyed astronomical instruments and records to resist imperial claims to their territories, see Giselle Byrnes, *Boundary Markers: Land Surveying and the Colonisation of New Zealand* (Wellington, 2001), 106; Julie Cruikshank, *Do Glaciers Listen?: Local Knowledge, Colonial Encounters, and Social Imagination* (Vancouver, B.C., 2005), 136; and Simon Schaffer, "'On Seeing Me Write': Inscription Devices in the South Seas," *Representations,* XCVII (2007), 113–116.

ods to facilitate marine empires, U.S. astronomers formulated techniques for determining longitude that did not rely on advanced technologies and were specially suited to the needs of land-based expansion.[23]

William Dunbar and President Thomas Jefferson refined a method of finding longitude on land that could be performed by a solitary astronomer without access to a timepiece. Their correspondence demonstrates some of the ways that early U.S. astronomy was oriented toward territorial expansion and suggests that Jefferson was not, as historians have proposed, a mastermind who "programmed" the scientific aspects of western exploration and expansion. Rather, Jefferson was a pragmatic experimenter who encouraged scientific solutions to some of the many challenges of continental imperialism. The president looked to the developing national astronomical community to solve the United States' longitude problem, and, while he consulted several astronomers, he was particularly interested in Dunbar's advice. As in European empires, scientific solutions to the needs of U.S. expansion emerged out of exchanges between distant territories and metropoles.[24]

The primary goal of the first wave of exploratory expeditions into the lands the United States had bought in the 1803 Louisiana Purchase was to determine the latitude and longitude of significant locales, particularly those along major rivers. The first and most detailed task in Jefferson's instructions to Meriwether Lewis in 1803 and Thomas Freeman in 1804 was thus to make astronomical observations "with great pains and accuracy." As Jefferson explained to Dunbar in 1804, astronomical observations will "enable us to prepare a map of Louisiana, which in it's contour and main waters will be perfectly correct, and will give us a skeleton to be filled up with details hereafter." Samuel Mitchill, the chair of the House of Representatives' Committee on Commerce and Manufactures, predicted that geo-

23. On longitude at sea, see Dava Sobel, *Longitude: The True Story of a Lone Genius Who Solved the Greatest Scientific Problem of His Time* (New York, 1995); Andrewes, ed., *Quest for Longitude;* and González González, *Astronomía y navegación.* Ephemerides (ephemeris, sing.) were tables that listed the coordinates of celestial bodies to assist with astronomical calculations.

24. On Jefferson as programmer, see William H. Goetzmann, *Exploration and Empire: The Explorer and the Scientist in the Winning of the American West* (New York, 1966), 5 (quote); Michael Adas, *Dominance by Design: Technological Imperatives and America's Civilizing Mission* (Cambridge, Mass., 2006), 24; and Greene, *American Science in the Age of Jefferson,* 196–198. On how imperial science developed through transatlantic exchanges, see James Delbourgo and Nicholas Dew, eds., *Science and Empire in the Atlantic World* (New York, 2008).

graphic information about the interior of the continent would do more for the United States' power and pride than maritime explorations had for Britain, Spain, or France. He wrote that "an expedition of discovery up these prodigious streams and their branches might redound as much honour, and [contribute] more to the interest of our government, than the voyages by sea round the terraqueous globe have done for the polished nations of Europe who authorized them." Astronomical observations in the continental interior were both vital to the United States' interests and important steps toward being counted among the powers of the earth.[25]

Yet performing these observations in the field was difficult, and Dunbar experienced several of these challenges firsthand during his 1804–1805 expedition up the Washita River. Despite the presence of Philadelphia chemist George Hunter, the Washita expedition lacked a qualified second astronomer who would have improved and eased Dunbar's observations. As he wrote to Jefferson during the expedition, "The Doctor has never been in the habit of using [astronomical] instruments . . . , otherwise his assistance in the lunar observation would be of great advantage," since Hunter could have measured the lunar altitude at the same moment that Dunbar observed the moon's distance from a reference star. Hunter's shortcomings as an astronomical assistant meant that "at present we must depend totaly upon calculation of the altitude of the two bodies regulated by the time kept by the Watch." Dunbar knew that relying on his watch was risky: not only was it constantly losing time, but any damage to this fragile instrument would deprive the expedition of a vital technology for determining longitude.[26]

The president's experience with the Lewis and Clark expedition inspired him to develop a clockless method of finding longitude. He had pondered "substituting a meridian at land, instead of observations of time, for ascertaining longitude by the lunar motions" during the planning stages of Lewis and Clark's journey, but Jefferson revisited this issue in earnest in May 1805

25. Th[omas] Jefferson to Meriwether Lewis, June 20, 1803, TJP [1] ("great pains"), Th[omas] Jefferson to [William] Dunbar, Mar. 13, 1804, [4] ("enable"); [Samuel Latham] Mitchill, "Louisiana," *Medical Repository,* 2d Hexade, I, no. 4 (1804), 407 ("expedition"); Jefferson to Thomas Freeman, Apr. 14, 1804, in Dan L. Flores, ed., *Southern Counterpart to Lewis and Clark: The Freeman and Custis Expedition of 1806* (Norman, Okla., 2002), 321.

26. William Dunbar to [Thomas Jefferson], Nov. 9, 1804, TJP ("Doctor," [3] "at present," [4]); Trey Berry, Pam Beasley, and Jeanne Clements, eds., *The Forgotten Expedition, 1804–1805: The Louisiana Purchase Journals of Dunbar and Hunter* (Baton Rouge, La., 2006), 216–217.

as he and Dunbar began planning an expedition, led by surveyor Thomas Freeman, to explore the Red River and Spanish Texas. "While Capt. Lewis's mission was preparing," Jefferson told Dunbar, "it was understood that his reliance for his longitudes must be on the Lunar observations taken, as at sea, with the aid of a timekeeper." Since "a thousand accidents" might harm a timepiece "and thus deprive us of the principal object of the expedition," Jefferson came to think that a clock was not in fact necessary because "we can always have a meridian at land, that would furnish what the want of it at sea obliges us to supply by the time keeper." Astronomical methods of finding longitude through lunar observations typically incorporated a timepiece but did not rely on accurate timekeeping to the same extent as the maritime technique made possible by Harrison's chronometer; the methods Jefferson suggested, however, would not use any sort of mechanical timepiece. Jefferson proposed three possible solutions, the basic idea of all three being that a terrestrial explorer could find the difference in distance between his own meridian and that of Greenwich, and thus his own longitude, by comparing the moon's right ascension and its distance from his current meridian with the moon's distance from the meridian at Greenwich at the moment when it had that same right ascension. Jefferson requested Dunbar's feedback on these astronomical techniques because he hoped to apply them "advantageously in our future expeditions, and particularly that up the Red river."[27]

Even as he was consulting Dunbar, Jefferson was already drawing on other connections to explore the feasibility and novelty of these methods. Isaac Briggs, surveyor general of the Mississippi Territory, told Jefferson that the method was indeed practicable, and Alexander von Humboldt agreed that Jefferson's "idea was correct, but not new" since a similar method had been published in Jean-Étienne Montucla and Joseph-Jérôme de Lalande's *Histoire des mathématiques* (1802) and had been first pro-

27. Th[omas] Jefferson to [Meriwether] Lewis, Apr. 30, 1803, TJP ("substituting"), Jefferson to [William Dunbar], May 25, 1805 ("While Capt.," [2], "advantageously," [3]). The term meridian refers to a "great circle on the celestial sphere passing through the poles and the zenith [the point directly overhead]," not a line of longitude on the earth. See Jacqueline Mitton, *A Concise Dictionary of Astronomy* (New York, 1991), 242. Right ascension is "one of the coordinates used to define position on the celestial sphere in the equatorial coordinate system." "It is the equivalent of longitude on the Earth [declension is the equivalent of latitude] but is measured in hours, minutes and seconds of time eastwards from the zero point, which is taken as the intersection of the celestial equator and the ecliptic." A right ascension of one hour "is the angle through which the celestial sphere appears to turn in one hour of sidereal time, as the Earth rotates." See ibid., 320.

posed in the 1600s. Far from being fazed by this preemption, Jefferson claimed Europeans had never sufficiently experimented with this method because it was "of no use at sea where a meridian cannot be had, and where alone the nations of Europe had occasion for it." Jefferson also believed that the seventeenth-century cosmographers who pioneered this method had relied on timekeepers.[28]

The intellectual challenges of perfecting this expansion-promoting scientific method became a national enterprise for astronomers and inventors in the United States. Jefferson asked several astronomers for advice who, in turn, circulated this longitude problem among their own scientific friends. For example, Jefferson requested Freeman's input on "the method of determining Longitude by the Right Ascension of the Moon, without the aid of Time," and Freeman furnished his own response while also forwarding the problem to Philadelphia mathematician Robert Patterson. Dunbar benefited from the advice of Benjamin Smith Barton in developing his clockless longitude method and published his writings on the subject in the American Philosophical Society's *Transactions.* Some U.S. inventors tinkered with workarounds. In 1807, a treasury official and astronomer named Joshua I. Moore sent Jefferson the description of a "device for finding the Longitude" envisioned by one Timothy Kirk, "the Inventor of some ingenious machineries." Kirk's device would use gears and weights to fix a plumb line's angle at a prime meridian and, as the machine was transported east or west, the difference in the angle of the plumb would determine longitude gravitationally. Though Moore assured Jefferson that Kirk's invention had no merit, it does confirm the degree to which the issue of longitude without a clock engaged the attention of mathematical practitioners in Jefferson's America.[29]

Formulating a new way to determine longitude on land was more than a theoretical exercise for Jefferson: he sought a practical method that would be applied in exploratory expeditions. He was thus glad to receive Moore's criticism even though Moore found that none of Jefferson's three proposed

28. Jefferson to [Dunbar], May 25, 1805, TJP (quotes, [3]).

29. Tho[mas] Freeman to [Thomas Jefferson], July 13, 1805, TJP ("method," [1]), Joshua I. Moore to [Jefferson], July 8, 1807 ("device" [3], "Inventor," [1]). On Dunbar's network and contributions, see William Dunbar to [Jefferson], Mar. 18, 1806, TJP; and William Dunbar, "On Finding the Longitude from the Moon's Meridian Altitude," in American Philosophical Society, *Transactions,* VI (1809), 277–278. On "mathematical practitioners" in early America whose work demonstrated the blurriness of the boundaries between artisanship and science, see Silvio A Bedini, *Thinkers and Tinkers: The Early American Men of Science* (New York, [1975]).

techniques worked. Moore's own clockless method of finding longitude could not be performed by a single observer, but he stressed that it did solve actual challenges of continental expeditions, especially mechanical failures and interference from Indians. In the example he sent Jefferson, Moore imagined: "I am on an exploring Expedition . . . and want to ascertain my longitude. My timepiece is lost, stolen, stopped, broken,—or it may be I think it unadviseable to let the Indians hear it's *[sic]* magical noise." Jefferson thanked Moore for his efforts to assist "our missionaries, if their time pieces failed" and forwarded the treatise to Dunbar in hopes that he could share it with Freeman before the Red River expedition set out.[30]

Freeman's Red River voyage would not begin until April 1806, giving Dunbar ample time to develop a new technique of his own. In his first letter on the subject, penned in July 1805, Dunbar deferentially mentioned that "there can be no doubt as to the principle" of Jefferson's proposed method, but he was nevertheless quick to point out some of its shortcomings. He noted that Jefferson's approach assumed astronomers in the field who relied on sturdy sextants could determine their meridian with the same ease that Jefferson could with his sophisticated universal equatorial instrument. Moreover, Jefferson's method would require "two good observers (three would be better)," and Dunbar realized that it was nearly impossible to outfit the Red River expedition with this many skilled astronomers. He also claimed that perfecting Jefferson's method would require "a very intricate calculation" and double altitude observations of stars that were "extremely difficult for inexperienced observers (and for others)." In short, making the necessary adjustments that would ensure the precision of Jefferson's method would make the operation too complex for the underskilled and underequipped observers of Freeman's team.[31]

To solve these problems, Dunbar proposed a method based on lunar altitude observations that could be performed by a single observer. Lunar altitude observations measure the direct angular distance along a meridian between the moon and the horizon and could be performed with a sextant, octant, or other portable instruments. Not only could astronomers measure the moon's meridian altitude more precisely than its distance from stars, but this technique satisfied Jefferson's other desiderata: it was "much

30. [Joshua I.] Moore, [Sept. 7, 1805], "To Find the Longitude Without a Timepiece," TJP ("I am on an exploring Expedition," [4]), Th[omas] Jefferson to [Moore], Sept. 19, [18]05 ("our missionaries").

31. William Dunbar to [Thomas Jefferson], July 9, 1805, TJP ("there can be no doubt," "very intricate," [3], "two good," "extremely," [4]).

better calculated for travellers by land than voyagers by sea," and "a Single observer with a good altitude instrument, altho' deprived of the use of a time keeper, may still make useful observations for the advancement of geographical knowledge." Dunbar proved the mathematical principals of this method with astronomical data from the Washita expedition and found that his calculations of longitude based on the moon's altitude had been so accurate "that I feel a disposition to rely upon them more than upon the Customary lunar distance." Dunbar had, moreover, formulated this method with a specifically Anglo-American kind of scientific explorer in mind. He told Jefferson that "it is probable that this method will be found chiefly useful to Scientific Gentlemen traveling by land, who are unprovided with a Chronometer and without the aid of an assistant or second observer." These gentlemen would be trained in mathematics and astronomy and "find no difficulty in making those calculations." Rugged yet astronomically competent men who explored the continental interior (such as Meriwether Lewis, Zebulon Pike, and, eventually, Stephen H. Long) were becoming significant and heroic figures in the United States, and Dunbar's technique supported and promoted this idealized Anglo-American man of science.[32]

Jefferson's and Dunbar's efforts to perfect and apply this method were, however, more or less wasted on the Freeman expedition. Dunbar had expressed doubts about Freeman's enthusiasm for astronomy, telling Jefferson that "he did not seem fond of astronomical observation when I knew him on the line of demarkation." Dunbar was right to be suspicious: although Freeman carried a telescope, chronometer, and three sextants, he seems to have made only eleven astronomical observations during more than four months in the field and only conducted observations for determining longitude four times. Despite instructions to experiment with Dunbar's clockless lunar altitude method, Freeman made little if any effort to test its efficacy. The expedition ended prematurely when Spanish troops forced Freeman's team to turn back, depriving Jefferson of the geographical knowledge he desired. Just as the threat of Creek violence had circumscribed Ellicott's astronomical work during the Florida boundary survey, armed Spaniards during the Red River expedition reminded U.S. officials

32. William Dunbar to [Thomas Jefferson], Oct. 8, 1805, TJP ("much better," [1], "Single," [1–2]), Dunbar to [Jefferson], Dec. 17, 1805 ("I feel," [1]). For Dunbar's lunar altitude observations during the Washita expedition, see William Dunbar to Thomas Jefferson, Nov. 10 1805, TJP. On lunar altitude instruments, see J. A. Bennett, *The Divided Circle: A History of Instruments for Astronomy, Navigation, and Surveying* (Oxford, 1987), 133.

that while astronomical fieldwork was itself an assertion of power, its practice relied on actual geopolitical control. The clockless longitude method had little impact on fixing geographic points of interest in the continental interior but it was not insignificant: it grew out of exchanges between the nation's metropole and its borderlands and was a practicable solution to a problem specific to the imperial United States.[33]

OBSERVATION, CIRCULATION, AND IMPERIALISM

In a 1793 "Essay on Those inquiries in Natural Philosophy, Which at Present Are Most Beneficial to the United States," Dr. Nicholas Collin encouraged experts in the United States to form a "tolerably accurate map" of the continent by conducting astronomical observations at sites that were most important "in a political, and oeconomical view." The lower Mississippi Valley clearly fit this description, and astronomical observations from this region were valuable to officials seeking to secure national power and commercial interests. The astronomers most responsible for making stargazing in the region useful to U.S. expansion—William Dunbar, Andrew Ellicott, and the Spanish naval officer José Joaquín de Ferrer y Cafranga—had extensive experience with the Spanish Empire. They conducted and shared local astronomical observations, took the lead in consolidating and analyzing astronomical data, and pushed the federal government to co-opt key aspects of Spanish imperial astronomy.[34]

The observatory on Dunbar's Natchez plantation was central to astronomical knowledge production in the lower Mississippi Valley. The circulation and comparison of astronomical data, particularly from sites far from each other, was necessary for improving geographical knowledge of the continent, and Dunbar's was the most southwestern observatory in the United States. He began planning to construct an "observatory better furnished than any to the west of 80° long[itude]" in 1799. And although it was not financially supported by the federal government, his observatory would serve as a crucial site for generating nationally beneficial geographical information. Dunbar stressed the usefulness and uniqueness of his observatory when he told Ellicott that "it will I hope be in my power to ex-

33. Dunbar to [Jefferson], Mar. 18, 1806, TJP (quote, [1]); William Dunbar to Thomas Freeman and his associates, Apr. 28, 1806, Letter Book, 1805–1812, 9, WDP; Flores, ed., *Southern Counterpart to Lewis and Clark*, 64–65, 110, 119, 141, 181, 190, 195, 335–342.

34. Nicholas Collin, "An Essay on Those Inquiries in Natural Philosophy, Which at Present Are Most Beneficial to the United States of North America," American Philosophical Society, *Transactions*, III (1793), xv.

change with you some few observations, which perhaps may be of use." "It is probable," he observed, "that I shall be the only person to the West of the mountains on this Continent, who will be employed in such pursuits." Dunbar was eager to remind eastern officials and experts that observations in the borderlands and, thus, Dunbar himself, were integral to mapping and ordering the Mississippi Valley.[35]

Much of the reason why Dunbar's observatory proved so successful was that its plantation setting enabled Dunbar to reproduce, and probably even surpass, the supposed detachment of metropolitan observatories despite the lower Mississippi Valley's environmental challenges and political instability. Dunbar consulted with Ellicott while designing the observatory so that it would resemble Philadelphia's (which David Rittenhouse had built in the 1780s), and he wanted the space to serve as both an astronomical observatory and experimental laboratory "where I shall place my library, microscopic apparatus, Electrical Machine, Chemical apparatus etc. etc." Ellicott embraced this opportunity to export metropolitan design to the borderlands, and he proposed that Dunbar should follow the Philadelphia model by constructing "an octagon of 8 feet radius," though he warned that even this might not be large enough "to embrace the several objects you have in view." It seems likely that Dunbar's plantation observatory would have even fostered astronomers' idealized detachment from political and social interference more effectively than its model in Philadelphia. Though the astronomers using Philadelphia's observatory could do little to force their fellow citizens milling around town to respect the separation between observatory and society, Dunbar had more than enough coercive power to enforce tranquility around his plantation (and considering Dunbar's brutality as a master, his slaves were probably more than happy to leave him closed up in his observatory as long as possible). Strictly controlled spaces were a prerequisite for conducting credible astronomical observations in the service of imperialism, and no spaces in the lower Mississippi Valley promoted hierarchy, order, and discipline better than plantations.[36]

35. William Dunbar to John Swift, June 15, 1799, "Extracts," 54 ("observatory"); Dunbar to Andrew Ellicott, May 4, 1799, AEP, II, Letters Received, 1798–1800, 430 ("it will I hope").

36. William Dunbar to Andrew Ellicott, Mar. 1, 1801, AEP, V, Correspondence, 1026 ("where I shall"), Ellicott to Dunbar, June 13, 1801, AEP, V, Correspondence, 1044 ("octagon"). On the Philadelphia observatory, see Brooke Hindle, *David Rittenhouse* (Princeton, 1964), 226. On how observatory design could affect knowledge production, see Mari E. W. Williams, "Astronomical Observatories as Practical Spaces: The Case of Pul-

Like Dunbar, Ferrer achieved distinction within the United States' scientific community because he applied astronomical data from the lower Mississippi Valley and wider Caribbean toward refining North American geography. Metropolitan astronomers in Europe had long collected and processed astronomical observations from colonies that promoted commercial, military, and political goals. Ferrer, who had performed a similar function at Spain's Royal Observatory, was central to this process in the early republic. He had received astronomical training as an officer in Spain's Royal Navy and furthered his education in Britain after being taken as a prisoner of war during the American Revolution. He served at Spain's Royal Observatory at Cadiz and conducted observations in Peru, Mexico, and throughout the Caribbean in the 1780s and 1790s. Ferrer resigned his Spanish commission around 1800 and established himself in New York City, where he focused on scientific pursuits and profiting from mercantile ventures. He was part of an international astronomical network: he shared his observational data and calculations of geographical coordinates with astronomers in Spain, France, Germany, Britain, and the United States. In 1801, he travelled down the Mississippi River and conducted longitude-finding observations at several key points, and these observations proved authoritative and directly useful to U.S. officials: Ellicott, Dunbar, and Thomas Freeman all incorporated them into the geographical charts and expedition reports that, as Ellicott noted, were "made for the use of our government." Ferrer also provided geographical knowledge directly to the American Philosophical Society (APS), which welcomed him as a member in 1801. He contributed seven articles to the *Transactions*, deposited manuscript tables of his longitude calculations for the Mississippi Valley and Caribbean at the

kowa," in Frank A. J. L. James, ed., *The Development of the Laboratory: Essays on the Place of Experiment in Industrial Civilization* (London, 1989), 118–136. One example of Dunbar's brutality occurred in 1777, when he locked an enslaved man named Adam in "the Bastile" (his plantation jail), ordered him whipped five hundred times, and then sold him away from whatever family and friends he might have had, all for the heinous crime of drinking rum. See Dunbar, Journal, 1776–1780, WDP, 27. It is not clear if Dunbar employed slaves in his observatory's scientific work, yet he had relied on slaves for the boundary survey, the Washita expedition, and natural history collecting around his plantation, so it seems likely that Dunbar would not have eschewed the aid of enslaved blacks in his observatory. See Minor to Gayoso, Aug. 24, 1798, PC-LLMVC, leg. 215-B, reel 75, Ser. 106, 20; Dunbar, Journal, 1804–1805, WDP, 145; and Dunbar, Journal, 1776–1780, WDP, 7. On the relationship between observatory space and imperialism, see Pang, *Empire and the Sun*, 121–143.

APS library, and drew a map of the coasts of northern Cuba and southern Florida for the society's collections.[37]

Despite receiving scant attention from historians, contemporaries acknowledged Ferrer's importance. Ellicott, who used Ferrer's observations in Venezuela to improve his own calculations for the longitude of Natchez, promoted the talents of this "ingenious Spanish Gentleman" in his publications and letters to Jefferson. French astronomer Joseph-Jérôme de Lalande told Ferrer in 1806 of his "great happiness that there is now an astronomer as useful as yourself in the United States, where there has not been another since [David] Ritinhouse *[sic]* died." And Franz von Zach, editor of a German scientific journal, published some of Ferrer's compilations from the *Transactions* as an example of astronomy in a "rapidly and powerfully burgeoning empire." European experts recognized that the United States was imperial and that a Spanish astronomer facilitated its expansion.[38]

Ferrer acted as an astronomical observer and calculator for both Spain and the United States and, in the process, circulated knowledge between them. For one, he calculated longitude for "Miller's Place on the river Coenecuch" because its "vicinity to Pensacola and the head of the river Perdido" made it "interesting to Geography and Navigation." Ellicott had derived a

37. Ellicott, *Journal*, 137 (quote); Flores, ed., *Southern Counterpart to Lewis and Clark*, 101; Jose Joaquin de Ferrer, "Astronomical Observations ... Chiefly for the Purpose of Determining the Geographical Position of Various Places in the United States, and Other Parts of North America," American Philosophical Society, *Transactions*, VI (1809), 158–164, "Observations on the Comet Which Appeared in September 1807, in the Island of Cuba," 345–359, and "Memoir on the Occultation of Aldebaran by the Moon on the 21st of October 1793," 213–221; "Tabla de las posiciones geográficas de varios puntos de la América," Feb. 6, 1801, APS Archives, III: Communications, APS; Ferrer, Chart of the Coast of Southern Florida, the Bahamas, and Cuba, circa 1800, Miscellaneous Manuscript Maps, 1747–1948, APS. On Ferrer, see Antonio de Alcalá Galiano, *Biografía del astrónomo español Don José Joaquín de Ferrer y Cafranga* (Madrid, 1858), esp. 12. On the circulation of astronomical data, see Portuondo, *Secret Science*, 211–256. On the Royal Observatory in Cádiz, see González González, *Astronomía y navegación en España*, 170–182.

38. Ellicott, *Journal*, "Appendix," 29 ("ingenious"); Joseph-Jérôme de Lalande to José Joaquín de Ferrer y Cafranga, Sept. 27, 1806, quoted in Galiano, *Biografía*, 14 ("great happiness"); *Monataliche correspondenz zur beförderung der erd-und Himmels-Kunde*, 11 (January–June, 1805), quoted in Greene, *American Science in the Age of Jefferson*, 142 ("rapidly"). For Ellicott praising Ferrer, see And[re]w Ellicott, "Astronomical, and Thermometrical Observations, Made at the Confluence of the Mississippi, and Ohio Rivers," American Philosophical Society, *Transactions*, V (1802), 189; Ellicott, *Journal*, 22, 118; and And[re]w Ellicott to Th[omas] Jefferson, Jan. 31, 1802, TJP.

longitude for this spot by observing a transit of Mercury, and Ferrer made its location more precise by incorporating observations of the same transit from Spain. He received data on the occultation of Jupiter's first moon from Spain's Royal Observatory (which had just moved from Cadiz to the Isle of Leon) and, by comparing these with Ellicott's same observations in New Orleans, recalculated the longitude of that city. Ferrer also brought knowledge from the heart of Spanish America to the APS. He donated several books on Mexican languages and history to the APS library and suggested that U.S. navigators incorporate the observations of Peruvian pilots—who, he admitted, "do not understand much of astronomy"—to correct the tables of magnetic variation published by Alexander von Humboldt and other European savants. He provided Spain with observational data from spots throughout the United States' contested borderlands, including Natchez, New Orleans, and the conjunction of the Ohio and Mississippi Rivers.[39]

The early republic's multinational community of astronomers collaborated to observe the total solar eclipse of June 16, 1806, and apply their findings toward improving geographical knowledge of the United States. As Dunbar put it, "This eclipse was to be seen all over Europe and North America" and was therefore "a very important phenomenon for settling comparative longitudes." Ferrer was the center of calculation for the eclipse observations that Dunbar and other astronomers sent to the APS from sites throughout the United States and Spain, and Ferrer integrated this data to produce a chart of geographical coordinates in both countries. As the only site west of the Appalachians to furnish observations on the eclipse, Dunbar's plantation observatory was particularly valuable to Ferrer's calculations. Ferrer incorporated the data from Dunbar's plantation to better determine the longitudes of eastern cities and used the plantation as a reference point from which to derive coordinates for important sites in the lower Mississippi Valley, including Natchez and New Orleans. The role of Dunbar's observatory as a fixed point from which the coordinates of nearby locations could be ascertained was similar to British astronomers' efforts to use their observatory in Parramatta, Australia, to calculate longitudes for sites throughout the Indian and Pacific Oceans. Slavery, and not the state,

39. Jose Joaquin de Ferrer, "The Geographical Position of Sundry Places in North America and in the W. Indies . . . ," American Philosophical Society, *Transactions,* VI (1809), 225–229 ("Miller's Place," 226); José Joaquín de Ferrer y Cafranga to John Vaughan, July 15, 1802, APS Archives IIa: Papers, APS, Ferrer to Robert Patterson, Dec. 3, 1805 ("do not"); "Nota de las observaciones y memorias remitidas al depósito hidrográfico de Madrid desde el año 1796 hasta 1807," in Galiano, *Biografía,* Appendix 9.

had funded Dunbar's observatory, but the facility nevertheless served U.S. expansion in many of the same ways that colonial observatories benefitted European empires.[40]

U.S. astronomers resented that their government did not financially support astronomy or astronomers as liberally as European empires. Complaints about federal support for astronomy gravitated around two main issues: the miserly funding for individual astronomers and the lack of national observatories that would facilitate geography, expansion, and maritime commerce. Dunbar summed up these concerns and the many benefits of governmental support for science in an 1802 letter to APS treasurer John Vaughan:

> There is no example of any encouragement being held out by Governmt; no spirit of inquiry set on foot at the public expense. What is the reason we have no State observatory to which individuals might send their contributions and from which they might receive astronomical intelligence. There is no want of Instruments and eminent astronomers. No naturalist travels at the public expense to explore our immense country and make us acquainted with the infinite ressources it Contains upon its surface, in its waters and within its bowells, from whence national advantages would result.... I conceive great hopes that the influence of our present illustrious President will correct and enlarge the views of our public men, and that under his auspices and protection, Arts, Science, and Literature may take a flight, which will at length carry them as far beyond those of European brethren, as we soar above them in the enjoyment of national liberty.

Vaughan forwarded this letter to President Jefferson, who might have taken its message to heart as he began planning exploratory expeditions into the Louisiana Territory.[41]

40. William Dunbar, "Observations of the Eclipse of the Sun, June 16th, 1806: Made at the Forest, Near Natchez ...," American Philosophical Society, *Transactions*, VI (1809), 262 (quotes). On Ferrer's eclipse calculations, see Jose Joaquin de Ferrer, "Observations of the Eclipse of the Sun, June 16th, 1806, Made at Kinderhook, in the State of New-York," ibid., 271–273, and Ferrer, "Further Observations on the Eclipse of 16th June, 1806 ...," 293–297. On the Parramatta Observatory, see Schaffer, "Keeping the Books at Parramatta Observatory," in Aubin, Bigg, and Sibum, eds., *Heavens on Earth*, 138.

41. Extract of a letter from W[illia]m Dunbar to John Vaughan, Mar. 21, 1802, TJP (quote, [3–4]). On the dearth of federal funding for science before the mid-nineteenth century, see A. Hunter Dupree, *Science in the Federal Government: A History of Policies and Activities to 1940* (Cambridge, Mass., 1957), 1–43.

Ellicott used Spain and other European empires as examples as he attempted to negotiate better compensation for his services during the Florida boundary survey. On his return from the expedition, Ellicott found that President John Adams's administration would not compensate him for back pay and expenses. Forced to sell his books and instruments to support his family, Ellicott tried loosening the government's purse strings by emphasizing the disparity between Spanish and U.S. support for astronomy. In an 1801 letter to the secretary of the treasury, Ellicott lamented that the federal government left him impoverished while the Spanish commissioners "[Stephen] Minor and [Thomas] Power divided about twenty six thousand dollars exclusive of their pay have been complimented by the court of Madrid and set up their carriages." A week later, Ellicott appealed to recently inaugurated Thomas Jefferson by comparing his own work in the Gulf South with more famous and well-funded French, Spanish, and British astronomical expeditions. He wrote that "my astronomical observations done are more numerous than those made by [Pierre-Louis Moreau de] Maupertuis and his associates under the arctic circle, D. [Antonio de] Ulloa and his companions under the equator and [Charles] Mason's and [Jeremiah] Dixon's for nearly four years in this country all added together." Ellicott told Jefferson that he had disregarded all the hardships of the survey "because I thought I was serving my country and expected an adequate compensation." Ellicott, moreover, believed that he did not accrue the honor his astronomical pursuits merited. He told Dunbar that not only was his pay being withheld but "it is likewise believed that the work was carried on and executed principally by the Spanish party and neither reasoning nor documents could yet change this opinion." Ellicott had often boasted about the U.S. commission's superiority to Spain's, and the idea that Minor and Power had outdone him in both fame and fortune through their work on the line no doubt added insult to Ellicott's sense of injury. It also provided more evidence for his argument that the prestige of the United States and its astronomers hinged on the federal government's willingness to match European empires' funding for scientific pursuits.[42]

Ellicott promoted restructuring federal support for astronomy to more closely resemble the example of European empires. Part of this agenda was the establishment of a national observatory and national prime meridian

42. Andrew Ellicott to Secretary of the Treasury [Samuel Dexter?], Apr. 8, 1801, AEP, V, Correspondence, 1037 ("[Stephen] Minor"), Ellicott to [Thomas Jefferson], Apr. 13, 1801, AEP, V, Correspondence, 1039 ("my astronomical"), Ellicott to Dunbar, June 13, 1801, AEP, V, Correspondence, 1045 ("it is likewise").

that would further sever the United States' ties with Britain. While calculating longitudes from the Florida boundary expedition, Ellicott suggested to Jefferson that "we appear yet to be connected to Great Britain by a number of small ligaments, which, tho' apparently unimportant, are nevertheless a draw-back upon that absolute independence we ought as a nation to maintain." He thus pushed Jefferson to authorize a series of observations that would establish a prime meridian at the Capitol building "from which it appears, (in the language Americans ought to use), that Greenwich is ... 76° 56′ 6″ east from the City of Washington." At least some Anglo-Americans believed that breaking away from British influence and achieving equality with Europe hinged on developing a national geographical perspective and, thus, astronomical institutions.[43]

More specifically, Ellicott thought the United States should emulate Spain's national observatory. In May 1801, he sent Jefferson his recommendations that building observatories in "sea-port towns" would be "immediately useful to the navigation." "As commerce is a subject of primary importance," Ellicott wrote, "and depends upon navigation ... it must appear singular, that the legislature of our country has wholy *[sic]* neglected institutions for the improvement of astronomy, and navigation" even though "in Europe we find observatories erected, and supported, by petty Dukes, and Princes." To remedy this, Ellicott suggested that Jefferson eschew the examples of Britain and France and, instead, follow the precedent set by Spain. "The immediate connection of observatories in Europe with Navigation," Ellicott argued, "appear to be neglected except at Cadiz, where I am informed by Don Jose Joaquin de Ferrer, an ingenious Spanish gentleman, they are connected with great advantage to the commerce of that City." Ellicott's encounters with the Spanish commission had inspired boasts of nationalistic pride during the Florida boundary survey, but his relationship with Ferrer revealed that Ellicott also admired Spain as a model for astronomical practices and institutions that could bolster the United States' wealth and power.[44]

43. Ellicott to [Jefferson], Apr. 13, 1801, TJP (quotes, [2]). On efforts to establish a national prime meridian, see Charles O. Paullin, "Early Movements for a National Observatory, 1802–1842," *Records of the Columbia Historical Society, Washington, D.C.*, XXV (1923), 36–56; Silvio A. Bedini, *Jefferson Stone: Demarcation of the First Meridian of the United States* (Frederick, Md., 1999); Matthew H. Edney, "Cartographic Culture and Nationalism in the Early United States: Benjamin Vaughan and the Choice for a Prime Meridian, 1811," *Journal of Historical Geography*, XX, no. 4 (1994), 384–395.

44. And[re]w Ellicott to [Thomas Jefferson], May 16, 1801, TJP ("sea-port towns," [1], "immediate," [2]).

Such proposals would have appealed to Jefferson. Not only did he promote science in a general sense, but he, too, had a specific interest in co-opting valuable aspects of Spanish and Spanish American astronomy. He pushed for the U.S. Navy to use Spanish astronomer José de Mendoza y Rios's "tables of navigation and Nautical astronomy" as a standard reference on all public ships and, in 1809, praised the APS for procuring a "pamphlet on the astronomy of the ancient Mexicans [because] If it be ancient and genuine, or modern and rational, it will be of real value." This pamphlet was an extract from Antonio de León y Gama's 1792 *Descripción histórica y cronológica,* and Jefferson seems to have thought that such a text could inform Anglo-American knowledge of ethnography and astronomy alike. Ferrer had donated this pamphlet to the APS in 1802.[45]

The officials and men of science who developed the U.S. Coast Survey, the primary institution through which Jefferson sought to use astronomy to strengthen national power and commerce, were also eager to offer leading positions to astronomers with experience serving the Spanish Empire. Following Jefferson's recommendation, Congress approved the Coast Survey in February 1807, and historians have asserted that the appointment of Swiss astronomer Ferdinand Rudolph Hassler as its first director was an obvious decision. Yet prominent members of the APS suggested that Dunbar and Ferrer were at least as qualified for the astronomical aspects of the survey. In his letter to Albert Gallatin, whom Jefferson had put in charge of planning the project, Ellicott wrote that "for this [astronomical] part of the business, I know of no man, in this country, as competent as Mr. Dunbar, of Natchez, neither do I know one other to whom I would entrust it, were I President of the United States." "But," he continued, "Mr. Dunbar, being no *quack,* and in easy circumstances, I do not expect he would undertake it, unless it was made an object worth his attention." In other words, Ellicott recognized that Dunbar would only act as the survey's astronomer if the federal government paid him extremely well. Mathematician Robert Patterson did note Hassler's astronomical skills but also recommended "Mr. Ferrer, of New York, [as] perfectly qualified, I presume, for every part of the

45. Thomas Jefferson to Jos[é] de Mendoza [y] Rios, May 4, 1806, ("tables") reproduced in "Science and Technology: Astronomy," Information File, Jefferson Library Special Collections, Robert H. Smith International Center for Jefferson Studies, Charlottesville, Va., Jefferson to Benjamin Smith Barton, Sept. 21, 1809 ("pamphlet"); Ferrer to Vaughan, July 15, 1802, APS Archives IIa: Papers. On how European men of science judged ancient civilizations based on perceptions of their astronomical competency, see Schaffer, "Asiatic Enlightenments of British Astronomy," in Schaffer et al., eds., *Brokered World,* 49–104.

work, and, as an astronomer ... second to no one in the United States—whether he may not still be a Spanish subject, I cannot tell." The question of Ferrer's national attachment proved to be a moot point, however, since his business interests rendered him unwilling to participate in the survey.[46]

Even though business kept Ferrer from being the Coast Survey's astronomer, his experience with Spanish imperial astronomy could still influence its scientific practices. At the time the survey was in the planning stage, APS treasurer John Vaughan sent Gallatin his English translation of "Mr. Ferrer's ideas, as to the mode he concieves *[sic]* most eligible for carrying into execution the Survey of the Coast." Ferrer's plan included a list of the astronomical instruments that the federal government would need to provide and suggested that the first step was to establish thirty observation stations at seaports and lighthouses. The observatory in Philadelphia would be the primary station. Ferrer argued that his "method admits of much more exactitude" than the plan that Gallatin had initially suggested, and, although Vaughan excluded it from his translation, Ferrer criticized Gallatin's methods because they "would require costs and a space of time 4 times greater" than his own approach. Appointing a Spaniard to design and lead the United States' largest and most well-funded astronomical project was a very real possibility in 1807; but, by the time the Coast Survey finally began in 1816, Ferrer had returned to Europe, and Hassler implemented his own methods for carrying out the survey. Anglo-Americans were not anxious that hiring Ferrer would reflect their nation's scientific backwardness. They were proud to count him as a member of their scientific community.[47]

46. Andrew Ellicott to Albert Gallatin, Apr. 17, 1807, in *American State Papers: Documents, Legislative and Executive of the Congress of the United States,* ed. Walter Lowrie and Matthew St. Clair Clarke, VII (Washington, D.C., 1832), 830 ("for this [astronomical] part"), emphasis in original, Robert Patterson to Albert Gallatin, Mar. 31, 1807, in *American State Papers,* VII, 829 ("Mr. Ferrer"); John Vaughan to Gallatin, Apr. 28, 1807, APS Archives, III: Communications, APS. On Hassler and the U.S. Coast Survey, see Florian Cajori, *The Chequered Career of Ferdinand Rudolph Hassler ...* (Boston, [circa 1929]), 44–45; Steven J. Dick, *Sky and Ocean Joined: The U.S. Naval Observatory, 1830–2000* (Cambridge, 2003), 7; Hugh Richard Slotten, *Patronage, Practice, and the Culture of American Science: Alexander Dallas Bache and the U.S. Coast Survey* (Cambridge, 1994), 42–46; and Paullin, "Early Movements for a National Observatory," *Records of the Columbia Historical Society, Washington, D.C.,* XXV (1923), 37. Although it is not entirely clear from the sources, it seems the Ferrer resigned his commission in the Spanish navy when he moved to New York but remained a Spanish subject. See Galiano, *Biografía,* 12–13.

47. Vaughan to Gallatin, Apr. 28, 1807, APS Archives, III: Communications ("Mr.

Officials and experts in the United States continued to look to Spanish America as an astronomical exemplar even as independence movements threatened the Spanish Empire. Giovanni Antonio Grassi, the Italian-born president of Georgetown College, particularly admired Francisco José de Caldas and his observatory, which was built in Bogota in 1804 and funded through José Celestino Mutis's Royal Botanical Expedition, as models for U.S. astronomy. In an 1812 letter, Grassi applauded Caldas for organizing "the first monument erected to Uranus in the New World" and for challenging stereotypes about the "ignorance" of Catholics by "joining the humble practice of the true religion with the study of the most sublime science." He added that it was "surprising, but nevertheless true," that there were no observatories in the United States and asked Caldas to send a description of his South American observatory. It would, however, be another thirty years before the United States built its own national observatory—the 1842 Naval Observatory—and, by then, Spain's American empire had mostly crumbled.[48]

• • •

Imperialism and the encounters it inspired were vital to the practice and application of astronomy in early America. Geopolitical competition motivated the work of the Spanish and U.S. commissions of the Florida boundary survey, violence against slaves enabled astronomers like William Dunbar to perform disciplined observations, and interimperial exchanges of data made José Joaquín de Ferrer a prominent figure in the United States' scientific community. But U.S. officials found that actualizing political order through astronomy depended on having real coercive power on the ground, and, as Andrew Ellicott and Thomas Freeman realized during their expeditions, native groups and Spain still called the shots throughout much of the Floridas and the lower Mississippi Valley. The persistence of geopolitical competition for the Gulf South would continue to affect the pursuit of knowledge well into the 1810s, not least because Spanish, Anglo, and French men of science in the region could still choose from an array of potential allegiances.

Ferrer's ideas"), José Joaquín de Ferrer y Cafranga to John Vaughan, Apr. 25, 1807 ("method").

48. Giovanni Antonio Grassi to Josef Francisco *[sic]* de Caldas, Mar. 12, 1812, APS Archives III: Communications (quotes). On Caldas and the observatory, see John Wilton Appel, "Francisco José de Caldas: A Scientist at Work in Nueva Granada," in American Philosophical Society, *Transactions*, LXXXIV, no. 5 (1994), esp. 72–76.

{ 4 }

ALLEGIANCE, IDENTITIES, AND NATIONAL SCIENTIFIC COMMUNITIES

• • •

The lower Mississippi Valley remained a contested place as the United States expanded into the region between the 1790s and 1810s. The United States established the Mississippi and Orleans Territories in what had been West Florida and lower Louisiana, but Spain, France, and Britain worked to weaken the United States' tenuous grip by attracting the allegiance of its white, black, and native inhabitants. Geopolitical competition was, moreover, hardly limited to imperial powers: native groups, adventurers, and pirates also strove for influence and territory.[1]

European men of science in the lower Mississippi Valley professed attachment to, and promoted the interests of, several empires, nations, and nonstate entities. All kinds of people throughout the world's borderlands maintained multiple allegiances and identities. But individuals who aspired to be recognized as men of science faced the added challenge of fashioning themselves as credible sources of natural knowledge, a reputation that had as much to do with personal and political attachments as intellectual prowess. Cultivating an identity as a man of science was a careful social performance in early modern Europe, and it was even trickier in an American borderland caught in the throes of the Age of Revolutions. Patronage networks—the relationships through which powerful individuals supported the work of learned men by offering them money, status, and access to social

1. On geopolitical competition and allegiance in the Gulf South, see Andrew McMichael, *Atlantic Loyalties: Americans in Spanish West Florida, 1785–1810* (Athens, Ga., 2008); Gene Allen Smith and Sylvia L. Hilton, eds., *Nexus of Empire: Negotiating Loyalty and Identity in the Revolutionary Borderlands, 1760s–1820s* (Gainesville, Fla., 2010); Kathleen DuVal, *Independence Lost: Lives on the Edge of the American Revolution* (New York, 2015); David Narrett, *Adventurism and Empire: The Struggle for Mastery in the Louisiana-Florida Borderlands, 1762–1803* (Chapel Hill, N.C., 2015); and Eric Herschthal, "Slaves, Spaniards, and Subversion in Early Louisiana: The Persistent Fears of Black Revolt and Spanish Collusion in Territorial Louisiana, 1803–1812," *JER*, XXXVI (2016), 283–311.

networks—were unstable, and the lines between science, fortune seeking, and intrigue were unclear.[2]

Local experts in the lower Mississippi Valley tried to benefit from circulating information among a variety of actual and potential patrons, and, in the process, they manipulated and blurred the boundaries between the United States' scientific community and those of other polities competing for the borderlands. Even after the United States claimed the lower Mississippi Valley, connections with U.S. officials were just a fraction of the web of patronage relationships and knowledge exchanges connecting the Gulf South to other parts of the hemisphere and the Atlantic world. The weakness of U.S. rule, the federal government's parsimonious support for science, and the lack of opportunities for non-Anglos all ensured that U.S. officials and institutions were not necessarily the most desirable or accessible sources of patronage for local men of science.[3]

Historians have mostly ignored scientific practitioners in the United States' vast contested territories, and this has engendered a false sense that the early United States had a well-defined scientific community that cohered around shared patriotism. However, territorial expansion into the borderlands brought Spanish, Anglo, and French men of science into the United States who—instead of being U.S. patriots inspired by a postindependence context of liberty and democracy—sought to exploit the imperial United States' regional weakness and rivalries. The Spanish naturalist and spy Thomas Power opposed U.S. expansion by sharing natural and political knowledge with Spanish governors and rogue U.S. officers. The Scottish planter and astronomer William Dunbar maintained attachments to both Spain and the United States, leveraged his power over enslaved blacks

2. On the challenges of fashioning oneself as a man of science, see Mario Biagioli, *Galileo, Courtier: The Practice of Science in the Culture of Absolutism* (Chicago, 1993); Hugh Richard Slotten, *Patronage, Practice, and the Culture of American Science: Alexander Dallas Bache and the U.S. Coast Survey* (New York, 1994); E. C. Spary, "Of Nutmegs and Botanists: The Colonial Cultivation of Botanical Identity," in Londa L. Schiebinger and Claudia Swan, eds., *Colonial Botany: Science, Commerce, and Politics in the Early Modern World* (Philadelphia, 2005), 187–203; James Delbourgo, "Fugitive Colours: Shamans' Knowledge, Chemical Empire, and Atlantic Revolutions," in Simon Schaffer et al., eds., *The Brokered World: Go-Betweens and Global Intelligence, 1770–1820* (Sagamore Beach, Mass., 2009), 271–320; and Peter J. Kastor, *William Clark's World: Describing America in an Age of Unknowns* (New Haven, Conn., 2011), 130–155.

3. On the relationship between science and national identity, see Carol E. Harrison and Ann Johnson, eds., *National Identity: The Role of Science and Technology*, *Osiris*, XXIV (2009).

into intellectual and political influence, and took advantage of U.S. officials' dependence on his expertise to increase his wealth and status. And the French engineer and slave trader Barthélémy Lafon struggled for years to win the patronage of U.S. officials and institutions before opting to work against U.S. interests by offering his expertise to pirates and Spanish officials. The stories of these three men reveal how territorial expansion both added to, and exacerbated deep tensions within, the United States' scientific community.[4]

KNOWLEDGE AND (THOMAS) POWER

Thomas Power (1765–1825) was a naturalist and a spy who worked to undermine U.S. expansion into the lower Mississippi Valley. Science and spycraft were not distinct fields for Power, nor did they necessarily generate distinct results: spying enabled him to do scientific fieldwork that, in turn, could legitimize his presence in sites of intrigue and generate useful political intelligence and natural knowledge. As the information he produced was usually meant to remain secret, Power's patronage relationships with Spanish governors and Anglo adventurers were highly personal affairs based on backroom meetings and uneasy trust. Even though Power was an educated and active naturalist, these shady associations ultimately limited his ability to fashion an identity as a credible man of science. Being recognized and remembered as a man of science depended on forging the right

4. On the scientific community in the early United States, see George H. Daniels, *American Science in the Age of Jackson* (New York, 1968); Nathan Reingold, "Definitions and Speculations: The Professionalization of Science in America in the Nineteenth Century," in Alexandra Oleson and Sanborn C. Brown, eds., *The Pursuit of Knowledge in the Early American Republic: American Scientific and Learned Societies from Colonial Times to the Civil War* (Baltimore, 1976), 33–69; Simon Baatz, "Philadelphia Patronage: The Institutional Structure of Natural History in the New Republic, 1800–1833," *JER*, VIII (1988), 111–138; Elizabeth Keeney, *The Botanizers: Amateur Scientists in Nineteenth-Century America* (Chapel Hill, N.C., 1992); Daniel Goldstein, "'Yours for Science': The Smithsonian Institution's Correspondents and the Shape of Scientific Community in Nineteenth-Century America," *Isis*, LXXXV (1994), 573–599; and Conevery Bolton Valencius et al., "Science in Early America: Print Culture and the Sciences of Territoriality," *JER*, XXXVI (2016), 73–123. On patriotism and science, see Brooke Hindle, *The Pursuit of Science in Revolutionary America, 1735–1789* (New York, 1974), 380–385; John C. Greene, *American Science in the Age of Jefferson* (Ames, Iowa, 1984), esp. 10–12; and Andrew J. Lewis, *A Democracy of Facts: Natural History in the Early Republic* (Philadelphia, 2011). On how expansion into the Mississippi Valley exposed tensions and multiple allegiances within the United States, see François Furstenberg, "The Significance of the Trans-Appalachian Frontier in Atlantic History," *AHR*, CXIII (2008), 647–677.

connections, but, as Power knew better than most, it was far from clear between the 1790s and 1810s that allying with the United States would prove the most promising path to fame and fortune in the Gulf South.[5]

Power's nationality was ambiguous, a valuable but potentially risky asset for a borderlands spy. Historians have tended to assume that Power identified as Irish or British and that Spain was merely his "adopted country" because his parents were Irish Catholics. But he was born in Tenerife on the Canary Islands, firmly within the Spanish Empire and at the center of the Atlantic world. Spanish officials usually labeled him a *canario* (Canary Islander); Power described himself (in English) as a "Spaniard." His Irish appearance did facilitate missions into U.S. territory, but, as Louisiana governor Manuel Gayoso de Lemos observed, Power sometimes met with bad treatment when Anglo-Americans did "acknowledge [him] as a Spaniard." There were other dangers in not looking Spanish. While working on the Florida boundary survey in 1798, Power fretted that it was "impossible to convince [the Choctaws that] I am a Spaniard; and in case any attempt is made against the Americans, I shall have but a bad chance."[6]

Power was already a learned citizen of the world when he came to Louisiana in the early 1790s, and, within a few years of his arrival, Spanish offi-

5. For other individuals from this period whose spying and scientific work overlapped, see Joseph G. Tregle, Jr., "Introduction," in Thomas Hutchins, *An Historical Narrative and Topographical Description of Louisiana and West-Florida* (Gainesville, Fla., 1968), xxiv–xxxv; Neil Safier, "A Courier between Empires: Hipólito da Costa and the Atlantic World," in Bernard Bailyn and Patricia L. Denault, eds., *Soundings in Atlantic History: Latent Structures and Intellectual Currents, 1500–1830* (Cambridge, 2009), 265–293; and Thomas J. Schaeper, *Edward Bancroft: Scientist, Author, Spy* (New Haven, Conn., 2011).

6. Charles Gayarré, *History of Louisiana*, III, *The Spanish Domination* (New Orleans, 1885), 345, 346 ("adopted"); "Captain Power's Testimony," *Misissippi [sic] Herald and Natchez Gazette*, Dec. 3, 1807 ("Spaniard"); Manuel Gayoso de Lemos to Stephen Minor, Sept. 5, 1797, Gayoso de Lemos Letters, LLMVC ("acknowledge"); Thomas Power to Gayoso, July 7, 1798, PC-LLMVC, leg. 215-B, reel 75, Ser. 106, 52 ("impossible"); Sebastián Calvo de la Puerta, marqués de Casa-Calvo, to Pedro Cevallos, July 18, 1805, MPA, SD, VIII, reel 61x, 509, PKY; Minor to Gayoso, July 29, 1798, PC-LLMVC, leg. 50, reel 111, Ser. 143, 1110; Jack D. L. Holmes, *Documentos inéditos para la historia de la Luisiana, 1792–1810* (Madrid, 1963), 248; François-Xavier Martin, *The History of Louisiana: From the Earliest Period*, 2 vols. (New Orleans, 1827–1829), II, 123; Royal Ornan Shreve, *The Finished Scoundrel: General James Wilkinson, Sometime Commander-in-Chief of the Army of the United States, Who Made Intrigue a Trade and Treason a Profession* (Indianapolis, Ind., [circa 1933]), 63; Andro Linklater, *An Artist in Treason: The Extraordinary Double Life of General James Wilkinson* (New York, 2009), 146.

cials recognized him as an expert on the nature and politics of an enormous swath of North America. As Governor Gayoso remarked, Power was a man "of much understanding and education" who had "acquired much knowledge during his extensive travels in the four parts of the world." Gayoso added, accurately enough, that Power was "very mischievous": Power seemed to delight in fomenting intrigue, and his writings expressed a sarcastic outlook and a lack of patience with those he considered to be his mental inferiors. By the first years of the nineteenth century, his scientific work and spycraft had taken him throughout the eastern United States, the Mississippi and Ohio River valleys, Louisiana, and the Floridas, and he had somehow found time to acquire "vast local knowledge of the Internal Provinces" west of the Mississippi. Although Power had initially stood out in Louisiana for his trilingualism and been recruited as a translator, Francisco Luis Hector, the baron de Carondelet, governor of Louisiana from 1791 to 1797, recognized Power's other talents and made the young man his confidential agent for expeditions into the United States. According to Power, these missions—which included an overland journey from New Orleans to Canada—were "discreet, extremely important for the Royal Service, and as arduous and dangerous as they were thorny and delicate." The overarching objective of his assignments was to incite Kentuckians and Tennesseans to break away from the Atlantic states in the hope of lessening the looming threat that the expanding United States posed to Spanish America.[7]

Power identified as a man of science to justify his travels through Tennessee, Kentucky, Ohio, and Chickasaw territory. He sometimes went under the name "Doctor Powers," and he claimed to be researching a natural history of the region. Writing in 1796, U.S. secretary of war James McHenry noted succinctly that "Powers *[sic]* is about thirty five years of age; of Irish descent; born in one of the Canary Islands; educated at St. Omer; bred a physician; a man of Science; seemingly versatile; and speaks French Spanish and English fluently." Although Power probably hoped that self-identifying as a naturalist would add a trustworthy veil to his geopolitical schemes, it also seems that "the ardour of [his] curiosity" was genuine. When the U.S. Army tried barring him from Kentucky in 1796, Power pro-

7. Manuel Gayoso de Lemos to Francisco Saavedra, Feb. 26, 1799, AHN, leg. 3901, apartado 4, no. 3, 1209 ("much understanding," "very mischievous"), 1212 ("acquired"); Casa-Calvo to Cevallos, July 18, 1805, MPA, SD, VIII, reel 61x, 509 ("vast local"); Thomas Power, "Ynstancia del agrimensor Thomas Power . . . solicitando el abono de sus haberes," [n. d. 1805?], ANC, Fondo Floridas, leg. 2, no. 18, reel 2, 1 ("discreet"); Holmes, *Documentos inéditos,* 248.

tested that "to leave this country, without obtaining some knowledge of its topography, soil, improvements, population, etc ... is to me a very painful disappointment." This lament could be read as an excuse for intrigue, but Power made time for scientific investigations even when they were not part of his service to Spain, and researching and publishing about the natural history of the Ohio Valley could have allowed Power to acquire broader recognition as an expert on interior North America. Power's career underlines both the inadequacy and flexibility of the term "man of science" for individuals whose investigations of nature were just one of several methods for achieving their social, political, and intellectual goals.[8]

Power's persona as a man of science by no means gave him a reputation for trustworthiness or disinterestedness. Nearly everyone who had heard of Power knew that he was, as U.S. astronomer Andrew Ellicott put it, "intreauging *[sic]* for the Spanish government in the state of Kentucky ... fomenting the discontent but too prevalent among our Citizens." Power himself would later declare that his status as a spy was "so generally known, that it would be idle and absurd, and altogether useless, for me to deny it." Standards of credibility were specific to particular communities and varied depending on one's role within them, and the same cunning that made Power so untrustworthy to some U.S. citizens put him in "the highest confidence" of his exclusive network of Spanish officials, Kentuckian separatists, and

8. Deposition of Samuel Allen, July 8, 1797, PC-LLMVC, leg. 213, 619 ("Doctor"); James McHenry to Anthony Wayne, May 25, 1796, NTC, M0367, box 2, folder 15, http://images.indianahistory.org/u?/ONWT,1548 ("Powers"); James Wilkinson, *Memoirs of My Own Times ...*, 3 vols. (Philadelphia, 1816), II, 203, Thomas Power to James Wilkinson, May 20, 1796, II, app. XLIII ("ardour"), Power to Wilkinson, May 26, 1796, II, app. XLIII ("to leave this country"); Martin, *History of Louisiana*, II, 123. Saint-Omer was a secularized Jesuit college in Liège with predominantly English faculty and students; the secondary curriculum emphasized physics, logic, languages, natural history, and mathematics. See Charles R. Bailey, "French Secondary Education, 1763–1790: The Secularization of Ex-Jesuit Collèges," American Philosophical Society, *Transactions*, LXVIII, no. 6 (January 1978), 88–92. For occasions on which Power pursued science independently of his duties, see Thomas Power to Juan Ventura Morales, Apr. 29, 1801, PC-HNOC, leg. 538-B, reel 78, Ser. 78, 241; and Power to Manuel Gayoso de Lemos, June 12, 1798, PC-LLMVC, leg. 215-B, reel 75, Ser. 106, 50. Several other naturalists did achieve notoriety in the United States and Europe by publishing on the Ohio Valley, and many of these men—including André Michaux, Thomas Hutchins, and Gilbert Imlay—were connected with international intrigue. See Frederick J. Turner, "The Origin of Genet's Projected Attack on Louisiana and the Floridas," *AHR*, III (1898), 666; Tregle, "Introduction," in Hutchins, *Historical Narrative and Topographical Description*, xxiii–xxxv; and Wil Verhoeven, *Gilbert Imlay: Citizen of the World* (London, 2008), 204.

U.S. double agents. Power was well aware of this, and he invariably played up the inseparability of credibility and loyalty in communications with his patrons. He told Carondelet in 1797 that he "acted with prudence, with honour, and the disinterestedness of an honest man, as well as with the zeal and fidelity which the King's service requires." In response, the governor promised to reward him with "an appointment, which will render you independent of hatred and jealousy." Official patrons believed the loyalty of their informants was vital and had to be bought.[9]

The travels of Power and other Spaniards in the trans-Appalachian West have received little historical attention and almost no recognition as scientific endeavors. Scholarship has perpetuated the Anglocentric argument that U.S.-initiated expeditions were "uniquely American" in that they combined scientific, commercial, and political goals while Spanish exploration was, in contrast, "secretive and nonscientific." Yet Power's missions were similar to, and entangled with, those of more famous French and Anglo-American naturalist spies operating throughout the region. For example, the French botanist André Michaux tried to use his 1793 western expedition, performed on behalf of the American Philosophical Society (APS), to forward Edmond Charles Genêt's plot to incite Kentuckians against Spain's Gulf South colonies and win westerners' backing for a French Louisiana. Spanish officials also wanted to secure the allegiance of westerners, so they sent Power to Kentucky in 1794 "to keep an eye on the movements, and progress, of . . . Genet's expedition, against Louisiana." Power's expeditions were also comparable to those of Zebulon Pike, the Anglo-American explorer who integrated spying, natural history, and astronomy in his 1806 to 1807 venture into the Southwest. Although Pike worked for the United States and Power served Spain, they shared a common patron in James

9. Andrew Ellicott to Timothy Pickering, June 5, 1797, AEP, III, Letters Sent, 1797–1800, 545 ("intreauging"); "Captain Power's Testimony," *Misissippi [sic] Herald and Natchez Gazette,* Dec. 3, 1807 ("so generally known"); Daniel Clark, *Proofs of the Corruption of Gen. James Wilkinson, and of His Connextion with Aaron Burr . . .* (Philadelphia, 1809), 22 ("highest"), Thomas Power to Francisco Luis Hector, the baron de Carondelet, May 9, 1797, note no. 36, 74 ("acted"), Carondelet to Power, May 26, 1797, note no. 38, 83–84 ("appointment," 84). See also the depositions collected by John Gordon, July 8, 1797, PC-LLMVC, leg. 213, 619–623; Clark, *Proofs,* 39; and Linklater, *Artist in Treason,* 153. On Spanish spying in the late eighteenth-century trans-Appalachian West, see William H. Goetzmann, *Exploration and Empire: The Explorer and the Scientist in the Winning of the American West* (New York, 1978), 43; and A. P. Whitaker, "Spanish Intrigue in the Old Southwest: An Episode, 1788–89," *Mississippi Valley Historical Review,* XII (1925), 155–176.

Wilkinson, the U.S. Army general who, among other schemes, plotted the secession of Kentucky, sent intelligence to Spanish officials, and supported Aaron Burr's imperial ambitions.[10]

Wilkinson was an independent power broker who sought to acquire geographic and political knowledge and use it to solidify his own influence in the continental interior. Power thus made an ideal ally. After they first "accidentally met" near Gallipolis, Ohio, in 1793, Wilkinson recalled that Power "appeared to be a man of travel and information; his conversation was interesting, and he represented himself, as engaged in collecting materials, for the purpose of forming a natural history of the country." Power became Spain's liaison to Wilkinson and, by 1796, had undertaken three daring missions to exchange intelligence with the general. He developed an attachment to Wilkinson that reached beyond the scope of serving Spain. Power tended to reserve his personal loyalties for exceptionally intelligent leaders, including Spanish officials with scientific leanings like Gayoso and the brilliant if overly ambitious Wilkinson. In 1797, a U.S. soldier who met Power in Natchez observed that he "spoke of Wilkinson's great superiority of understanding and that he was sorry for my ignorance in not perceiving it." Just as Power performed the role of a devoted client, so, too, did Wilkinson play his part as patron by using his influence with Spanish officials to advocate for Power's promotion. After only a few years in North America, Power's political and intellectual networks transcended imperial boundaries.[11]

10. John Logan Allen, "Pike and American Science," in Matthew L. Harris and Jay H. Buckley, eds., *Zebulon Pike, Thomas Jefferson, and the Opening of the American West* (Norman, Okla., 2012), 84 ("uniquely"), 90 ("secretive"); "Mr. Power's Narrative and Deposition, *Respecting His Mission, in 1795*," in Wilkinson, *Memoirs*, II, app. XLV ("to keep an eye on"). On Michaux, see Gilbert Chinard, "André and François-André Michaux and Their Predecessors: An Essay on Early Botanical Exchanges between America and France," American Philosophical Society, *Proceedings*, CI (1957), 352; and Elizabeth Hyde, "André Michaux and French Botanical Diplomacy in the Cultural Construction of Natural History in the Atlantic World," in Sue Ann Prince, ed., *Of Elephants and Roses: French Natural History, 1790–1830*, American Philosophical Society Memoir, CCLXVII (Philadelphia, 2013), 89–100. On how Wilkinson's plots overlapped with Pike's expedition, see Goetzmann, *Exploration and Empire*, 43–50.

11. Wilkinson, *Memoirs*, II, 203 ("accidently"); Robert Newman to James McHenry, June 18, 1797, NTC, M0367, box 2, folder 20, http://images.indianahistory.org/u?/ONWT,1790 ("spoke"); James Wilkinson to Manuel Gayoso de Lemos, Sept. 22, 1796, in Clark, *Proofs*, 40; David E. Narrett, "Geopolitics and Intrigue: James Wilkinson, the Spanish Borderlands, and Mexican Independence," *WMQ*, 3d Ser., LXIX (2012), 104; Ralph E. Ehrenberg, "'Forming a General Geographical Idea of a Country': Mapping

Power's involvement with Wilkinson led U.S. General Anthony Wayne to accuse him of being *"a spy,"* which, in turn, prompted Power to reflect on the meaning of "the term *spy.*" Though the "blockhead" Wayne had meant it to convey "a person employed by one nation to ... detect the schemes of a rival and hostile nation," Power equated spies with a type of philosopher whose penetration into the workings of human nature gave them a higher place in the chain of being than officials like Wayne. As he told Wilkinson:

> To an intelligent and reflecting mind, how mortifying and humiliating it must be, on taking a philosophical survey of society, to behold in the front ranks,... wretched bodies without a soul, whose sluggish intellects and contracted views, reduce them, in the scale of beings, nearly to the level of the most torpid of the brute creation. To contemplate these creatures, decorated with all the insignia of power,... exacting the tribute of homage and submission from men, possessing fifty times their understanding, and adorned with virtues,... [I] blush to think, that I belong to the same species .... High rank, honours, exalted posts, dignified offices, confer not any mental prerogative .... [Such leaders] will ever be an object of scorn, and derision, for the philosopher. I will take my leave of General Wayne, by assuring him, that he need be under no kind of apprehension of being taken for a *spy* in any part of the world.

This rant was an early instance in which Power revealed the Jacobin streak that increasingly influenced his politics, an ideological shift that he shared with both other Louisianans and other Spanish men of science in the Americas.[12]

Power was hardly above self-praise in his pursuit of recognition and promotion, and he cited the danger of his work, loyalty to Spain, and hatred of the United States to solicit greater financial compensation for his services. He explained to Governor Gayoso that "the hazards I have encountered, and the narrow escapes I have made in the accomplishing of whatever may have been the objects of my missions" had earned him so little pay that he remained "a loser" in 1798. He recognized that there was no longer a chance

---

Louisiana from 1803 to 1820," in Alfred E. Lemmon, John T. Magill, and Jason R. Wiese, eds., *Charting Louisiana: Five Hundred Years of Maps* (New Orleans, 2003), 124–125.

12. James Wilkinson to Thomas Power, Nov. 11, 1795, in Wilkinson, *Memoirs,* II, Appendix XL *("a spy"),* Power to Wilkinson, Nov. 12, 1795, Appendix XLI ("term *spy,*"), emphases in originals; Antonio Lafuente, "Enlightenment in an Imperial Context: Local Science in the Late-Eighteenth-Century Hispanic World," *Osiris,* XV (2000), 159–160; Ernest R. Liljegren, "Jacobinism in Spanish Louisiana, 1792–1797," *LHQ,* XXII (1939), 47–97.

that Kentucky would break away from the United States, but, rather than attributing this unfortunate result to the 1795 Treaty of San Lorenzo, which had eased Spanish commercial restrictions against U.S. citizens, Power blamed the incompetence of his coconspirators. "Without incurring the reproach of vanity," he boasted, "had every person concerned, displayed the same activity, firmness, constancy, fidelity, secrecy, honor, and disregard of personal interest, that have invariably marked my conduct, the Politicks of the Western Country would at this day wear a widely different aspect." These were not merely professional concerns for Power. His hatred of "our Northern neighbours" was a sentiment that "originate[d] in an inveterate rancour, too deeply rooted ever to be conquered." Power was (especially when asking for a raise) emotionally invested in challenging the expansion of the United States.[13]

The patronage Power received from Carondelet, Gayoso, and Wilkinson helped him land positions that blended science and spycraft, most notably that of surveyor and secretary for the Florida boundary commission of 1798 to 1800. He participated in the scientific and managerial aspects of the expedition and penned a "historical diary of occurrences on the line," an account of "topographical, botanical, and physical observations," and a journal devoted to astronomical observations. He also gathered intelligence on "the operations of our false neighbors," particularly Ellicott's own intriguing and the movements of U.S. troops.[14]

Yet Power's reputation as a spy limited the types of scientific employment officials would give him and, therefore, also limited his wider public recognition as a man of science. Gayoso had proposed that Power should take over as Spain's astronomer after William Dunbar quit the boundary survey in 1798, but Ellicott raised such a fuss at this prospect that Gayoso agreed to give the less capable but more trusted Stephen Minor the prestigious job of astronomer. Ellicott, however, had played right into Gayoso's hands: the governor took advantage of Ellicott's trust in Minor, who had been Ellicott's childhood friend, to facilitate Spanish intelligence gathering during the expedition. Gayoso asked Minor to procure "a continued and early information of all their politics and of all their movements and actual

13. Thomas Power to Manuel Gayoso de Lemos, Sept. 29, 1798, PC-LLMVC, leg. 215-B, reel 75, Ser. 106, 62.

14. Power to Morales, Apr. 29, 1801, PC-HNOC, leg. 538-B, reel 78, Ser. 78, 241 ("historical"); Gayoso to Saavedra, Feb. 26, 1799, AHN, leg. 3901, apartado 4, no. 3, 1211 ("operations"). For Power gathering and communicating intelligence on the United States and Ellicott, see Power to Gayoso, July 7, 1798, PC-LLMVC, leg. 215-B, reel 75, Ser. 106, 52, Power to Gayoso, Sept. 20, 1798, and Power to Gayoso, Oct. 14, 1798, 68.

object of the American forces" and added that he was "glad of the intimate harmony that I find subsists between you and Mr. Power, whom I look upon to be a man of penetration and discernment." This plan would only succeed if Power and Minor were not "looked upon as Spyes," and, since Power was already notorious for intrigue, Minor made a better front man. Ellicott and Dunbar achieved international and historical recognition as astronomers through their work on the boundary. Power did not.[15]

After the survey ended in East Florida in 1800, Power traveled to the Chesapeake and Pittsburgh on his way back to New Orleans and, at least according to Ellicott, planned to make a stop in Philadelphia. Whether he went to Philadelphia is unknown, but Ellicott worried that Power's plan was to stoke international discord. He told the secretary of state that "Mr. Power has been so long in the habit of intrigue and duplicity that he is only at home when in the midst of confusion—his former residence in Philadelphia procured him an extensive acquaintance with the partizans of France." It is possible that, like other Spanish men of science active in the greater Caribbean, Power headed to Philadelphia to consult the city's libraries and scientific resources, collections that would have helped him finalize the diary of geographic and botanical observations that he completed while travelling home down the Ohio and Mississippi Rivers. Philadelphia was a shared hub of learning for the United States and the Caribbean through which information often crossed imperial boundaries. For one, Anglo-Americans in Philadelphia were deeply interested in natural knowledge from throughout the Spanish Empire: Philadelphia publishers produced some three hundred imprints relating to the Hispanic world before 1830, and Charles Wilson Peale, one of the city's leading naturalists, claimed that Spanish officials had encouraged him to explore South America and ship specimens to Madrid's Royal Cabinet of Natural History. Some Spanish naturalists working in the Caribbean also relied on Philadelphia's institutions and resources. In 1798, the botanist Baltasar Boldo interrupted his plant collecting in Cuba to visit Philadelphia and read "the most modern botanical works that can be acquired, and that cannot be found in Spain, in order to rectify the manuscripts of [his] general history of [Cuban] plants." During his stay in the city, Boldo acquired seeds from North America, Botany Bay, China, and Siberia

15. Manuel Gayoso de Lemos to Stephen Minor, Oct. 16, 1798, Gayoso de Lemos Letters (quotes). For Ellicott's complaints, see "Extracts from A. Ellicott's Communication to Secretary Pickering, Darling's Creek, November 8th, 1798," in Wilkinson, *Memoirs,* II, app. XXXI. On Ellicott's childhood friendship with Minor, see Andrew Ellicott to Timothy Pickering, Sept. 12, 1797, AEP, III, Letters Sent, 1797–1800, 576.

that he forwarded to Madrid's Royal Botanical Garden. If Power had similar plans for his visit to Philadelphia, they were thwarted (or perhaps overshadowed) by his reputation for intrigue and his responsibility to fulfill his primary duty of reporting political intelligence to his patrons.[16]

By the final years of Spanish rule in Louisiana, Power had begun to prosper. His superiors gave him employment as surveyor of the Feliciana District, social prestige as a militia captain, a large cash reward for his work on the boundary line, and forty thousand arpents of land on the Tickfaw River. He had acquired enough money by 1798 to purchase his first slave and bought five more over the next ten years. Power also furthered his connections within Spanish Louisiana's social and scientific circles in 1799 by marrying the niece of the colony's surveyor general. He had, in other words, every reason to suspect that embracing allegiance to Spain had been the right decision.[17]

The 1803 transfer of Louisiana from Spain to France and, then, to the United States threatened to disrupt Power's connections with Spanish officials, but it also offered a wealth of potential allegiances and employment opportunities for someone with his access to political and natural knowl-

16. Andrew Ellicott to Timothy Pickering, Mar. 23, 1800, AEP, IV, Correspondence, 1004 ("Mr. Power"); Power to Morales, Apr. 29, 1801, PC-HNOC, leg. 538-B, Ser. 78, reel 78, 241; Merle E. Simmons, "Bibliography of Pre-1830 Philadelphia Imprints Pertaining to the Hispanic World and Hispanic Culture," unpublished manuscript, APS; Charles Willson Peale to Thomas Jefferson, Oct. 11, 1801, in James P. McClure and J. Jefferson Looney, eds., The Papers of Thomas Jefferson Digital Edition, Main Ser., XXXV, http://rotunda.upress.virginia.edu/founders/TSJN.html. My thanks to Christopher Heaney for the Simmons "Bibliography" and Peale letter. On Boldo, see Baltasar Manuel Boldo to Joaquín de Santa Cruz y Cárdenas, conde de Mopox, Aug. 14, 1798, copy, Colección Guillén, MXXXI, MS 2243, Mopox tomo 4, doc. 20, AMNM ("most modern"), Mopox to Francisco Saavedra, Dec. 29, 1798, Colección Guillén, MXXIX, MS 2241, Mopox tomo 2, doc. 23. On how Portuguese naturalists also sought botanical knowledge of the Americas in Philadelphia, see Safier, "Courier Between Empires," in Bailyn and Denault, eds., *Soundings in Atlantic History*, 268, 273, 275–277.

17. For Power's positions and rewards, see Power to Gayoso, Sept. 29, 1798, PC-LLMVC, leg. 215-B, reel 75, Ser. 106, 63; Vicente Sebastián Pintado to Carlos de Grand Pre, Dec. 2, 1803, Papers of Vicente S. Pintado, reel 2, MSS 17, 619, 315, PKY; Andrew Ellicott to Thomas Jefferson, Apr. 13, 1801, AEP, V, Correspondence, 1039; and "Relativo al derecho de media asunata por ventas de tierras realengas que debe satisfacer el Sr. Tomás Power por cuarenta mil arpanes de tierra en la Mobila," (1805), ANC, Fondo Floridas, leg. 8, no. 15, 1, reel 11. For Power's slaves, see Gwendolyn Midlo Hall, *Afro-Louisiana History and Genealogy, 1718–1820*, http://www.ibiblio.org/laslave/. On Power's marriage, see Francisco Broutin, Historical Notaries' Indices, Index, I, 10, New Orleans Notarial Archives.

edge. Wilkinson offered one tempting option in which Power would pose as a French agent to forward Wilkinson's vision of seizing Louisiana and Texas, either as his own empire or under the aegis of the United States. But the imminent French takeover of Louisiana also promised Power the greatest scientific opportunity of his career when Sebastián Calvo de la Puerta, the marqués de Casa-Calvo, appointed Power to lead an expedition to the source of the Red River and run the boundary between French Louisiana and Spanish Texas. Dunbar, for one, was certain that the expedition would make Power an international celebrity. "I congratulate you my good friend," he told Power, for "nothing is more gratifying to the inquisitive mind of a man of genius than the discovery of new objects hitherto concealed from the persevering researches of the Philosopher and Man of Science." Dunbar wrote excitedly about the "hidden treasures" of botanical, ethnographic, and geographic information that Power would surely reveal, adding that "in Europe the result of your researches will be looked for with a degree of impatience." "If the notices brought by [James] Bruce from the Sources of the diminutive Nile were a subject of eager Curiosity to the learned in Europe," Dunbar explained, "with how much more avidity will they devour the intelligence to be expected from the sources of this Great Father of Rivers and from a new world." This was Power's chance to reshape his public identity from a spy who dabbled in science to a world-renowned explorer, but France ruined this vision by selling Louisiana to the United States and leaving Power with no expedition to lead.[18]

Power did get his chance to explore Texas in 1805 and 1806, but not in a way that would earn him wide scientific recognition. Casa-Calvo, who had experience with Spanish scientific expeditions in Cuba, made himself the head of a sixty-three man team that mapped much of the Texas interior and sought to settle the boundaries of the Louisiana Purchase. He relegated Power to his usual role as surveyor / spy because "his experience and known talent will provide us with necessary information ... as much in the realm of science as in politics." Casa-Calvo and Nicolas deFiniels, a former engineer in both the French and U.S. armies, had the leading roles in the expedition's geographic research, though Power did have opportunities to break off from the main party and explore the Atchafalaya and Sabine River regions. U.S.

18. William Dunbar to Thomas Power, July 8, 1803, WDP, Letter Book, 1802–1805, 11 (quotes). On Wilkinson's plot, see "Extract of a Letter, in Cypher, from Thomas Power, in New Orleans, to Brigadier-General James Wilkinson ...," May 5, 1803, in Wilkinson, *Memoirs,* II, 253; and Narrett, "Geopolitics and Intrigue," *WMQ,* 3d Ser., LXIX (2012), 120.

officials grew suspicious that this expedition was a cover for inciting western Indians and Louisianans to reject U.S. rule and, when the members returned, ordered Spanish officials to leave the Orleans Territory. Power, however, was one of several Spanish agents who remained in New Orleans and continued to pose a threat to the United States' fragile authority.[19]

Power's attachment to Spain grew more dubious and his Jacobin leanings more pronounced as Spanish influence in the lower Mississippi Valley—and Power's patronage networks—slowly dissolved. In one private letter, he voiced approval of Aaron Burr's plot to conquer northern New Spain because it might relieve the "miserable wretches groaning under the galling chains of Religion, Political, and Civil Tyranny ... and raise them to the rank of men." Power, a Spanish agent, used the language of the Black Legend to rationalize the invasion of Spanish America. This private letter, so uncharacteristic of his official communications, raises doubts about the sincerity of Power's oft-professed attachment to the Spanish crown.[20]

Power's loyalties would be put on trial during some of the most dramatic court cases in the early United States: Burr's treason trial of 1807 and Wilkinson's court-martial of 1811. Power was still friends with Wilkinson (and owed him money) during the Burr trial in 1807. So, when Burr sought to deflect attention from himself by asking Power to confirm that Wilkinson was a double agent, Power avoided implicating the general by emphasizing his own allegiance to Spain. Power claimed that he "had the honor to be, and still [was], an officer in the service of his C.M. [Catholic Majesty]" and was "bound by the strongest ties that can fasten on a man of principle and honor, not to answer any interrogatories" that might compromise the interest of "the King my Master." Nevertheless, Power's reputation and ambiguous nationality called his credibility into question in matters of law as much as matters of science. Ellicott, a longtime enemy of Power, attempted to dis-

19. Carlos Martínez de Yrujo, marqués de Casa Yrujo, to Sebastián Calvo de la Puerta, marqués de Casa-Calvo, July 20, 1805, MPA, SD, VIII, 531, PKY (quote). On Casa-Calvo's experience in Cuba, see Joaquín de Santa Cruz y Cárdenas, conde de Mopox, "Documentación relativa a los individuos que tomaron parte en la comisión de Cuba," July 2, 1798, Colección Guillén, CCCXCV, MS 1578. On the Texas expedition, see Sebastián Calvo de la Puerta, marqués de Casa-Calvo, to Felix Tala, Apr. 12, 1806, PC-HNOC, leg. 142-A, reel 79, Ser. 144, 867; and Jack D. L. Holmes, "The Marqués De Casa-Calvo, Nicolás DeFiniels, and the 1805 Spanish Expedition through East Texas and Louisiana," *Southwestern Historical Quarterly*, LXIX (1966), 327–329, 331–333, 335–337. On Spanish officials in New Orleans, see McMichael, *Atlantic Loyalties*, 70.

20. Thomas Power to Stephen Minor, Feb. 6, 1807, Minor (Family) Papers, 1774–1891, LLMVC.

credit his testimony by declaring that "he knows Thomas Power, and that his general character is bad, . . . that he was an Englisman *[sic]* by descent, an African by birth, a Jesuite by education, a French Jacobin in principle, and one of the first agents of the massacres at the Cape." Ellicott was right to identify a radical streak in Power, but it is unlikely that Power had any role in Jean-Jacques Dessalines's massacre of French whites at Cap-Français in April 1804, since he was busy performing his duties as surveyor of the Feliciana District. There was, however, great anxiety in the United States in the wake of the Haitian Revolution, and Ellicott might have leveled this accusation to turn public opinion against Power by casting him as a threat to white power.[21]

Power would find himself back on the stand in 1811, this time testifying against Wilkinson at the general's court-martial in Frederick, Maryland. Wilkinson had attracted Power's ire by breaking his promise not to publish the certificate Power had written to exculpate him from charges of treason in 1807, and Power now devoted himself to bringing about Wilkinson's downfall. "The well known Don Thomas Power, a Spanish subject," was notorious when he arrived in Frederick, but, as usual, he was known for being a spy and not a man of science. The court-martial was a moment that determined Power's and Wilkinson's futures and legacies; as Power warned Wilkinson in court, this "is a fatal moment for me sir—I must stab or be stabbed." And Power got stabbed. The silver-tongued Wilkinson used Power's reputation for intrigue to convince his accusers that "Power is altogether unworthy of credit" because he had been *"employed by me, as a spy, for the United States,* to watch over the very government, of which, he has sworn on colonel Burr's trial [that] 'he was an officer.'" This accusation was almost certainly false, but Power knew he had lost and gave up in dramatic fashion. On his third day on the stand, he showed up drunk to court and made a great show of surreptitiously pocketing some documents that apparently proved he was deliberately aiming "to destroy Wilkinson." When the court ordered him to share these papers, Power "threw them up with tears in his eyes and left the place." It is hard to believe that someone so comfortable with intrigue would have made such a blunder; he was prob-

21. "Captain Power's Testimony," *Misissippi [sic] Herald and Natchez Gazette,* Dec. 3, 1807 ("had the honor"), emphasis in original; "From the Washington Monitor," *Public Advertiser* (New York, N.Y.), June 11, 1808 ("he knows"); Frederick Walther to Vicente Sebastián Pintado, Feb. 9, 1804, Papers of Vicente S. Pintado, reel 2, MSS 17,619, 405, Thomas Power to Pintado, May 4, 1804, 485; Thomas Perkins Abernethy, *The Burr Conspiracy* (Gloucester, Mass., 1968), 254–256.

ably just tired of all these trials concerning his adventures as a youth and wanted to go home to his family. Wilkinson was acquitted of all charges that he had been a double agent, and Power returned to New Orleans, where he died in 1825.[22]

Imperial competition shaped and circumscribed Power's patronage relationships, scientific activities, goals, identity, and legacy. During his fifteen years of gathering political intelligence and studying nature, Power's attempts to fashion himself as a man of science were constantly overshadowed by his reputation as a spy. Ironically, it was Power's patrons who most limited his opportunities for wider recognition: Carondelet, Gayoso, Casa-Calvo, and Wilkinson all valued Power more for his intriguing than his science, so they relegated him to inconspicuous appointments and prevented his observations from circulating beyond their exclusive community of conspirators.

Power's overlapping pursuits of natural and political knowledge were ideally suited for the contested borderlands in which he lived. But a life dedicated to resisting U.S. expansion and flouting boundaries that only seem clear in retrospect—including boundaries between Spanish and U.S. information networks, legitimate governments and conspiratorial factions, and scientific research and political intrigue—ensured that Power has been all but entirely excluded from U.S. history and the history of science. This exclusion has skewed scholars' understandings of intellectual life in early America. Power was just one of the many white, black, and native inhabitants of America's contested spaces who applied natural knowledge toward challenging the United States, and these men and women were just as much a part of the republic's intellectual world as patriots whose scientific work promoted its aggrandizement. Imperialism simultaneously enlarged the United State's learned community and ensured it was rife with enemies.

WILLIAM DUNBAR AND "THE SCIENCE OF POLITICS"

In contrast to Thomas Power's obscurity, contemporaries and historians alike have considered William Dunbar (1750–1810)—a Scottish-born astronomer with plantations near Natchez and Baton Rouge—to be the most excellent scientific practitioner in the lower Mississippi Valley. Dunbar successfully fashioned himself as a man of science because he leveraged his

22. *Alexandria Herald* (Alexandria, Va.) Nov. 22, 1811 ("well known"); *American Watchman and Delaware Republican* (Wilmington, Del.), Nov. 27, 1811 ("fatal," "destroy," "threw"); Wilkinson, *Memoirs of My Own Times,* II, 88 *("employed"),* 107 ("Power"), emphasis in original; Shreve, *Finished Scoundrel,* 237–242.

wealth, which depended on his power over enslaved blacks, and position between competing powers to secure patronage across political boundaries. Both Spanish and U.S. officials looked to him for the expertise, instruments, and national prestige that they considered necessary to ordering the lower Mississippi Valley. Dunbar capitalized on this trust to dictate the conditions under which his expertise would facilitate Spanish and U.S. imperialism and to shape the networks of knowledge and patronage that connected him with experts and officials in the United States.[23]

Soon after arriving in British West Florida in 1773, Dunbar began researching agricultural technologies and environmental conditions in the hope of establishing a Caribbean-style plantation. He visited local plantations "to study the art" of making indigo, kept meteorological observations, and experimented with several potentially valuable commodities. Dunbar also drew heavily on laborers and specimens from the Caribbean. He purchased most of his first slaves from Jamaica and asked slave traders on the island to send "a small collection of seeds etc. of your various kinds of fruits and curious trees" that he could test on his plantation; he was particularly interested in introducing the "fine pine apple." The routes that supported slavery and science were often one and the same throughout the greater Caribbean.[24]

Dunbar's first shift in political attachment coincided with the American Revolution, but it was the loss of wealth and security, and not any political principles, that led him to affiliate with Spain. In 1778, patriot raiders led by

23. Studies of Dunbar include Franklin L. Riley, "Sir William Dunbar—The Pioneer Scientist of Mississippi," Mississippi Historical Society, *Publications,* II (1899), 85–111; Eron Rowland, ed., *Life, Letters, and Papers of William Dunbar . . .* (Jackson, Miss., 1930), esp. 10–11; John Francis McDermott, "Philosophic Outpost on the Frontier: The Library of William Dunbar of the Forest" (1965), unpublished MS, MDAH; Robert J. Malone, "Everyday Science, Surveying, and Politics in the Old Southwest: William Dunbar and the Influence of Place on Natural Philosophy" (Ph.D. diss., University of Florida, 1996); Arthur H. DeRosier, Jr., *William Dunbar: Scientific Pioneer of the Old Southwest* (Lexington, Ky., 2007); and Andrew McMichael, "William Dunbar, William Claiborne, and Daniel Clark: Intersections of Loyalty and National Identity on the Florida Frontier," in Smith and Hilton, eds., *Nexus of Empire,* 273–275. On a more famous European astronomer who also built a scientific identity by circulating knowledge and building patronage relationships across national boundaries, see Mary Terrall, *The Man Who Flattened the Earth: Maupertuis and the Sciences in the Enlightenment* (Chicago, 2002), esp. 7–12.

24. William Dunbar to John Ross, July 12, 1775, in "Extracts," 11 ("study"); Dunbar, Journal, 1776–1780, WDP, 2–3; Dunbar to Messrs Thompsons of Kingston, Jamaica, June 18, 1775, "Extracts," 9 ("small collection").

Captain James Willing robbed Dunbar and many other West Florida loyalists "of everything that could be carried away." But all was not lost. Dunbar had hidden his slaves and "obtained permission from the Spanish Govt" to buy a plantation in Spanish Louisiana "and soon after brought down all [his] Negroes." He also retained his lands on the eastern bank of the Mississippi and, in 1779, divided his twenty-one slaves between his "Spanish Plantation" in Louisiana and his "English Plantation" near Baton Rouge. After the Treaty of Paris in 1783, both of these were within Spain's dominion. Dunbar was hardly enthusiastic about becoming a Spanish subject, but he mused in 1781 that "I should have no objection to live under Spanish or any other well regulated government." He added "there is no need of the name American to make my residence agreeable." Like Thomas Power, Dunbar thrived under Spain: his wealth, landholding, and social standing all improved, and he found a profitable outlet for his astronomical expertise by acting as a land surveyor for Spain and accepting payment in land.[25]

Dunbar became increasingly, though not exclusively, attached to the United States after the 1795 Treaty of San Lorenzo. This treaty placed his Natchez plantation—his family's primary residence since 1792—within the United States' borders, though he still retained lands in Spanish West Florida. At least initially, Dunbar's shift in political affiliation did not earn him the support and recognition he had expected. Shortly after Andrew Ellicott arrived in Natchez to claim the district for the United States in 1797, Dunbar asked the U.S. secretary of state for an appointment as surveyor general of the Mississippi Territory, a position that was as close as one could get to being a federally funded astronomer in the early republic. Dunbar told the secretary that his zeal for the United States, access to Spanish maps, and "local information" made him the perfect man for the job. But Dunbar recognized that he needed more than expertise to attain this appointment. He needed to prove his allegiance. He thus requested a letter of recommendation from Ellicott and asked him to emphasize that Dunbar was not, despite rumors spread by his "secret enemies," a Spanish agent. Dunbar also asked Ellicott to highlight his supposedly warm feelings for the U.S. government, and, accordingly, Ellicott told the secretary of state that Dunbar was "a gentleman possessed of an uncommon share of Philo-

25. Dunbar, Journal, 1776–1780, WDP, 31 ("of everything"), 33 ("obtained"), 35 ("Spanish"); Dunbar to John Ross, Dec. 29, 1781, "Extracts," 19–20 ("I should"); Dunbar to Manuel Gayoso de Lemos, Apr. 4, 1797, WDP, Correspondence, 1778–, Gayoso, Grant of land to Dunbar, Apr. 5, 1797, Plat for survey of granted lands, Apr. 11, 1797; DeRosier, *William Dunbar*, 66, 69.

sophical information" and that "I have no doubt of his zeal for the U.S. and attachment to our excellent constitution." Dunbar had covered all his bases, but U.S. officials, much like the Spanish governors who employed Power, opted instead to appoint a surveyor who was both an astronomer and intelligence agent, Isaac Briggs.[26]

Dunbar had not pinned all his hopes for patronage and recognition on the United States. During the same week that he asked Ellicott for a letter of recommendation, he was also considering Louisiana governor Manuel Gayoso de Lemos's offer of a potentially more lucrative position as Spain's astronomer for the Florida boundary survey. Although he did ponder if his "scruples ... respecting American Citizenship" should preclude him from accepting this job, he reasoned that a person could be "employed for a time as a Geometrician, by a nation, with which he may have no connection, either as Citizen or Subject." Although the thirty-five hundred peso salary itself was no small reward, he told a friend that "by going upon this trip I shall have an opportunity of discovering valuable lands and *perhaps minerals of a precious nature.*" "I am possessed of some chemical knowledge," he asserted, "and shall carry with me a portable laboratory for making assays." Dunbar probably ended up taking the job with the Spanish boundary commission because it seemed like a chance to get filthy rich in plantation lands and mineral treasures.[27]

Dunbar was proficient in several scientific pursuits, but he most proudly identified as an astronomer. He established a regional reputation for astronomical expertise in the early 1790s while working as district surveyor of Spanish Natchez, and his work on the Florida boundary line made his name more widely known. By 1800, Dunbar was corresponding with leading U.S. and British astronomers, including the famed discoverer of Uranus, William Herschel.[28]

26. William Dunbar to Timothy Pickering, [n.d. (1797?)], "Extracts," 28 ("local"); Dunbar to Andrew Ellicott, Aug. 27, 1797, AEP, II, Letters Received, 1798–1800, 105 ("secret"), Ellicott to Pickering, Sept. 12, 1797, AEP, III, Letters Sent, 1797–1800, 573 ("gentleman"). On federal surveyors as astronomers, see Hindle, *Pursuit of Science,* 337. On Briggs as a political agent, see Silvio A. Bedini, *Thomas Jefferson: Statesman of Science* (New York, 1990), 351.

27. William Dunbar to Andrew Ellicott, Mar. 27, 1798, AEP, II, Letters Received, 1798–1800, 211 ("scruples"); Dunbar to John Ross, Aug. 21, 1797, "Extracts," 31–32 ("by going"), emphasis in original.

28. Dunbar to Dr. [William] Herschel, June 16, 1799, "Extracts," 55. On Dunbar's work as a surveyor in the Spanish colonies, see Malone, "Everyday Science, Surveying, and

Dunbar's success as an astronomer in the Spanish American borderlands largely depended on maintaining connections with British scientific networks. In 1796, soon after he had learned that the 31° boundary would be run, Dunbar asked an English friend, John Swift, to supply him with the instruments and observational techniques needed to ensure that Gayoso would choose him as Spain's astronomer. Dunbar requested that Swift forward an "Astronomical Circle armed with its acromatic *[sic]* Telescope" as soon as possible because "I conceive there will be no instruments in this country so well calculated for this business." He added that "if you have any learned Geometrical friend ask him to put down in a few words the most approved and the most correct method of tracing a parallel of Lat[itude" since "some new thought may have been started on this subject which has not yet reached us in this country." In exchange for astronomical instruments and methods, Dunbar offered to supply British astronomers with "Celestial news from this quarter." Dunbar was happy enough to play the colonial data provider in his long-distance relationships with Britons while he used the instruments and techniques they provided to bolster his reputation among officials vying for influence in the lower Mississippi Valley.[29]

Dunbar acted as an astronomical instrument broker for both Spain and the United States. He had begun accumulating scientific instruments in the 1770s and soon became a source for local residents and officials who wanted to procure or repair instruments, such as when he "enjoyed the pleasure of making Coupet happy by putting a New glass in his Telescope" in 1776. Slavery and the international market in plantation-grown commodities made his instrument collection possible: he sent purchase orders to English instrument maker Edward Troughton concurrently with shipments of cotton bound for England and used the profits to purchase apparatus, including a six-foot Gregorian reflecting telescope worth £150. As he wrote in 1803, "The small fortune which I have acquired by cultivating the Earth alone, enables me to procure any instruments of moderate expence which might facilitate my researches." It was a risky, costly, and slow process for Spanish and U.S. officials in the region to purchase and receive scientific instruments, so they looked to Dunbar as an established local supplier, and, as Dunbar soon discovered, this could be a profitable enterprise.

Politics in the Old Southwest," 179; and Riley, "Sir William Dunbar," Mississippi Historical Society, *Publications,* II (1899), 93.

29. William Dunbar to John Swift, June 30, 1796, "Extracts," 25–26 ("Astronomical"); Dunbar to Swift, Oct. 14, 1804, WDP, Letter Book, 1802–1805, 26 ("Celestial").

He provided nearly all the Spanish boundary commission's instruments at a price "a hundred per cent above the first cost in London" and would continue to sell instruments to Spain at a 100 percent markup well after becoming a U.S. citizen. U.S. officials also bought astronomical instruments from Dunbar, but Dunbar was outraged to discover that the United States imposed duties on shipments of instruments into New Orleans. He told the treasurer of the APS that "science which ought from the honor of our Country to be fostered and encouraged by our Gov is hereby oppressed, certainly the man who imports rare and valuable scientific instruments ... for the express purpose of advancing the philosophical and geographic knowledge of our Country ought not to be punished by a fine." Like countless other rich white men throughout U.S. history, Dunbar was quick to offer patriotic justifications for why he deserved a tax break.[30]

Dunbar used his access to astronomical instruments to court patronage. When supplying the Spanish boundary commission in 1798, Dunbar told Governor Gayoso that "Shou'd your Excellency have the curiosity to view the [astronomical] Circles or any other of the articles" Dunbar would "have [them] moved with care to the Governor's house." Dunbar knew Gayoso would be particularly pleased with instruments that promised to reflect the dignity of Spain and outshine the U.S. astronomers in the field, so he assured the governor that "our instruments will not only be much more eligant but better adapted to the purpose than those of Mr. Ellicott, his Great [zenith] Sector only excepted." Dunbar also provided London-made instruments for Gayoso's personal collection: in April 1799, he offered to import a "new improved Achromatic telescope" for Gayoso and added that if there were "any other object your Excelly may want, I hope you will do me the honor to put in my power to procure them for you." Dunbar further curried Gayoso's favor by sharing his own most sophisticated instruments. "My great ambition is to have a grand telescope of extraordinary powers," he told Gayoso, and, "judging of your Excelly's curiosity by my own[,] ... it

30. Dunbar, Journal, 1776–1780, WDP, 13 ("enjoyed"); William Dunbar to [torn], Jan. 17, 1803, WDP, Letter Book, 1802–1805, 6 ("small fortune"); Manuel Gayoso de Lemos to Dunbar, June 26, 1798, PC-LLMVC, leg. 215-B, reel 75, Ser. 106, 152 ("hundred"); Dunbar to John Vaughan, Oct. 20, 1805, WDP, Letter Book, 1805–1812, 4 ("science"); Dunbar to [John Swift?], Apr. 1, 1802, WDP, Letter Book, 1802–1805, 1, Dunbar to Power, July 8, 1803, 12–13. On Dunbar's instruments, see Riley, "Sir William Dunbar," Mississippi Historical Society, *Publications*, II (1899), 107–108. For Dunbar providing instruments for U.S. officials, see William Dunbar to Henry Dearborn, July 13, 1805, WDP, Letter Book, 1802–1805, 37; and Isaac Briggs to Albert Gallatin, Sept. 8, 1803, Surveyor General's Files, Outgoing Correspondence, Ser. 1153, box 3198, MDAH.

may be set up in your Excelly's garden, where we will have the satisfaction of viewing together ... the Planets and other heavenly bodies." Gayoso would, however, die of yellow fever in July of 1799, leaving Dunbar to seek the patronage of another powerful official with an interest in science, Thomas Jefferson.[31]

As in his dealings with Spanish patrons, Dunbar's eventual success with receiving appointments from U.S. officials—positions that included territorial judge, justice of the peace, legislator, and cotton-gin inspector—was interconnected with his wealth and scientific expertise. Mississippi's first territorial governor, Winthrop Sargent, observed that Dunbar's "Integrity, science and Wealth" made him "Respectable" as a judge. Ellicott once told the secretary of state that "the best informed, most wealthy, and virtuous part of the inhabitants of the Mississippi Territory" were governed by "the most worthless part of the community" but suggested that, "if good order, virtue, and reputation were the objects [of government], why not elect such men as Wm Dunbar[?]" As a federalist and closet monarchist, Dunbar found the popular basis of Jeffersonian republicanism distasteful. Still, he was not above deceiving potential patrons, including the territory's second governor, William C. C. Claiborne, into believing that he shared their principles. One Mississippi official noted in 1803 that the "deep designing" Dunbar was "very capable of imposing upon those who are fond of flattery" and had "so far imposed upon Governor C[laiborne]" that he lauded Dunbar as a good republican. The official added that "Dunbar is a man possessing great qualifications; and all he wants to qualify him to become an eminent public character, is ... a better knowledge of practical politics." "But he is too old, and too scientific a character," he surmised, "to become a sudden convert in the science of politics." What this official failed to see was that Dunbar converted science into politics to establish himself as a leading figure in Natchez no matter which party or nation held sway.[32]

31. William Dunbar to Manuel Gayoso de Lemos, [date illegible (1798?)], PC-LLMVC, leg. 215-B, reel 75, Ser. 106, 121 ("Shou'd"), Dunbar to Gayoso, Apr. 13, 1799, leg. 212-B, 376 ("new improved"). On Gayoso's education and scientific interests, see Irving A. Leonard, "A Frontier Library, 1799," *HAHR*, XXIII (1943), 21–51. Zenith sectors measure the angular distance between a star and the meridian to determine geographical positions on earth.

32. Winthrop Sargent to John Marshall, Aug. 25, 1800, in Dunbar Rowland, ed., *The Mississippi Territorial Archives, 1798–1803* ... (Nashville, Tenn., 1905), 274 ("Integrity"); Andrew Ellicott to [James Madison], Dec. 29, 1801, in Clarence Edwin Carter, ed., *The Territorial Papers of the United States*, V, *The Territory of Mississippi, 1798–1817* (Washington, D.C., 1937), 133–134 ("best informed," 133, "if good order," 134), Edward Turner

Dunbar also mastered the politics of science: he expertly forged the patronage networks necessary to become an esteemed participant in the United States' scientific community. Dunbar's first move was impressing and flattering Ellicott, who became a vocal advocate of Dunbar's talents. Dunbar praised Jefferson's scientific achievements as well but, more importantly, won Jefferson's patronage by presenting him with detailed information about Gulf South nature that he had acquired in the Spanish service. Dunbar had written a manuscript on "the face of the Country, vegetable and animal productions" during the 1798 boundary survey and presented a copy to his then patron Governor Gayoso. In a clever manipulation of global scientific networks, Dunbar sent this manuscript to a friend in London via Jefferson two years later. This was his first letter to Jefferson, and Dunbar saturated it with the deferential language needed to cement a patron-client relationship. He wrote, for example, that since "men of learning and genius are indulgent to those of inferior talents, I have suffered my notes and observations to appear before you." Dunbar asked Jefferson to forward the notes to London; instead, Jefferson sent them to the APS, asked its savants to "judge of [Dunbar's] degree of science," and recommended that they welcome Dunbar as a member. The APS promptly offered Dunbar membership—a testament to both his scientific abilities and elite social status—and he remained an active member of the society until his death. The APS's obituary remembered Dunbar as "the self-taught Astronomer of the woods, whose communications have so often enriched our volumes, and reflected credit on the Society." It was a stretch to call Dunbar self-taught: he had private tutors as a child and, as a young man, studied science and mathematics while earning a master of arts degree from King's College, Aberdeen. Yet, all of the most revered scientific figures in the early republic—particularly Benjamin Franklin, John Bartram, and David Rittenhouse—were considered autodidacts, and describing Dunbar as self-taught was another signal of the APS's eagerness to integrate him, even in memory, into the national scientific community.[33]

---

to John C. Breckinridge, Nov. 2, 1803, 267, 274 ("deep"), 275 ("Dunbar is"). On Dunbar's appointments, see DeRosier, *William Dunbar,* 181–188.

33. William Dunbar to Manuel Gayoso de Lemos, Nov. 7, 1798, PC-LLMVC, leg. 215-B, reel 75, Ser. 106, 167, LLMVC ("face"); Dunbar to Thomas Jefferson, July 14, 1800, TJP, [1] ("men of learning"), Th[omas] Jefferson to [Caspar Wistar], Dec. 16, 1800 ("judge"); "Obituary Notice," American Philosophical Society, *Transactions,* I, New Ser. (1818), xviii ("self-taught"); William Dunbar to Andrew Ellicott, May 25, 1798, AEP, I, Letters Received, 1796–1798, 244; Ellicott to Pickering, Sept. 12, 1797, AEP, III, Letters Sent, 1797–1800, 576, Ellicott to Pickering, July 12, 1798, 680. On how individual initiative and inter-

Dunbar took advantage of both Gayoso's and Jefferson's national pride and reliance on his astronomical expertise to garner more pay and prestige. Gayoso and Jefferson had comparable traits that made them valuable patrons for an aspiring man of science: both had wide-ranging scientific interests, governed enormous territories, and had the power to make official appointments. Even so, the similarities between the arguments Dunbar made for higher pay to Gayoso in 1798 and Jefferson in 1804 are striking. Dunbar told Gayoso during the planning stages of the Florida boundary expedition that he ought to be paid more than the U.S. commission's astronomer because "the high dignity of a great and ancient monarchy when but in competition with that of a new born republic demands, it would seem, some distinction." Dunbar elaborated that this monarchical distinction should include more money and "4 male servants and a laundress" to attend him during the expedition. Careful to preserve an air of deference, Dunbar blamed his "parsimonious or even niggardly" pay on Juan Ventura Morales, the intendant of Louisiana and one of Gayoso's most hated rivals. Dunbar's rhetoric worked: Gayoso raised his pay to thirty-five hundred pesos and provided him with excellent accommodations during the survey. As one Louisianan planter observed, Dunbar experienced few of "the inconveniences and hardships" typical of "astronomical operations" because he was "not only abundantly provided of all necessities of life but even of luxury." Six years later, Dunbar found himself in a very similar situation, only now it was President Jefferson who sought to hire him as the leader of a scientific expedition, one that would explore the Washita River. Jefferson informed Dunbar that he would only be given three thousand dollars for both this and a later expedition, arguing that the poor funding was immaterial considering "your attachment to science and attainment in it, and the dispositions to aid it necessarily flowing from these." As Dunbar forthrightly replied, however, "the Patriotism of men of Science and Genius" was not sufficient incentive to devote time and energy to scientific research. "When a great Empire talks of compensation," Dunbar told Jefferson, "this ought

---

personal relationships were essential to securing patronage in the early republic, see Howard S. Miller, *Dollars for Research: Science and Its Patrons in Nineteenth-Century America* (Seattle, Wash., 1970), ix. On the importance of social standing to membership in the APS, see Sally Gregory Kohlstedt, "Savants and Professionals: The American Association for the Advancement of Science, 1848–1860," in Oleson and Brown, eds., *Pursuit of Knowledge in the Early American Republic,* 308–309. On Dunbar's education, see DeRosier, *William Dunbar,* 18–20. On Franklin, Bartram, and Rittenhouse as "an American triumvirate of natural genius," see James Delbourgo, *A Most Amazing Scene of Wonders: Electricity and Enlightenment in Early America* (Cambridge, Mass., 2006), 144.

to be adequate to the importance of the service and honorable both to govt and to the selected individuals." Like Gayoso, Jefferson knew that he depended on Dunbar's expertise and acquiesced to his demands, offering him an additional five thousand dollars.[34]

Such negotiations were indicative of how Spanish imperialism had worked in the Gulf South, a borderland where relatively weak officials depended on local power brokers and experts. Individuals like Dunbar had thrived in this system of negotiated authority, and, as Dunbar recognized, they could continue to do so because the United States was also an overextended "Empire" in which control over the borderlands was just as tenuous—and just as reliant on the cooperation of locals—as Spain's had been. Unlike naturalists in revolutionary France who renegotiated patronage relationships during the shift from monarchy to republic by appropriating the rhetoric of their new republican leaders, Dunbar saw no need to alter the language he had used to address an imperial governor when writing to a republican president. Since the United States was a "great Empire" like Spain, Dunbar could draw on his experience as an astronomer in the Spanish service to teach Jefferson how the region's scientific experts could and would serve U.S. expansion. The negotiated relationships between local experts and U.S. officials were both essential to U.S. efforts to order the borderlands and a legacy of Spanish colonialism that local individuals perpetuated for self-interested reasons.[35]

Imperial competition structured how Jefferson and Dunbar envisioned each other's role in national networks of knowledge, power, and patronage.

34. William Dunbar to Manuel Gayoso de Lemos, Apr. 15, 1798, PC-LLMVC, leg. 215-B, reel 75, Ser. 106, 135 ("high dignity"), 136 ("4 male"), LLMVC, Dunbar to Gayoso, Apr. 23, 1798, leg. 215-B, reel 75, Ser. 106, 145 ("parsimonious"); Julien Poydras to Dunbar, June 25, 1798, "Letterbook of Private and Commercial Correspondence of an Indigo and Cotton Planter (Julien Poydras), 1794–1800, Pointe Coupée Parish, Louisiana," LSM, film 18, 335, Ser. H, reel 1, 45, PCL ("inconveniences"); Thomas Jefferson to Dunbar, Apr. 15, 1804, WDP, Correspondence, 1800–1804 ("your attachment"); Dunbar to Jefferson, May 13, 1804, WDP, Letter Book, 1802–1805, 16 ("Patriotism"); Jefferson to Dunbar, Mar. 14, 1805, WDP, Correspondence, 1805–1812; Jack D. L. Holmes, *Gayoso: The Life of a Spanish Governor in the Mississippi Valley, 1789–1799* (Baton Rouge, La., 1965), 216–222. On Gayoso's and Jefferson's scientific interests and activities, see Gayoso to Stephen Minor, Nov. 29, 1798, Gayoso de Lemos Letters; Leonard, "A Frontier Library," *HAHR*, XXIII (1943), 21–51; and Silvio A. Bedini, *Jefferson and Science* (Chapel Hill, N.C., 2002), esp. 35–40.

35. Dunbar to Jefferson, May 13, 1804, WDP, Letter Book, 1802–1805, 16 (quotes); E. C. Spary, *Utopia's Garden: French Natural History from Old Regime to Revolution* (Chicago, 2000), 9, 190.

Jefferson recognized that Dunbar's local expertise could help secure U.S. control over its territorial claims, writing in January 1801 that "I hope I shall not . . . lose the benefit of your communications." "A Philosophical vedette at the distance of 1000. miles, and on the verge of the terra incognita of our continent," he explained, "is precious to us here." "Vedette" is a term for a sentinel stationed in advance of a military force, and it seems Jefferson viewed Dunbar as a vanguard in international contests for North America. Such geopolitical competitions also inspired Jefferson to patronize other scientific pursuits and institutions, such as Lewis and Clark's expedition and the short-lived Military and Philosophical Society. Dunbar, for his part, hoped Jefferson's administration would provide greater funding and honor to men of science in the United States, particularly those in the nation's borderlands. As he told Ellicott later in 1801, he had expected "that under a president who is a Philosopher and a person of general science, Learning and the arts [would] be patronized, Invention and discovery encouraged and rewarded, but the cry of economy by the votaries of your present Government alarms me." Not only was Dunbar still referring to the United States as someone else's government (despite being a citizen and official of it), but he suggested that U.S. officials had better step up their support for the sciences if they wanted to triumph over other powers and win control of the continent. As he put it, "Other governments and even societies of private individuals . . . Carry on discoveries to the most distant corners of the Globe and into the interior of Continents of perilous approach, not excepting our own frontiers." The Jeffersonian expeditions of 1803–1806, including Dunbar's own voyage up the Washita River, briefly promised to satisfy Dunbar's concerns. But funding ran dry for these expeditions in 1807, and federal policymakers could only offer Dunbar an "apology" that the "frequent drafts . . . made upon your time and patriotism" would receive no further compensation than the government's "most grateful acknowledgments." Dunbar had long since recognized that the U.S. government was not a reliable source of patronage. Luckily for him, he had hedged his bets and maintained ties with Spain.[36]

36. Thomas Jefferson to William Dunbar, Jan. 12, 1801, TJP, [2] ("I hope"); Dunbar to Andrew Ellicott, Oct. 3, 1801, AEP, V, Correspondence, 1069–1070 ("under a president"), 1070 ("Other governments"); Henry Dearborn to Dunbar, Mar. 30, 1807, in Rowland, ed., *Life, Letters, and Papers of William Dunbar,* 197 ("apology"); Deborah Allen, "Acquiring 'Knowledge of Our Own Continent': Geopolitics, Science, and Jeffersonian Geography, 1783–1803," *Journal of American Studies,* XL (2006), 205–232; Sidney Forman, "The United States Military Philosophical Society, 1802–1813: *Scientia in Bello Pax,*" *WMQ,* II (1945), 273–276. On federal funding for science, see A. Hunter Dupree, *Science in the*

In July of 1803, as Dunbar was busy sending details about the Mississippi and Louisiana Territories to Jefferson, he was also helping Spain plan and execute its own scientific projects. Thomas Power, who had been appointed to lead a Spanish expedition up the Red River, looked to Dunbar as a source for advice and instruments, and Dunbar was more than happy to oblige, for he "presume[d] that the Great Monarch who employs you" would fund the expedition generously. Dunbar listed the instruments he was willing to sell to Spain—including astronomical apparatus, timepieces, and a portable "laboratory for chemical experiments"—at a 100 percent markup. More importantly, Dunbar demanded knowledge and land as payment, rewards that he claimed to merit owing to his ongoing attachment to Spain. He informed Power that "I would solicit a copy of the [expedition's] Journal relating to Science and discovery" and promised that he would "certainly preserve Silence" with any sensitive information because he "still retain[ed] a predilection" for the Spanish government. Dunbar also continued his old practice of exchanging his scientific services for Spanish land, asking for "a tract of land of some magnitude" in West Florida that was large enough to divide among his eight children. Although Power's expedition was cancelled after the Louisiana Purchase, Dunbar's access to instruments and professed affection for Spain had promised to provide him a substantial reward.[37]

Despite his supposed predilection for Spain, Dunbar lost no time in betraying his promise not to share its secrets with U.S. officials. Soon after Jefferson had learned of the Louisiana Purchase, he sent questionnaires to several men in Louisiana and the Mississippi Territory, including Dunbar, to amass data on the geography, natural resources, boundaries, and population of the newly acquired region. In his reply, written one month after his letter to Power, Dunbar divulged that "I was applied to by the Spanish commission to furnish instruments; I shall probably be able to discover upon what principles this line was intended to be ran." Dunbar then asked Power directly about the boundary, and Dunbar quoted "the Gentleman who was to have been entrusted with the scientific part of the Spanish Commission" in his next letter to Jefferson.[38]

Federal Government: A History of Policies and Activities (Baltimore, 1986), 22–43. On Dunbar's Washita River expedition, see Trey Berry, Pam Beasley, and Jeanne Clements, eds., *The Forgotten Expedition, 1804–1805: The Louisiana Purchase Journals of Dunbar and Hunter* (Baton Rouge, La., 2006).

37. Dunbar to Power, July 8, 1803, WDP, Letter Book, 1802–1805, 11–13.

38. William Dunbar to Thomas Jefferson, [date torn, but probably Aug. 19, 1803], WDP, Letter Book, 1802–1805, 13 ("I was applied"); Dunbar to Jefferson, Sept. 30, 1803,

In the early 1800s, Dunbar constructed a persona that fit with, and contributed to, the vision of patriotic national science then developing in Philadelphia and other northeastern cities. He began ordering scientific books "peculiar to the U.S." from Philadelphia and asking members of the APS about "which Books are the most usefull for an American Botanist." Dunbar's requests satisfied the pretensions of Philadelphians who were then fashioning themselves as metropolitan natural philosophers, men who were self-consciously trying to take over the responsibilities of organizing and interpreting American nature that had, during the colonial era, been the purview of savants in England. Dunbar played the role of provincial collector flawlessly: he sent naturalist Benjamin Smith Barton a list of Gulf Coast quadrupeds, provided samples of "dubious Plants" to Philadelphian botanists who promised "to examine them and return the Linnaean name," and submitted his observations on geography, astronomy, hydrology, agronomy, and linguistics for publication in the society's *Transactions*. Dunbar and the Philadelphians who aspired to lead the United States' scientific community relied on each other to fashion, respectively, the provincial and metropolitan scientific identities that furthered their personal ambitions. The networks that integrated and made up the nation's scientific community evolved as much from the initiative of individuals in the borderlands as those in eastern cities.[39]

Dunbar made the local knowledge of diverse French, Spanish, and Anglo residents of the lower Mississippi Valley available to eastern men of science. He was pivotal to a regional scientific network that was centered in the lower Mississippi Valley but reached into the Caribbean and far West, and scientific practitioners in the region's U.S. and Spanish territories sought Dunbar's patronage because he could make them known in wider scientific and political spheres. Martin Duralde, the commandant of Opelousas in Spanish Louisiana, was one of these men. Dunbar had asked Duralde and other learned men in the region to compile Indian language vocabularies

---

in Clarence Edwin Carter, ed., *The Territorial Papers of the United States*, IX, *The Territory of Orleans, 1803–1812* (Washington, D.C., 1940), 68 ("Gentleman").

39. William Dunbar to John Vaughan, June 22, 1803, WDP, Letter Book, 1802–1805, 10 ("peculiar"); Henry Muhlenberg to Dunbar, July 5, 1808, in Rowland, ed., *Life, Letters, and Papers of William Dunbar*, 200 ("which Books"), 203 ("dubious"), 203–204 ("examine"); Dunbar to Benjamin Smith Barton, Mar. 1, 1801, "Extracts," 64. Several of Dunbar's writings were published in American Philosophical Society, *Transactions*, VI, parts 1 and 2 (1809). On Philadelphian naturalists refashioning themselves, see Lewis, *Democracy of Facts*, esp. 42–71.

and send them to his Natchez plantation, from whence Dunbar would forward them to Jefferson and the APS. Duralde sent Dunbar two vocabularies, one for the Atakapas, who lived near Duralde's outpost, and another for the Chitimachas, a group whose distance required Duralde to rely on, and extol the credibility of, his own extended network of informants. Duralde seized this chance to demonstrate the breadth of his scientific interests: he described human and mammoth bones unearthed by well diggers in Spanish Louisiana—including how a planter used "the hollow of [a mammoth's] hip bones ... to press his indigo"—and propounded theories about these fossils and regional geology based on his own observations and native stories. He lamented that "I have been obliged, (after a course of studies of little use to me) to sacrifice a real desire for natural and abstract science" and added that it would give him "real pleasure" to perform further scientific work on Dunbar's behalf. Dunbar sent Duralde's writings to Jefferson, who forwarded them to the APS, which published them in its *Transactions*. The incorporation of one man, Dunbar, gave the United States' scientific community access to information circulating within the multinational Gulf South.[40]

Although anti-Scottish prejudice—what Stephen Minor pinned down as Dunbar's "shuffling, evasive, Scotch sincerity and honesty"—led some Spanish and U.S. officials to distrust Dunbar, prominent leaders in both nations maintained confidence in his allegiance and credibility. Indeed, Spanish and U.S. leaders tried taking advantage of each other's trust in Dunbar. Jefferson appointed Dunbar to lead a scientific expedition up the Red River in 1805 and asked Governor Claiborne to assure Sebastián Calvo de la Puerta, the marqués de Casa-Calvo, "that the object of the journey is merely geographical and scientific, having in view nothing unfriendly to Spain." Casa-Calvo observed that "it is well known that the President, under the

40. Martin Duralde to William Dunbar, Apr. 24, 1802, Historical and Literary Committee, American Indian Vocabularies Collection, APS (quotes); Thomas Jefferson to Dunbar, Mar. 3, 1803, TJP. For Dunbar's regional connections, see [Barthélémy] Lafon to William Dunbar, Aug. 19, 1805, in Rowland, ed., *Life, Letters, and Papers of William Dunbar*, 181, John Sibley to Dunbar, Apr. 2, 1805, 162–173; Dunbar to Messrs Thompsons of Kingston, Jamaica, June 18, 1775, "Extracts," 9; Dunbar to Peter Walker, June 10, 1804, WDP, Letter Book 1802–1805, 18–19; Dunbar, Journal, 1804–1805, WDP, 173; William Dunbar, "On the Language of Signs among Certain North American Indians," American Philosophical Society, *Transactions*, VI, no. 1 (1809), 1–8; and Dunbar, "Description of a Singular Phenomenon Seen at Baton Rouge," American Philosophical Society, *Transactions*, VI, no. 1 (1809), 25. On how Dunbar was the leading figure in the Mississippi Territory's fledgling scientific institution, see *Constitution of the Mississippi Society, for the Acquirement and Dissemination of Useful Knowledge* (City of Washington, 1804).

pretext of promoting the sciences, makes use of philosophy ... in his ambition to procure knowledge of the Country that we possess." To assuage Casa-Calvo's fears, Claiborne offered to employ a Spanish agent to accompany the U.S. expedition; Dunbar chose Power, a decision that attested to Dunbar's respect for Power's scientific talents and, perhaps, Dunbar's intention to once more profit from serving Spain's interests. Despite the threat that Jefferson's blend of science and imperialism posed to Spain, it was Dunbar's participation that almost led the two governments to cooperate. Claiborne and Casa-Calvo agreed, despite all previous evidence to the contrary, that Dunbar's scientific interests made him politically disinterested. However, the main reason why Casa-Calvo was satisfied with Dunbar's appointment was that he believed, as he told Claiborne, that "even though [Dunbar] lived in Natchez he has always been devoted to the Spanish Government." That Spanish and U.S. officials simultaneously thought Dunbar was allied with their respective governments nearly enabled the Red River expedition to go forth and generate new knowledge about the West. By the time the expedition was eventually launched in 1806, however, Dunbar had quit his role as leader, and Spanish soldiers forced the U.S. explorers out of Texas.[41]

Dunbar successfully fashioned himself as an Anglo-American man of science: even the U.S. House of Representatives praised him for "reflecting lustre on his adopted country by his profound science and great public worth." But Dunbar never fully reconciled himself with being an American. As his public and scientific work began to wind down in 1806, Dunbar sent a friend in Philadelphia an order for a customized carriage. In the copy of the purchase request that Dunbar penned in his own letter book, he wrote, and soon crossed out, that he would like to have his initials inscribed on

41. Stephen Minor to Manuel Gayoso de Lemos, [n.d.], PC-LLMVC, leg. 215-B, reel 75, Ser. 106, 49 ("shuffling"); Thomas Jefferson to [William C. C.] Claiborne, May 26, 1805, in Carter, ed., *Territorial Papers of the United States,* IX, 451 ("object"); Casa-Calvo to Cevallos, July 18, 1805, MPA, SD, VIII, reel 61x, 509 ("well known"); Casa-Calvo to Claiborne, July 15, 1805, MPA, SD, VIII, reel 61x, 527, PKY ("even though"); Claiborne to Dunbar, July 29, 1805, in Dunbar Rowland, ed., *Official Letter Books of W.C.C. Claiborne, 1801–1816,* 6 vols. (1917; rpt. New York, 1972), III, 141; Holmes, "Marqués De Casa-Calvo," *Southwestern Historical Quarterly,* LXIX (1966), 326. For Claiborne describing Dunbar as disinterested, see Claiborne to Casa-Calvo, July 11, 1805, in Rowland, ed., *Official Letter Books of W.C.C. Claiborne,* III, 119. For a U.S. official deriding Dunbar's Scottish heritage, see Turner to Breckinridge, Nov. 2, 1803, in Carter, ed., *Territorial Papers of the United States,* V, 267, 274. On the 1806 Red River expedition, see Dan L. Flores, ed., *Southern Counterpart to Lewis and Clark: The Freeman and Custis Expedition of 1806* (Norman, Okla., 2002).

the carriage's door because "~~[I] presume it not being fashionable among you Republicans to have your coat of arms painted upon your Carriages~~." Seeming to remember that he, too, was now an American who professed Republican political leanings, he changed his copy to read that he wanted a simple "WD" on the carriage because "we Republicans must not think of Coats of arms."[42]

Dunbar's only real allegiance was to himself. He secured a lasting reputation as both a loyal American and a distinguished scientific practitioner by manipulating transnational webs of patronage, promoting the interests of competing nations, and helping define the networks that connected experts in the borderlands with prominent officials. These strategies had enabled Dunbar to thrive under British and Spanish rule, and they continued to serve him well in the imperial United States. Far from marking a clear split between colonial and national eras, U.S. expansion into the lower Mississippi Valley perpetuated a colonial context in which violence against enslaved blacks, transnational exchanges, and (at least until the 1810s) geopolitical competition were vital to how men of science pursued knowledge and constructed identities. Men like Dunbar—planters who exploited their power over slaves to study nature, gain political influence, patronize their own networks of experts, and build connections with eastern officials and institutions—would continue to dominate the region's intellectual life throughout the antebellum era.

PIRATING KNOWLEDGE: BARTHÉLÉMY LAFON'S GULF SOUTH

Like William Dunbar, Barthélémy Lafon (1769–1820) worked to craft a reputation for scientific excellence by cultivating patrons among both Spanish and U.S. officials. Born in France, Lafon migrated to Spanish Louisiana in 1790 and embarked on a career as an engineer, mapmaker, and slave trader that lasted well into the era of U.S. rule. But, even as it became clear that the United States would emerge triumphant in geopolitical competitions for the lower Mississippi Valley, Lafon sought opportunities for advancement among Spanish officials and, increasingly, in the more shadowy realms of the circum-Caribbean, where pirates and filibusters offered extra-state opportunities for wealth and status. The success of U.S. imperialism

42. "REPORT; Presented on the 26th of January to the House of Representatives," *National Intelligencer, and Washington Advertiser* (Washington, D.C.), May 28, 1804, ("reflecting"); William Dunbar to John Vaughan, [1806], WDP, Letter Book, 1805–1812 ("~~[I] presume~~").

in the Gulf South neither solidified the porous boundaries of the United States' scientific community nor precluded the possibility that men of science in the borderlands would continue to work against U.S. interests.

During his thirteen years as a Spanish subject in Louisiana, Lafon won support from Spanish officials and gained recognition as a competent engineer and cartographer. He seems to have received his first building contracts from the New Orleans cabildo after the devastating fire of 1794, and he spent the next six years designing public and private buildings, surveying lands, constructing levees, and cutting drainage canals to fight yellow fever. Lafon earned the esteem of Louisiana's premier patron of the sciences, Governor Manuel Gayoso de Lemos, who gave Lafon a piece of prime New Orleans real estate at no cost with the expectation that he would use it to establish a foundry. Lafon also proposed several other projects that promised to make him an important figure in New Orleans, including erecting a public bath and a "Hall of Spectacle" that would showcase plays by his own theater company. He began acquiring a wider reputation as a cartographer around 1800 when James Pitot presented two of Lafon's maps of Louisiana to officials in France.[43]

Lafon was also building his fortune as a slave trader. He purchased his first slave, Isabel, in 1794, but entered the trade in earnest from 1799 to the summer of 1803. During these years, he purchased ten slaves and sold twenty-eight, a discrepancy that suggests he was funneling smuggled slaves to New Orleans buyers. Lafon dealt in blacks born in Africa and Louisiana as well as Caribbean blacks set adrift during the tumultuous 1790s. He sometimes kept families together, as with seven "mulatto rouge" slaves that he sold as a unit in July 1803. This was his last sale until 1807, and it seems that the transition to U.S. governance and consequent prohibition of the foreign slave trade in Louisiana in 1804 disrupted his business. Nonetheless, as with many New Orleans slave owners, Lafon maintained bonds with blacks that extended beyond the slave economy: he had a lasting relation-

43. "Salle De Spectacle, Lafon Au Public," *Moniteur De La Louisiane* (New Orleans), Sept. 4, 1802 (quote); Harriet Pierpoint Bos, "Barthelemy Lafon" (Master's thesis, Tulane University, 1977), 22–30, 36–41; Ehrenberg, "'Forming a General Geographical Idea of the Country,'" in Lemmon, Magill, and Wiese, eds., *Charting Louisiana*, 126; Jack D. L. Holmes, "Some French Engineers in Spanish Louisiana," in John Francis McDermott, ed., *The French in the Mississippi Valley* (Urbana, Ill., 1965), 142; "City Council Records and Deliberations of the Cabildo," book 4, IV, Sept. 19, 1800, to July 9, 1802 (transcribed by the Works Projects Administration), 182, PCL; James Pitot, *Observations on the Colony of Louisiana from 1796 to 1802*, trans. Henry C. Pitot (Baton Rouge, La., 1979), 88.

ship with Modeste Foucher, a free woman of color from Saint Domingue, and left a legacy to her and their children in his will.[44]

As Lafon's involvement with the slave trade waned following the United States' incorporation of Louisiana, he put greater emphasis on profiting from his scientific projects and cultivating patronage ties with officials and experts in the United States. Lafon, like other remotely stationed men and women in European empires and the United States, sent the material products of his work—including maps, architectural designs, and scientific publications—to metropolitan officials and natural philosophers who, he hoped, would value this otherwise difficult-to-access information. Lafon's written and illustrated texts stood in for him in the eastern United States, but they had only limited success forging social bonds.[45]

Three weeks after selling off the enslaved mulatto rouge family in 1803, Lafon began circulating his maps through the emerging networks of governance connecting New Orleans with Washington, D.C. He forwarded a "Manuscript map . . . of the Western part of this Country" to Secretary of State James Madison via Daniel Clark, the New Orleans merchant who first introduced Dunbar to Thomas Jefferson. The map was an advertisement that Lafon was willing and able to provide federal officials with valuable local knowledge from their new territorial claims: as Clark told Madison, Lafon "has in his possession materials to compleat almost a perfect map of the Country, and would undertake the work . . . if it suited our Government to give him encouragement." Although this manuscript map did not lead to any federal funding, it did bring Lafon's work and, to some extent, his reputation as a geographic expert, to the very center of U.S. politics. Writing from Washington, D.C., one "gentleman" noted in 1804 that he "had the pleasure of examining the *Plan and View of New Orleans* and its vicinity, together with the *Map of Louisiana*" drafted by "an artist of that country, who . . . was Geographer and Engineer to the Spanish government." These maps were "suspended in the Senate Chambers," and congressmen and the president alike "agree[d] that they are *chef d'auvres [sic]*." Although some geographical works from the American West discouraged U.S. official's ardor for expansion, this gentleman believed Lafon's maps proved that

44. Hall, *Afro-Louisiana History and Genealogy,* http://www.ibiblio.org/laslave (quote); Bos, "Barthelemy Lafon," 2.

45. Susan Scott Parrish, *American Curiosity: Cultures of Natural History in the Colonial British Atlantic World* (Chapel Hill, N.C., 2006), 123; Daniela Bleichmar, *Visible Empire: Botanical Expeditions and Visual Culture in the Hispanic Enlightenment* (Chicago, 2012), 7, 101–103; Kastor, *William Clark's World,* 144–151.

Louisiana was "a country well worthy of the attention of the United States." The scientific productions with which Lafon sought to build patronage connections with eastern officials encouraged U.S. imperialism.[46]

In April 1805, Lafon aspired to solicit Jefferson's patronage in hopes of being appointed surveyor general of the Orleans Territory, the southern portion of hitherto French and Spanish Louisiana. Writing in French, Lafon flattered the president by praising his "desire to make knowledge prosper," and he stressed that his experience in the region made him uniquely qualified to be surveyor general. Not only had he lived in the province for fifteen years, but he was "constantly devoted to engineering, and the study of the country," and "local geography" had been his "dominant occupation for eight consecutive years." Lafon added that he spoke French, English, and Spanish, the latter being "indispensible to a surveyor general" because all of the province's titles had been "made in Spanish for forty years hence." He reminded Jefferson that one of his maps already hung in the halls of Congress and tried boosting his credibility by listing local men of science and political leaders as references, most notably Dunbar and Governor William C. C. Claiborne. But Jefferson did not write him back, and an eastern Anglo, Seth Pease, was appointed to the position. The same experience as a cartographer and engineer for Spain on which Lafon based his own value to Jefferson, combined with French Louisianans reputation for resenting U.S. rule, might have led the president to doubt Lafon's allegiance.[47]

Lafon failed to win Jefferson's patronage, but his *Carte Générale du Territoire d'Orléans* (1806) did earn praise from eastern reviewers and became a standard, if occasionally pilloried, map of the lower Mississippi Valley. U.S. geographers William Darby and H. M. Brackenridge both drew on Lafon's *Carte Générale* while drafting their own maps of the region, and, though Brackenridge called it "the best" map of Louisiana published before 1814, he added that, "from my own observation, and from what I have

46. Daniel Clark to James Madison, Aug. 17, 1803, in J. C. A. Stagg, ed., Papers of James Madison Digital Edition, http://rotunda.upress.virginia.edu ("Manuscript"); "Extract of a Letter from a Gentleman at Washington to His Friend in This City, Dated December 21, 1804," *Louisiana Gazette* (New Orleans), Mar. 5, 1805 ("gentleman"); Kastor, *William Clark's World.*

47. Barthélémy Lafon to Thomas Jefferson, Apr. 11, 1805, Letters of Application and Recommendation during the Administration of Thomas Jefferson, 1801–1809, RG 59, M418, reel 7, National Archives, Philadelphia (quotes). On French Louisianans' resentment of Anglo rule, see Joseph G. Tregle, Jr., "Creoles and Americans," in Arnold R. Hirsch and Joseph Logsdon, eds., *Creole New Orleans: Race and Americanization* (Baton Rouge, La., 1992), 131–185.

FIGURE 9. *Carte générale du territoire d'Orléans.* By B[arthélémy] Lafon. 1806. Library of Congress, Geography and Map Division, G4010 1806 .L3

learned from others, it is by no means to be relied on." The Spanish geographer José Antonio Pichardo, who had been commissioned by Spain to determine the boundaries of the Louisiana Purchase, was far more critical. Pichardo lambasted Lafon for both plagiarizing a Spanish hydrographical chart to produce the *Carte Générale* and for "spoiling and marring" the original "in order to appropriate to the United States the [Sabine River], making it nearer to Natchitoches, and giving it another course." Lafon probably hoped that such cartographic annexations would help win the support of expansion-minded U.S. officials.[48]

48. H. M. Brackenridge, *Views of Louisiana; Together with a Journal of a Voyage up the Missouri River, in 1811* (Pittsburgh, Pa., 1814), 164 ("best"); "Explanation of the Map," in William Darby, *A Geographical Description of the State of Louisiana, the Southern Part of the State of Mississippi, and Territory of Alabama . . . ,* 2d ed. (New York, 1817), [357]; Charles Wilson Hackett, ed., *Pichardo's Treatise on the Limits of Louisiana and Texas: An Argumentative Historical Treatise with Reference to the Verification of the True Limits of the Provinces of Louisiana and Texas . . . ,* 4 vols. (Austin, Tex., 1931–1946), I, 375, 376 ("spoiling"). For praise, see "Lafon's Map of the Orleans Territory," *Medical Repository,* V, 2d Hexade (1808), 300–301.

Lafon not only drew on Spanish charts when creating his maps. He relied heavily on a Caribbean community of experts, particularly geographer refugees who had migrated to New Orleans from Saint Domingue, the once-wealthy French colony that had recently been rocked by the French and Haitian Revolutions. Lafon both incorporated the skills of these geographers and used them as envoys to help forge stronger ties between New Orleans and eastern centers of power and print. As he was drafting his *Carte Générale* in early 1805, Lafon agreed to pay Joseph Pilier, who was "from the isle of St. Domingue and presently a resident of New Orleans," sixteen piastres a month to make copies of extant maps, assist him with "geographic or geodetic operations," and perform "other work of this nature." Lafon's arrangement with Jacques Tanesse, "former surveyor of St. Domingue," was even broader. Lafon and Tanesse agreed to work together on a map of the Mississippi River, and Tanesse promised to "furnish all the material in his power" about its geography. More importantly, Lafon hired Tanesse as his publishing agent in Washington, D.C., where he would present one copy of their map to the secretary of war and another to a printer as the basis for an engraving. Lafon and other French geographers in the greater Caribbean looked to the United States for economic opportunities after regime changes and revolutions destroyed the patronage networks on which they had relied.[49]

Lafon hoped to take a leading role in the lower Mississippi Valley's regional scientific community. Writing to William Dunbar from New Orleans, Lafon reflected that "it is greatly to be desired for the advancement of the sciences that there should be a scientific paper here in which the observations and discoveries of the country would be recorded." The paper would embrace the same breadth of topics as scientific journals published in Europe and the eastern United States, and Lafon was keen on using it to compile "an ancient map of America" that would show "the place from which [Indian groups] came and the gradations of the power of each of these nations." Even when contemplating native history, Lafon viewed maps as means of conveying political influence. Lafon also proposed that Dunbar should contribute meteorological news from Natchez that Lafon could integrate with his own observations from New Orleans and those of correspondents in the Caribbean so as to better predict regional weather

49. Engagement particular de M. Joseph Pilier et B. Lafon ingénieur, Feb. 17, 1805, Barthélémy Lafon Contract Book, 1804–1821, MSS 316, HNOC ("from the isle"), Entre nous . . . Lafon ingénieur géographe . . . et Jacques Tanesse ci-devant arpenteur à St. Domingue, July 5, 1805, Lafon Contract Book ("former").

patterns, especially hurricanes. Unbeknownst to Lafon, however, Dunbar looked down on him: Dunbar once told Jefferson that the Frenchman was not "much of an astronomer." As a self-consciously modern man of science, Dunbar would have scoffed at Lafon for advocating the meteorological theories of "ancient astrologers." (Lafon did acknowledge that astrology was "a science rejected to-day" but insisted that the predictions of the ancients fit "positively what we have been experiencing.") Dunbar was happy to sell Lafon instruments but was reluctant to compromise his own reputation by associating with him too closely.[50]

Lafon's eclectic methodology was most apparent in his *Annuaire Louisianais* (1808), a self-published outlet in which he presented his theories about nature and attachment to the United States. The *Annuaire* introduced the principles of astrology, displayed Lafon's meteorological charts, and illustrated a method for determining one's meridian on land without the aid of astronomical instruments. Lafon stressed that this astronomical method was ideally suited for continental exploration because it overcame the difficulties of carrying "good astronomical instruments" across "inhabited countries" and made it possible "to establish the course of rivers and fix the climates where different animal and vegetable species live." It was also, most likely, meant to attract the attention of President Jefferson and demonstrate Lafon's engagement with the ongoing discussion throughout the United States' scientific community concerning the development of astronomical methods specific to continental expansion.[51]

Unlike some learned Anglo-American men in the early republic, Lafon rarely hesitated to publish his own theories about natural phenomena, and he used the *Annuaire* to intervene in the Atlantic-wide debate about the nature and origin of meteorites. He declared: "I am going to try to explain the formation of these meteorites; people will doubtless find me very bold in giving a parallel explication to those of the great masters ... but the explanations of those savants are not found to be precisely satisfying." Drawing on reports from Languedoc and Connecticut, Lafon argued that meteorites were neither lunar ejecta nor rocks hurling through space but "that weather or the atmosphere produces them, by the aggregation of the matter

50. Lafon to Dunbar, Aug. 19, 1805, in Rowland, ed., *Life, Letters, and Papers of William Dunbar*, 178–181 ("ancient astrologers," "science," "positively," 180, "greatly," "ancient map," "place," 181); Dunbar to Thomas Jefferson, July 6, 1805, WDP, Letter Book, 1802–1805, 34 ("much of an astronomer").

51. B[arthélémy] Lafon, *Annuaire Louisianais pour l'ánneé 1809 ...* (New Orleans, 1808), 74.

that is kept in fluctuation at very great heights." Although the atmospheric theory of meteorite formation had many adherents, Lafon believed that "a more precise and expanded explanation" was in order because the idea of *"rocks that fall from the sky is not something easy to conceive."* He reflected that understanding how meteorites formed in the atmosphere meant that "we no longer find incomprehensible the words: GOD MADE THE WORLD FROM NOTHING" because *"the spirit of God (the creative fire, the invigorating light of spring)"* formed the world from a still unsolidified universe "inanis et VACUA" through a similar process. Far from settling for cataloging and describing specific aspects of nature, Lafon considered himself perfectly qualified to theorize about the very origin of the earth. Eastern Anglo-American naturalists might have eschewed grand theories in their own research, but their caution did not necessarily characterize the scientific community of the vast and multiethnic United States.[52]

In the *Annuaire,* Lafon presented himself as a natural philosopher who interpreted the observations and specimens he received from the peripheries of his own regional scientific network. His "Notice on an Insect-Plant found in Natchitoches," an essay based on "the astonishing Natural History specimens sent by my correspondents in the interior of Louisiana," described how the men who initially observed the insect all agreed that it had "metamorphosed *into a plant.*" Lafon, however, discredited this explanation as a result of these men's "imagination in the marvelous" because their hypothesis "did not accord with the natural order" in which living things always move from less perfect to more perfect states (a kind of evolutionary thinking also evident in his theory of meteorite formation). After examining the *insecte-plante* with a microscope, Lafon suggested that a fungus had permeated its body and given it a plant-like appearance. He regretted that there were no engravers in New Orleans skilled enough to represent this insect in the *Annuaire,* grumbling that "the lack of artists will long delay the progress of the sciences in New Orleans."[53]

The *Annuaire* was also a political text that asserted Lafon's and Louisiana's attachments to the United States. It included a twenty-one point chronology of the world that listed U.S. independence, "the presidency of

52. Ibid., 88 ("I am going"), 91 ("more precise"), 93 ("we no longer"), emphases in original. On early republican savants' reluctance to theorize, see Lewis, *Democracy of Facts.* For contemporary theories about meteorites, see John G. Burke, *Cosmic Debris: Meteorites in History* (Berkeley, Calif., 1986), 37–69.

53. Lafon, *Annuaire,* 81–84 ("Notice," 81, "metamorphosed," 82, "imagination," 83–84, "did not accord," 84), emphasis in original.

Th. Jefferson," and "the Union of Louisiana and the United States" among such notable dates as Genesis and "the invention of the Zodiac." Lafon also used his "Essay on the Limits of Louisiana" in the *Annuaire* to argue that the Louisiana Purchase actually included all of Spanish New Mexico, and Spanish officials considered this essay a serious enough threat to their territorial claims that they penned a detailed refutation of it.[54]

As with his maps, Lafon tried to attract patronage from eastern officials and experts by circulating his *Annuaire*. Lafon wrote Jefferson again in 1808, this time in English, and forwarded him a copy of "the Louisiana Annuary edited by me—I wish it may please you." Although he boasted that the *Annuaire* would "give an idea of the territory of Orleans" and "contribute to fix your opinion upon this country," Lafon quickly transitioned to reciting his own qualities. He stated that "Natural philosophy, Geography, Astronomy and natural history divide my Leisure" and offered Jefferson wider access to the material artifacts of his wide-ranging scientific pursuits, particularly a portfolio with nearly four hundred images "drown *[sic]* after nature." Lafon reminded the president that "amongt *[sic]* your public occupations you spare moments to cultivate sciences and arts" and suggested that patronizing Lafon would behoove such "a distinguished philosopher and Learned administrator." This time Jefferson did thank him for "the offer of your valuable observations on the country you inhabit" and commended his almanac as more "scientific" than any "on this side of our country." But Jefferson gave Lafon no cause to hope for any federal support. The very content of the *Annuaire* might have undercut Lafon's efforts to attract the president's patronage, since Jefferson rejected the atmospheric theory of meteorites and would have dismissed the astrological and Christian aspects of Lafon's work. More importantly, Jefferson was chary of appointing "foreigners" to federal offices, and his consistent unwillingness to support Lafon suggests that the president saw French Louisianans as un-American. Jefferson had argued in 1807 that "whether we consider the natural rights of the native citizens, his knolege of the affairs of his country, or the superior reliance on his attachments, the trusts of every country are safest in it's *[sic]* native citizens." If a foreign-born, French-speaking Louisianan was to have a place in Jefferson's vision of the American nation, it was primarily

54. Ibid., front matter (quotes), 131. The Spanish translation and refutation of Lafon's essay are "Tratado de límites por Bartolome Lafon, yngeniero geógrafo en la Nueva Orleans," 1809, Real Academia de la Historia, film 84–34-L. reel 2, 87–95, HNOC, and "Refutación succincta de … los papeles nombrados 'Tratado de lmites por B. Lafon,'" Real Academia de la Historia, film 84–34-L. reel 2, 96–100.

as a subject of philanthropy and upliftment, not as a leading figure in the national scientific community.[55]

In 1812, Lafon sent the *Annuaire* to the nation's foremost geographer, Jedidiah Morse. Lafon claimed to have some tips on how Morse could make his own maps more accurate and suggested that the two geographers could exchange mutually useful information, a more equal relationship than the patron-client ties he hoped to forge with Jefferson. Morse thanked Lafon for his "obliging communications" in his *American Universal Geography,* but, as was often the case in Lafon's career, the timing just did not work in his favor: Morse received the *Annuaire* too late to incorporate its contents into the 1812 edition of his geography, and, by the time he published the next edition in 1819, Lafon's data were outdated. Morse also seems to have preferred citing Anglo-Americans as the authorities on Louisiana.[56]

Though Lafon might not have earned the kind of national recognition that he desired, he did secure political and scientific appointments at the local level. Six months after Louisiana was transferred to U.S. control, Orleans territorial governor Claiborne authorized the trilingual Lafon to act as a surveyor. Over the next five years, Lafon became an officer in the territorial militia, a deputy surveyor of the Orleans District, and a district alderman. Part of the reason why Lafon did so well in New Orleans politics was that he did not play fair. In 1808, one New Orleans resident raised a fuss that Lafon had won his seat as alderman by lying about his residential address in local newspapers. It was the second time that Lafon had used this trick successfully. Yet this measure of political success did not necessarily bolster Lafon's credibility: officials in the region suspected Lafon of being too concerned with his own advancement and too attached to Spanish surveying techniques to be fully reliable.[57]

55. Barthélémy Lafon to Thomas Jefferson, July 15, 1808, Special Collections Research Center, William and Mary Digital Archive, https://digitalarchive.wm.edu/handle/10288/15087 ("Louisiana Annuary"), Jefferson to Lafon, Aug. 28, 1808, https://digitalarchive.wm.edu/handle/10288/14785 ("offer"); Jefferson to Caspar Wistar, Feb. 25, 1807, in Florian Cajori, *The Chequered Career of Ferdinand Rudolph Hassler ...* (Boston, 1929), 43 ("foreigners"); Bedini, *Thomas Jefferson,* 190, 386–388; Burke, *Cosmic Debris,* 56–58; Brian Steele, *Thomas Jefferson and American Nationhood* (Cambridge, 2012), 169–174.

56. Barthélémy Lafon to Jedidiah Morse, Jan. 8, 1812, MSS 319, HNOC; Jedidiah Morse, *American Universal Geography,* 6th ed., I (Boston, 1812), iv (quote); Morse, *American Universal Geography,* 7th ed., I (Charleston, 1819), 658–671.

57. On Lafon's appointments, see "Authorization to Barthelemi Lafon to Act as Surveyor," June 23, 1804, in Carter, ed., *Territorial Papers of the United States,* IX, 244, "Return of Appointments in the Militia of the Territory of Orleans," Dec. 29, 1806, IX, 698;

One of Lafon's last attempts to ingratiate himself with prominent U.S. officials and experts came in 1812 when he finally sent himself to Washington, D.C., instead of his maps and publications. He went to petition Congress to support his claims to a tract of New Orleans property, but he found companionship and, in a minor way, patronage by befriending Benjamin Henry Latrobe, perhaps the most famous engineer and architect in the nation. Latrobe did promote Lafon's talents, proposing that the federal government hire Lafon to design the New Orleans lighthouse, but officials rejected this bid. Latrobe was actually less interested in finding employment for his new friend than in benefiting from Lafon's firsthand expertise in the intricacies of New Orleans politics. Latrobe told his son, Henry, who was then promoting the Latrobes' planned waterworks in that city, that "Lafon says that anything at N. Orleans may be done by intrigue." Latrobe was also surprised to find that Lafon was "an unassuming agreeable man ...." "I supposed I should find him riding a high horse," he confessed, "and rather a L'enfant than of plain practical common sense." He later added that "LaFon is still here, engaged in a work on the hieroglyphical science of the antients." "He is a Kind of enthusiast in that business and will be laughed at," he concluded, "otherwise he is *no* fool." In an era when the disciplinary boundaries of science were beginning to harden, Latrobe had to emphasize to the younger generation that Lafon's interests in esoteric wisdom and crafty politics was in no way incompatible with his capacity to produce and apply natural knowledge.[58]

Orleans Gazette; and Commercial Advertiser (New Orleans), Mar. 26, 1806; and Bos, "Barthelemy Lafon," 22. On the election, see [Anonymous] to Messieurs, Sept. 28, 1808, Raymond and Roger Weill Collection, MSS 340, HNOC. On how perceptions of Lafon's self-interestedness weakened his credibility, see John K. Smith to James Madison, May 15, 1810, The Papers of James Madison, Digital Edition, ed. J. C. A. Stagg, http://rotunda.upress.virginia.edu/founders/JSMN-03-02-02-0419. For how Lafon "pretend[ed]" to survey private lands using the method "pursued under the Spanish government" instead of the method that U.S. officials aimed to institute, see Seth Pease to [Albert Gallatin], June 30, 1807, in Carter, ed., *Territorial Papers of the United States*, IX, 746–748 ("pretend[ed]," 746, "pursued," 747); and Peter J. Kastor, *The Nation's Crucible: The Louisiana Purchase and the Creation of America* (New Haven, Conn., 2004), 81, 86.

58. John C. Van Horne and Lee W. Formwalt, eds., *The Correspondence and Miscellaneous Papers of Benjamin Henry Latrobe*, 3 vols. (New Haven, Conn., 1984–1988), III, 307n, Benjamin Henry Latrobe to Henry S. B. Latrobe, May 3, 1812, III, 287, B. H. Latrobe to H. S. B. Latrobe, Apr. 5, 1812, III, 276 ("an unassuming," "I supposed," "and rather"), 277 ("Lafon says"), B. H. Latrobe to H. S. B. Latrobe, June 10, 1812, III, 312 ("LaFon is still here," "He is a Kind," "otherwise"), emphasis in original. On the Latrobes' efforts to construct the New Orleans waterworks, see editors' note in John C. Van Horne

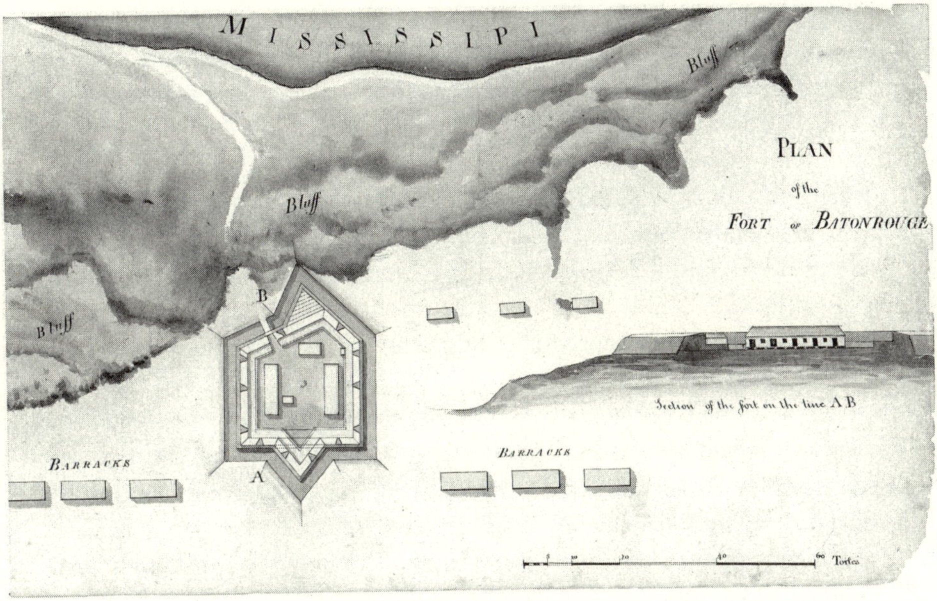

FIGURE 10. Plan of the Fort of Baton Rouge. By [Barthélémy Lafon]. 1814. The Historic New Orleans Collection. 1970.2.16 i, ii

Lafon's ongoing but largely unrecognized efforts to use science to help secure U.S. power in the lower Mississippi Valley climaxed from 1810 to 1815, when he worked with the U.S. Army as an engineer. He designed and built forts that, like so much of U.S. imperialism in the Gulf South, were layered on foundations previously established by the Spanish, French, and British empires. Lafon's plan for a fort overlooking the Mississippi River at Baton Rouge, for example, was designed by a French engineer working for the United States who intended to strengthen fortifications previously built by British engineers, occupied by Spanish soldiers, and seized by insurgents in West Florida's independence movement of 1810. General James Wilkinson also commissioned Lafon to prepare large-scale maps of the lower Mississippi Valley and, in 1812, appointed him chief engineer. Three years later, Lafon served with Andrew Jackson's forces against the British at the Battle of New Orleans, a conflict that brought an end to Britain's efforts to compete for influence in the lower Mississippi Valley.[59]

---

and Lee W. Formwalt, eds., *The Correspondence and Miscellaneous Papers of Benjamin Henry Latrobe*, II, *1805–1810* (New Haven, Conn., 1984), 808–811.

59. On Lafon's work for the U.S. Army, see Ehrenberg, "'Forming a General Geographical Idea of the Country,'" in Lemmon, Magill, and Wiese, eds., *Charting Louisiana*,

Lafon's fortunes began to decline in 1816 as creditors pressed him to pay off debts. Nearly bankrupt, Lafon decided to leave the United States and was forced to sell off his property, including a 450-volume library that included works in English, French, Spanish, and Latin on mathematics, astronomy, alchemy, and exorcism. Historians have argued that victory in the Battle of New Orleans solidified the attachment of Louisianans to the United States, but the case of Lafon—and other Louisianan engineers serving the U.S. military—suggests that even Louisianans who had been loyal to the U.S. could still reshape their political allegiances in the years after 1815. This was because Spain and other polities continued to vie for regional influence.[60]

Ironically, the U.S. scientific community chose this moment to reach out to Lafon. He had once more sent out a copy of his "Plan of the City and environs of New Orleans," this time to the APS. Although Lafon had probably given up on receiving any tangible reward for circulating his maps in the United States, a leading figure in the society noted in 1816 that he had "Written to B. Lafon, Esq., Geographer and Engineer . . . Soliciting Communications." The most prestigious scientific institution in the nation wanted Lafon's participation, but, after seeking recognition from the U.S. scientific community for thirteen years, Lafon had moved on.[61]

And that's when Lafon went a-pirating. In 1817, he became affiliated with the new extra-state polity led by Jean and Pierre Laffite at Galveston, a self-proclaimed government that professed its loyalty to, and purportedly acted under the authority of, the leaders of the Mexican independence movement. Complicating things further, the Spanish governor in Havana had hired the Laffite brothers as spies in 1815, and Spanish officials tolerated the Laffites' occasional plundering on the belief that they were in fact working to rid the Gulf of corsairs and infiltrate U.S.-based filibustering movements. Pierre

---

132; Bos, "Barthelemy Lafon," 104–105; A[rsène] Lacarrière Latour, *Historical Memoir of the War in West Florida and Louisiana in 1814–15; With an Atlas* (Philadelphia, 1816), 117, 179; and Stanley Faye, "Privateersmen of the Gulf and Their Prizes," *LHQ,* XXII (1939), 1068–1069.

60. On Lafon's debts and library, see Bos, "Barthelemy Lafon," 5, 113. For Louisianans' loyalty, see Kastor, *Nation's Crucible,* 153–180; and Gene Allen Smith, "'Motivated Only by the Love of Humanity': Arsène Lacarrière Latour and the Struggle for the Southwest," in Smith and Hilton, eds., *Nexus of Empire,* 298–320.

61. "Donations; Received by the American Philosophical Society, since the Publication of Vol. VI.—Old Series; For the Library," in American Philosophical Society, *Transactions,* I, New Ser. (1818), 451 ("Plan"); Peter S. DuPonceau, Historical and Literary Committee, Letter Books, I, APS Archives, VIII, Committees, APS ("Written").

and Jean Laffite were, however, serious about going through the motions of forming a "new government" at Galveston. They set up an admiralty court through which they legitimized the seizure of Spanish, U.S., and English vessels filled with calicoes, cash, and slaves. They advertised the commercial opportunities of trading with their settlement in New England newspapers and attracted ships flying under the Mexican, U.S., and Venezuelan flags. The Galvestonians also drafted a charter on board the schooner *Carmelita* that averred their fidelity to the Mexican republic and assigned official positions to some of the thirty-odd men who had established themselves on the island. The owner of the *Carmelita* and secretary pro tempore who penned the charter was Lafon.[62]

Lafon's political attachments shifted fluidly from 1815 to 1820. After the Battle of New Orleans, he had joined the filibustering New Orleans Associates and, probably in hopes of rebuilding his fortune by taking Spanish lands, planned to participate in an invasion of royalist Texas. After this scheme fizzled out, he devoted more of his energies to pirating and smuggling, using the *Carmelita* to trade in the prize goods that were moving through Galveston. He arrived at Galveston in April 1817 and officially associated himself with the pirate government by signing an "oath of fidelity to the Mexican republic." But Lafon, like the Lafittes, had a secret. Though the two pirating brothers had hidden their role as Spanish agents from the other Galvestonians, they revealed this connection to Lafon, and he joined in their conspiracy. By August 1817, the royal governor at San Antonio understood that Lafon, after more than thirteen years of attachment to the United States, was once more in the Spanish service. And, of course, Lafon

62. Testimony of Raymond Espanol, Oct. 7, 1817, in James Monroe, *Message from the President of the United States, Communicating Information of the Proceeding of Certain Persons Who Took Possession of Amelia Island and of Galvezton, during the Summer of the Present Year, and Made Establishments There* (Washington, D.C., 1817), 18 (quote); "Felipe Fatio's Report on the Plan of the Brothers Laffite to Capture the Corsairs on the Gulf," May 24, 1817, in Harris Gaylord Warren, "Documents Relating to the Establishment of Privateers at Galveston, 1816–1817," *LHQ*, XXI (1938), 1097–1100; Warren, *The Sword Was Their Passport: A History of American Filibustering in the Mexican Revolution* (Baton Rouge, La., 1943), 173–176; Jean L. Epperson, "Jean Lafitte and Corsairs on Galveston Bay," *Laffite Society Chronicles*, III, no. 2 (1997), 5–7. On the admiralty court, see "Extracts of a Letter from Beverly Chew … Aug. 1, 1817," in Monroe, *Message from the President of the United States*, 9, 11; and Warren, *Sword Was Their Passport*, 186. For commerce, see "NEW ORLEANS, *April 24*," *Independent Chronicle and Boston Patriot*, June 18, 1817. On the loyalty oaths, see "A Register of the Proceedings at Galveston, April 15th 1817," in Monroe, *Message*, 44–45, and Testimony of Raymond Espanol, Oct. 7, 1817, 18.

sent the governor a map. Spanish officials valued the written and illustrated texts in which Lafon packaged natural knowledge from the borderlands, including a hydrographic map of the Bay of Galveston and a geographical essay on the continental interior. Lafon, for his part, probably found Spanish officials' recognition and admiration ingratiating after years of feeling underappreciated by Anglo-Americans.[63]

Like Thomas Power, Lafon sent both political intelligence and scientific information to Spanish officials. But the military reports that Lafon shared with the Spanish consul at New Orleans were largely fabricated. He claimed to have intercepted dispatches regarding an alliance between U.S. filibusters, pirates, and Comanches that would overthrow Spanish rule in northern Mexico. Lafon suggested dividing the Indians among themselves, buying off the pirates, and undermining the filibusters through intrigue, a strategy that would simultaneously foil this plot and cripple the revolutionary campaign of Mexican patriots. The Spanish consul Felipe Fatio (son of the Florida planter) believed Lafon's tale and awarded him a retainer of seven hundred dollars for his service. At least part of the reason why Fatio considered Lafon to be a credible source of political intelligence was that Fatio believed him to be the most excellent cartographer and engineer in the region.[64]

Lafon was able to thrive during his final years (he bought seventeen slaves and sold none between 1818 and 1820) because Spain and other polities continued to compete for regional influence and offer viable patronage opportunities to Gulf South men of science. But, by the early 1820s, the Laf-

63. "Register of the Proceedings at Galveston, April 15th 1817," in Monroe, *Message,* 44 (quote). For Lafon's many allegiances, see Bos, "Barthelemy Lafon," 118; Faye, "Privateersmen of the Gulf and Their Prizes," *LHQ,* XXII (1939), 1070; Epperson, "Jean Lafitte and Corsairs on Galveston Bay," *Laffite Society Chronicles,* III, no. 2 (1997), 5; and Robert C. Vogel, "Who Were These Guys? Some of the Lesser Characters in the Story of the Lafittes," *Lafitte Society Chronicles,* IX, no. 1 (2003), 11. For Lafon and the Lafittes being Spanish agents, see "Jean Laffite's Diary of a Voyage to Galveston," quoted in Warren, "Documents Relating to the Establishment of Privateers at Galveston, 1816–1817," *LHQ,* XXI, no. 4 (1938), 1106n; Warren, *Sword Was Their Passport,* 180; and Faye, "Privateersmen of the Gulf and Their Prizes," *LHQ,* XXII (1939), 1070. On the map, see Antonio Martínez to Joaquín de Arredondo, Aug. 6, 1817, in Virginia H. Taylor, ed., *The Letters of Antonio Martínez, Last Spanish Governor of Texas, 1817–1822* (Austin, Tex., 1957), 34. On the geographical texts, see Faye, "Privateersmen of the Gulf and Their Prizes," *LHQ,* XXII (1939), 1069–1070. Barthélémy Lafon's "Entrada de la Bahia de Galvestown, 1816" is reproduced in Jean L. Epperson, "The Three Pirates Lafon," *Laffite Society Chronicles,* XI, no. 1 (2005), 10.

64. Warren, *Sword Was Their Passport,* 180–182.

fites' community of pirates had disappeared, and Spain had accepted Mexican independence and transferred Florida to the United States. As competition over the region declined, so, too, did the variety of patrons who could legitimize, and encourage the persistence of, a wide range of experts and investigative methods. The United States had become the only imperial power interested in ruling the Gulf South and, therefore, U.S. officials, men of science, and institutions gained more exclusive—but by no means complete—power to define the parameters of legitimate natural knowledge. But Lafon was spared the effort of once again realigning his political attachments and soliciting new patrons. He died during a New Orleans yellow fever epidemic in September 1820 before he had to face a new world in which U.S. domination severely limited the allegiances and opportunities available to the Gulf South's diverse inhabitants.[65]

• • •

Imperialism brought both new territories and new men of science into the United States. The cases of Thomas Power, William Dunbar, and Barthélémy Lafon suggest that many experts in America's borderlands were ambivalent about—or directly resisted—U.S. rule. These men sought to secure financial support and build reputations by attracting patrons from, and flouting the distinctions between, the various nations, empires, and upstarts competing for influence in the region. Power's access to political and natural knowledge made him valuable to both Spanish governors and Anglo-American adventurers, but his reputation as a spy and opposition to U.S. expansion helped ensure that he never achieved wider recognition as a credible scientific practitioner. Lafon found at least some patronage from Spain, the United States, and the pirates Laffite, but he eventually chose to distance himself from the U.S. officials and men of science who never quite welcomed him as one of their own. Far more than Power or Lafon, Dunbar did secure a reputation among contemporaries and historians alike as a significant man of science. This was largely because he convinced the right officials and institutions—especially Thomas Jefferson and the APS—that he was a patriot who would help them know and dominate the contested borderlands. But it was also because he was an Anglo, and, ultimately, it was Anglos who conquered the Gulf South and wrote the ethnographies and histories that defined which kinds of people counted as scientific Americans.

65. Hall, *Afro-Louisiana History and Genealogy,* http://www.ibiblio.org/laslave.

{ 5 }

ETHNOGRAPHY AND INTELLIGENCE IN THE TIME OF CONQUEST

• • •

In the early decades of the 1800s, several Anglo-Americans observed the diverse peoples of the Gulf South with an eye for determining if they were capable of participating in the political and intellectual life of the United States. To a striking extent, Anglos argued that the potential to do science—which meant studying nature in ways that they identified, in increasingly narrow terms, as legitimate—was a key criterion for evaluating which groups were intellectually fit for citizenship and which groups ought to be excluded, enslaved, or evicted. As Anglo-Americans analyzed the African Americans, natives, and European-descended creoles they encountered in the Gulf South, they also defined themselves as a scientific, experimenting people whose ability to understand and exploit the region's environment justified their rule.[1]

Anglo-Americans generally agreed that other American peoples were scientifically ignorant. But they disagreed over whether this ignorance could be ameliorated through so-called civilizing measures like education or whether it was somehow innate. This debate was, at its core, about the nature of human difference. This subject had fascinated Europeans and Anglo-Americans—as well as natives, Africans, and creoles—for hundreds of years, but Anglo-Americans' ideas about the causes and consequences of human variety were shifting dramatically during the late eighteenth and early nineteenth centuries. Put simply, the widely held theory that all humans were essentially similar though mutable through social and envi-

1. On the relationship between U.S. expansion and Anglo-Americans' perceptions of other ethnicities, see Reginald Horsman, *Race and Manifest Destiny: The Origins of American Racial Anglo-Saxonism* (Cambridge, Mass., 1981); D. W. Meinig, *The Shaping of America: A Geographical Perspective on 500 Years Of History*, II, *Continental America, 1800–1867* (New Haven, Conn., 1993), 176–196; and Michael Adas, *Dominance by Design: Technological Imperatives and America's Civilizing Mission* (Cambridge, Mass., 2006). On the challenges of incorporating non-Anglos into the United States' political community, see Peter J. Kastor, *The Nation's Crucible: The Louisiana Purchase and the Creation of America* (New Haven, Conn., 2004).

ronmental circumstances began losing ground to the notion that there were distinct races, which some ethnographers maintained were distinct species, with immutable mental capacities. The shift between these perspectives was gradual, messy, historically contingent, and never total: anatomical interpretations had been influential since at least the mid-eighteenth century, most white Americans refused to accept polygenesis (the separate creation of each race) because it contradicted scripture, and events like revolutions and slave uprisings inspired sharp turns in ethnographic discourse. Nevertheless, a more biologically deterministic explanation for human difference did take hold among many Anglo-Americans between the 1780s and 1840s, and the way people have interacted with one another has not been the same since.[2]

U.S. expansion, and the encounters among peoples it precipitated, did much to both encourage these changing perspectives and inspire resistance to them. On the one hand, several developments in the early nineteenth-century Gulf South gave Anglo-Americans ample reasons to expound and apply an ethnography that stressed the inferiority of some groups' brains. These developments included the United States' victory in geopolitical competitions for regional dominance, the entrenchment of slavery during the cotton and sugar booms, and the removal of southern Indians (this chapter focuses on the Choctaws, the first group to face removal after the 1830 Indian Removal Act). On the other hand, blacks, natives, creoles, and even some Anglos in the Gulf South opposed the hardening correlation between

2. For Anglo-Americans' changing views of human difference, see Winthrop D. Jordan, *White over Black: American Attitudes toward the Negro, 1550–1812* (Chapel Hill, N.C., 1968); Bruce R. Dain, *A Hideous Monster of the Mind: American Race Theory in the Early Republic* (Cambridge, Mass., 2002); and John Carson, *The Measure of Merit: Talents, Intelligence, and Inequality in the French and American Republics, 1750–1940* (Princeton, N.J., 2007), 11–95. For how some natives, Africans, and European-descended creoles viewed human difference, see, respectively, Daniel K. Richter, *Facing East from Indian Country: A Native History of Early America* (Cambridge, Mass., 2001), 189–236; Wyatt MacGaffey, "Dialogues of the Deaf: Europeans on the Atlantic Coast of Africa," in Stuart B. Schwartz, ed., *Implicit Understandings: Observing, Reporting, and Reflecting on the Encounters between Europeans and Other Peoples in the Early Modern Era* (Cambridge, 1994), 249–267; and Jorge Cañizares Esguerra, "New World, New Stars: Patriotic Astrology and the Invention of Indian and Creole Bodies in Colonial Spanish America, 1600–1650," *AHR*, CIV (1999), 33–68. On the rocky shift between eighteenth- and nineteenth-century visions of human difference, see Colin Kidd, *The Forging of Races: Race and Scripture in the Protestant Atlantic World, 1600–2000* (New York, 2006); and Andrew S. Curran, *The Anatomy of Blackness: Science and Slavery in an Age of Enlightenment* (Baltimore, 2011).

race, social status, and intelligence by drawing on the older ethnographic paradigm that education and environment made people what they were. Ethnography in the United States, in short, developed amid the violent and peaceful interactions brought about by U.S. imperialism and defined contested hierarchies of mental ability that Anglo-Americans used to legitimate their supremacy.

IMPROVABILITY, EDUCATION, INCORPORATION

From the 1780s to the 1810s, an era when the United States remained just one of the polities competing for influence in the Gulf South, most Anglo-Americans believed that intellectual talents, like other physical and moral faculties, varied according to one's experiences and environmental circumstances. Certain individuals might have been born with more talents than others, but, in general, education, climate, and social structure made some men and women smarter than others. Although a people's philosophical and technological accomplishments did not make them fixedly better or worse than any other group, eighteenth- and early nineteenth-century Euro-Americans did increasingly identify intellectual talent as a meaningful object of inquiry, one that could be evaluated independently of moral or spiritual faculties. These views undergirded Anglo-Americans' observations of the peoples of the Gulf South, including the region's Anglos, who, many suspected, would degenerate from exposure to the semitropical environment and a lack of educational opportunities. Some officials and experts built on the theory of mankind's essential sameness to trumpet education, uplift, and incorporation. Other Anglo-Americans, however, looked to this same theory to justify conquest and exclusion. Contests and collaborations with the peoples of the borderlands—combined with prejudices and uncomfortable doubts that ethnic differences might be more than skin-deep—shaped how Anglo-American philanthropists and oppressors alike produced and applied ethnographic knowledge about blacks, natives, and European-descended creoles during the early decades of U.S. expansion.[3]

3. On climate and education as roots of human difference, see William [Ragan] Stanton, *The Leopard's Spots: Scientific Attitudes toward Race in America, 1815–59* (Chicago, 1966), 7–11; Jordan, *White over Black*, 187–190; and Carson, *Measure of Merit*, 12–75. On the degeneration of Anglo-Americans, see Francis Baily, *Journal of a Tour in Unsettled Parts of North America, in 1796 and 1797*, ed. Jack D[avid] L[azarus] Holmes (Carbondale, Ill., 1969), 306–307; Andrew Ellicott, *The Journal of Andrew Ellicott . . .* (Philadelphia, 1803), v; Conevery Bolton Valenčius, *The Health of the Country: How American Settlers Understood Themselves and Their Land* (New York, 2002), 229–243; and

Some of the early republic's most prominent men of science enthusiastically promoted the scientific achievements of Gulf South blacks. Philadelphia physician and abolitionist Benjamin Rush used the example of James Derham, a "practitioner of physic" in Spanish Louisiana, to prove that black men could achieve distinction in the sciences. In a published 1788 letter to the Pennsylvania Abolition Society, Rush claimed that Derham knew many "new medicines" and that the two men had discussed the peculiarities of the Gulf South's disease environment. More importantly, Rush saw Derham's ability to make the most of the "opportunities of improving in medicine" offered by his master, New Orleans physician Robert Dow, as evidence that the civilizing influence of Euro-Americans could make black men into men of science. That Rush seized on Derham and other exemplars of black Americans' scientific competence to forward abolitionism indicates that intelligence was becoming both a distinct topic of study and a key variable in debates over human difference.[4]

Anglo-Americans intent on exploiting slave labor also acknowledged that Gulf South blacks were capable of intellectual attainment but used this obvious insight to justify oppressing blacks and denying them access to knowledge. In the turbulent years around 1800—when slave uprisings in Louisiana and Haiti, the Louisiana Purchase, and Aaron Burr's conspiracy punctuated a period of intense competitions around the lower Mississippi Valley—the region's slaveowners were prone to see the enslavement of blacks as part of an ongoing contest for power that they had to win. As Louisiana planter John Mills told his cousin in 1807, "You must know that unless there is order and subordination kept up amongst negroes, they would soon be masters, instead of slaves, for tho' they are black, they have as great a propensity, to command and be tyrants as white people." Since enslaved blacks might still prevail in this power struggle, they needed to be "kept in ignorance as much as possible" to ensure that whites maintained an edge in their rivalry. Mills also included some observations about Louisiana whites in his letter, noting, "Here you will ask what do [planters] want with so many negroes, the answer is, to make more money—again, you will ask what do they want with so much money, the answer is to buy more negroes."

Edward E. Baptist, *Creating an Old South: Middle Florida's Plantation Frontier Before the Civil War* (Chapel Hill, N.C., 2002), 27.

4. Benjamin Rush, "Rush's Testimony about Durham's *[sic]* Expertise," 1788, Letters from James Durham to Benjamin Rush, HSP / PACHS.net, http://www.pachs.net/web_of_healing/archives/Historical_Society/rush2.html (quotes); Jordan, *White over Black*, 448–455.

Anglo-American slaveowners found mastery to be its own reward, one for which they would torture people who, even according to many planters, were born with the same spark as themselves.[5]

Before the 1820s, most educated Anglo-Americans were optimistic that southern Indians, including the Choctaws, had the potential to become scientific. For example, a friend cautioned John McKee, a U.S. Indian agent who lived among the Choctaws, to bear in mind that "interest[,] education," and "want of a regular . . . exercise of useful arts" accounted for the differences between whites and "the Tawny sons of the Desart *[sic]*." But, he continued, "by comparing a savage with a well informed Christian we may see the superiority of the latter in moral science[,] a superiority that raises him as much above the former as he is above some of the more sagacious of the animal tribe." Natural knowledge and moral knowledge had become discreet faculties and bases for qualitative comparisons between peoples. Other Anglo-Americans interpreted linguistic evidence and the remains of ancient earthworks as proof that Gulf South Indians had once been scientifically accomplished, had declined, and could once again be uplifted. William Bartram suggested that this decline, which had made Gulf South natives less "ingenious and industrious in manufactures" than their ancestors, was a result of their increased dependency on Euro-Americans. In his response to a questionnaire by Philadelphia naturalist Benjamin Smith Barton, Bartram claimed, "We must seek for [natives'] arts and Sciences amongst nations far distant from the settlement of the White people—Or recover them by industry and experiment of our own." In other words, Bartram concluded that Anglo-Americans' best hope for acquiring native knowledge of North America was to either expand farther west or expand their own experimental agenda.[6]

In the early years of U.S.-Choctaw relations, some eastern men of science, including Barton, thought that natives' expertise could improve Anglo-Americans' scientific practices. Barton advocated investigating and incorporating the anatomical insights that Choctaws had acquired through

5. John Mills to Gilbert Jackson, May 19, 1807, John Mills Letters, MSS 1375, LLMVC (quotes). On some of the ways the enslaved blacks contested Anglo-American power, see Adam Rothman, *Slave Country: American Expansion and the Origins of the Deep South* (Cambridge, Mass., 2005), 95–101.

6. Samuel L. Campbell to John McKee, Jan. 12, 1795, John McKee Papers, circa 1792–1825, box 1, folder 2, no. 152, LOC ("interest"); William Bartram, Answers to Benjamin Smith Barton's Queries about the Indians [1789], ed. John Howard Payne, Bartram Family Papers, coll. 36, 29, HSP ("ingenious"); Amos Stoddard, *Sketches, Historical and Descriptive, of Louisiana* (Philadelphia, 1812), 347, 418, 430, 447–449.

extensive contact with their own peoples' remains. In 1804, as editor of the *Philadelphia Medical and Physical Journal,* Barton published a description of a mortuary rite in which Choctaws placed their dead on scaffolds. "After the flesh is sufficiently rotted," he reported, "they send for the 'Bone-pickers,' who are persons expressly appointed for this purpose." "They scrape, with their nails, the flesh very clean from the bones." Barton's editorial comments emphasized that this operation "cannot but be interesting to the historian of the science of Anatomy." "Some progress must necessarily have been made in a knowledge of the structure of the human body," he surmised, "particularly of the bony fabric, by a people who had adopted, and rigidly pursued [bone picking] for a long series of years." But Barton soon grew less interested in learning what Choctaws knew about bones and more committed to learning about Choctaws through their bones. In 1812, around the same time that some European physicians were starting to cultivate extensive collections of human skulls, Barton wrote to William C. C. Claiborne, governor of the Orleans Territory, requesting "some information concerning the genius, manners, etc. of the Southern Indians." He was especially "anxious to procure two or three Indian skulls" as well as details concerning "what nation of Indians the skulls belong." This early example of skull collecting suggests that the shift to more fixed and anatomical understandings of human difference, which would become especially pronounced among Philadelphia naturalists by the 1830s, was already under way by 1812.[7]

Anglo-Americans' views of natives' intellectual prowess were not isolated from wider, and often comparative, analyses of other American peoples. Writing in 1818, geographer William Darby described a two-week westward journey from New Orleans to the Sabine River that revealed "human beings from the most civilized to the most savage." Wealthy Mississippi Valley planters represented man's furthest advancement; the inhabitants of the Sabine frontier were stuck at "the earliest dawn of human improvement." Darby described the white, native, and ethnically mixed inhabitants of the border with Spanish Mexico as equally uncivilized, but he nevertheless berated Euro-Americans for uttering "disgusting expressions" about white frontiersmen being "worse than savages." He insisted that distinctions be-

7. James Boyd, "Extracts from a MS Journal of the Late Mr. James Boyd, of Lancaster, in Pennsylvania," *Philadelphia Medical and Physical Journal,* I, part 1 (1804), 98 ("After"); Benjamin Smith Barton, "Notes on the Preceding Paper by the Editor," *Philadelphia Medical and Physical Journal,* I, part 1 (1804), 101 ("cannot"); Benjamin Smith Barton to William C. C. Claiborne, Aug. 17, 1812, Violetta Delafield–Benjamin Smith Barton Collection, 1783–1817, Ser. 1, Correspondence, MSS B.B284d, APS ("some information").

tween natives and even the most savage whites made sense because "the frontier men of the United States, the pastoral creole of Louisiana, and the horsemen of the Spanish internal provinces, are in a much greater degree superior to the aboriginal savages of America, in point of improvement, than they are inferior in mental endowments to the most polished society in Philadelphia." Although he did not flatly deny the improvability of Indians, Darby did claim that philanthropists would do well to give up on natives and refocus their attention on Louisiana's backcountry whites, for "more knowledge could be instilled into the[ir] minds . . . in twenty years, than could be imparted in two centuries to all the savage tribes from the Mexican gulf to Hudson's bay." Darby's observations, penned when Anglo-Americans were emerging as the clear victors in the international struggle for the lower Mississippi Valley, are indicative of Anglo-Americans' growing pessimism that they could, or even should, make Indians fit to participate in national political and intellectual life.[8]

The Gulf South's European-descended creoles were also subject to ethnographic scrutiny. White, black, and mixed-race Louisianans all identified as creoles, but Anglo-Americans writing about the Gulf South generally used the term to denote, and blur distinctions between, the Spanish and French-descended peoples of Louisiana and Florida; indeed, Anglo-Americans sometimes used "French" to mean Louisianans from France, French creoles, and even Spanish creoles. Observing the manners and minds of creoles soon became a pastime for many Anglos in the Gulf South. Lieutenant George McCall was interested in the geography and natural history of Florida, for example, but also "had ample opportunity to see daily, and study hourly, the strange commixture of manners and habits of these descendants of the Spaniard, the Frenchman, and the Englishman, who make up the population." He wrote that most of Pensacola's white inhabitants, "who are termed *'Creoles,'*" had little interest in politics and "no love for knowledge." Despite this supposed ignorance, Anglo-Americans almost invariably agreed that creoles could become competent U.S. citizens and capable men of science if they received a proper education. The Florida chronicler John Lee Williams observed that, although "the creoles had, before the transfer of Florida to the United States, assumed something of a national character," he predicted that "all distinctions will soon be lost" be-

8. William Darby, *The Emigrant's Guide to the Western and Southwestern States and Territories* . . . (New York, 1818), 61–62 ("human," 61, "earliest," 61–62, "disgusting," 62). On philanthropists' pessimism, see Bernard W. Sheehan, *Seeds of Extinction: Jeffersonian Philanthropy and the American Indian* (Chapel Hill, N.C., 1973), 180–275.

tween "the old and new inhabitants" once creoles were exposed to the same education enjoyed by Anglo-Americans.[9]

Anglo-Americans partially derived their inclination to view creoles as scientifically underdeveloped from the old tradition of the Black Legend, the English stereotype that Catholics, particularly Spaniards, were unenlightened and morally cruel. But Anglos also acquired specific arguments about creoles' intellectual shortcomings from Spanish and French observers who had written about the Gulf South during the Spanish era. One French official attributed Louisiana's persistent backwardness to the early eighteenth-century Bourbons for populating it with "ignorant men," and Pierre-Louis Berquin-Duvallon, a cynical commentator on Spanish Louisiana whose work was translated and published in New York, believed the province's poverty derived from the character of its "Creoles," a "race [that] seem[s] to have degenerated" from the initial French settlers and whose "ignorance exceeds all human credibility." For Berquin-Duvallon, "the very air" of Louisiana was "mortal to the muses," and he was persuaded that "there are not ten men" in the province "capable of appreciating the merit of a Descartes and Newton ... a Buffon and a Linnaeus." Other French commentators eschewed environmental explanations and, instead, blamed the Spanish government's inadequate education system or the inhabitants' own apathy. Spanish and French stereotypes about creoles circulated to eastern Anglo-Americans through published texts, manuscripts, and, most likely, word of mouth: the way Anglo-Americans described creoles' ignorance was, in part, a product of the interimperial circulation of ideas.[10]

9. George McCall to H., Nov. 8, 1822, in McCall, *Letters from the Frontiers,* ed. John K. Mahon (Gainesville, Fla., 1974), 13–15 ("ample," 13, "who," 14, "no love," 15, emphasis in original); John Lee Williams, *The Territory of Florida* (1837), ed. Herbert J. Doherty, Jr. (Gainesville, Fla., 1962), 115–116 ("creoles," 115, "all distinctions," 116). On uses of the terms "creole" and "French," see Virginia R. Domínguez, *White by Definition: Social Classification in Creole Louisiana* (New Brunswick, N.J., 1986); Joseph G. Tregle, Jr., "Creoles and Americans," in Arnold R. Hirsch and Joseph Logsdon, eds., *Creole New Orleans: Race and Americanization* (Baton Rouge, La., 1992), 131–185; Peter J. Kastor, "'They Are All Frenchmen': Background and Nation in an Age of Transformation," in Kastor and François Weil, eds., *Empires of the Imagination: Transatlantic Histories of the Louisiana Purchase* (Charlottesville, Va., 2009), 239–267; and Shirley Elizabeth Thompson, *Exiles at Home: The Struggle to Become American in Creole New Orleans* (Cambridge, Mass., 2009), 7–10. On Louisiana creoles as objects of ethnographic study, see Stoddard, *Sketches, Historical and Descriptive, of Louisiana,* 319; Meinig, *Shaping Of America,* II, 189; and Christopher P. Iannini, *Fatal Revolutions: Natural History, West Indian Slavery, and the Routes of American Literature* (Chapel Hill, N.C., 2012), 258.

10. The Spanish translation of the French official's report is "Fragmento que deve

Like earlier Spanish and French observers, Anglo-American planters, officials, and yeomen had high expectations for the agricultural potential of the Gulf South and evaluated creoles on their ability to exploit the climate and soil for profit. Anglos found some creoles to be experimental agronomists. Jean Étienne de Boré, for instance, had established the first viable sugar plantation in lower Louisiana in 1795 with the help of white and black experts fleeing Saint Domingue, and U.S. officials began picking Boré's brain for details on growing and processing sugar well before the Louisiana Purchase. However, observers from the United States became increasingly prone to denigrate creoles' agricultural practices. One of their arguments was that creole planters lacked an experimental mindset. Ignoring experimenters like Boré, Connecticut-born traveler Amos Stoddard claimed that "the little knowledge of agriculture [creoles] possessed was never reduced to experiment," a failure that had "impede[d] the prosperity of that colony" from the French era to the 1810s. Another critique Anglo-Americans leveled at creole planters was that they improperly managed their slaves. Stoddard remarked that "the French and Spanish planters . . . treat[ed] their slaves with great rigor" and were "extremely ignorant of agricultural pursuits, and of the quantum of labor in the power of a slave to perform in a given time." Techniques for compelling or enabling slaves to be more productive were a primary focus of agronomic experimentation for planters, and, ac-

---

servir para la memoria, y reconocimiento de la Luisiana," AHN, leg. 3900, apartado 1, no. 82, 351 ("ignorant men"). See also Pierre-Louis Berquin-Duvallon, *Travels in Louisiana and the Floridas, in the Year, 1802, Giving a Correct Picture of Those Countries*, trans. John Davis (New York, 1806), 53 ("there are not," "capable"), 59 ("Creoles," "race," "ignorance"), 61 ("very air," "mortal"). On the Black Legend, see David J. Weber, *The Spanish Frontier in North America* (New Haven, Conn., 1992), 336. For explanations of creole ignorance that stressed education and apathy, see James Pitot, *Observations on the Colony of Louisiana from 1796 to 1802*, trans. Henry C. Pitot (Baton Rouge, La., 1979), 32; and Pierre Clément de Laussat, *Memoirs of My Life to My Son during the Years 1803 and After . . .*, ed. Robert D. Bush, trans. Agnes-Josephine Pastwa (Baton Rouge, La., 1978), 98. For creoles not in fact being as ignorant as writers like Berquin-Duvallon claimed, see Jack D. L. Holmes, "Educational Opportunities in Spanish West Florida, 1781–1821," *FHQ*, LX (1981), 77–87; *Pierre-Joseph Favrot's Education Manual for His Sons: A Précis of Knowledge for an Eighteenth-Century Louisiana Gentleman*, trans. Martha Scott Gruning, ed. Guillermo Náñez Falcón (Tulane, La., 1988); and Shannon Lee Dawdy, *Building the Devil's Empire: French Colonial New Orleans* (Chicago, 2008), 56. For an example of a French writer's observations about creoles circulating to eastern Anglo-Americans via manuscript, see Le Chevalier D'Annemours, "Mémoire sur le district du Ouachita dans la province de la Louisianne," [1803], MSS 917.6.Ex7, APS, and Thomas Jefferson to John Vaughan, May 5, 1805, Thomas Jefferson Papers, 1732–1828.

cording to most Anglo-American observers, creoles had failed to keep up with the times. Some Anglo-Americans deemed creoles to be such ignorant and cruel planters that they could not even keep their slaves alive. In an 1803 medical topography of the lower Mississippi Valley, physician Garrett Elliott Pendergrast blamed creole slaveowners for the shocking prevalence of neonatal tetanus, a fatal form of lockjaw in newborns, among Louisiana's slaves. He estimated that "at least five eighths of all [enslaved blacks] that are born die of this complaint within the first two weeks after birth" on the plantations south of Baton Rouge, while those born in Natchez, where most planters were Anglo-American, did not experience this disease at all. Pendergrast judged that the cause of this drastic discrepancy was "the greater inhumanity [that] the Spanish and French planters exercised over the parents of these unfortunate children."[11]

Most Anglo-Americans before the 1810s would have agreed that creoles, blacks, and Indians could acquire the ability to perform science through education, but the ways Anglo-Americans wrote about creoles suggest that they had started to equate whiteness with the potential for genius. Stoddard thought that "the Creole French are at least a century behind other civilized nations in the arts and sciences," yet he also stated that creoles "display a good share of genius and penetration." The notion of creoles' "genius" had spread among Anglo-Americans in the immediate aftermath of the Louisiana Purchase through Thomas Jefferson's compiled *Account of Louisiana.* Quoting New Orleans merchant Daniel Clark's reply to Jefferson's questionnaire about Louisiana, the *Account* asserted both that creoles were ignorant

11. Jean Étienne de Boré to Andrew Ellicott, Feb. 25, 1799, TJP; Stoddard, *Sketches, Historical and Descriptive, of Louisiana,* 294–295 ("little knowledge," 294–295, "impede[d]," 295), 332–333 ("French," 332, "extremely," 333); Garrett Elliott Pendergrast, *A Physical and Topographical Sketch of the Mississippi Territory, Lower Louisiana, and a Part of West Florida* (Philadelphia, 1803), 33–34 ("at least," 33–34, "greater," 34). On Boré, see Nathalie Dessens, *From Saint-Domingue to New Orleans: Migration and Influences* (Gainesville, Fla., 2007), 79; Charles Gayarré, "A Louisiana Sugar Plantation of the Old Regime," *Harper's New Monthly Magazine,* LXXIV (1887), 606–621; [Georges-Henri-] Victor Collot, *A Journey in North America . . . ,* trans. J. Christian Bay, [II] (Florence, 1924), 171–175; Carlyle Sitterson, *Sugar Country: The Cane Industry in the South, 1753–1950* (Lexington, Ky., 1953), 4–13, 119; and William C. C. Claiborne to Thomas Jefferson, July 10, 1806, in Dunbar Rowland, ed., *Official Letter Books of W.C.C. Claiborne, 1801–1816,* 6 vols. (1917; rpt. New York, 1972), III, 361–364. On techniques for extracting greater profit from slave labor, see Richard Follett, *The Sugar Masters: Planters and Slaves in Louisiana's Cane World, 1820–1860* (Baton Rouge, La., 2005), 151–194; and Lorena S. Walsh, *Motives of Honor, Pleasure, and Profit: Plantation Management in the Colonial Chesapeake, 1607–1763* (Chapel Hill, N.C., 2010), 613–623.

and that they were "endowed with a good natural genius, and an uncommon facility of learning whatever they undertake." The idea that certain individuals could be endowed with genius from birth instead of developing a genius in a particular field had only developed in the mid-eighteenth century, and to argue that an entire population, much less a group that was generally disparaged for its lack of intellectual accomplishments, was born with "natural genius" was a bold claim.[12]

Creoles' supposed ignorance was, in the eyes of some U.S. officials, an obstacle to their full incorporation into the nation. Governor Claiborne feared that the integration of creoles into the United States might fail because "sudden and total reformation is best calculated for enlightened minds;—the experiment may prove hazardous with Creole ignorance." Claiborne trusted that science, virtue, and republican citizenship were closely interconnected and was convinced that creoles' credulity—which stemmed from the Spanish government's failure to cultivate the sciences and, thus, the powers of discernment and reason—made them unfit for a representative government. He therefore sought to transform Louisiana's creoles into scientific and virtuous Americans through education. In 1804, Claiborne ordered the multinational commandants stationed throughout the territory, many of whom had served the Spanish or French Empires, to "impress upon the inhabitants, the propriety of educating their Children," stressing that education led to "virtue" and that U.S. rule would ensure that "commerce and agriculture [would be] promoted, and the arts and Sciences in Louisiana particularly cherished." Officials' efforts to cultivate promising citizens and to weed out undesirable ones were continuing experiments that emerged from and shaped U.S. imperialism.[13]

12. Stoddard, *Sketches, Historical and Descriptive, of Louisiana,* 175 ("display"), 308, 319, 320 ("Creole French"); *An Account of Louisiana: Being an Abstract of Documents, in the Offices of the Departments of State, and of the Treasury* (Philadelphia, 1803), 38 ("endowed"). For Clark's response to the questionnaire, see Daniel Clark to James Madison, secretary of state, Sept. 8, 1803, in Clarence Edwin Carter, ed., *The Territorial Papers of the United States,* IX, *The Territory of Orleans, 1803–1812* (Washington, D.C., 1940), 38. On changing ideas of genius in the eighteenth century, see Joyce E. Chaplin, *The First Scientific American: Benjamin Franklin and the Pursuit of Genius* (New York, 2006), 2–3; and James Delbourgo, "The Newtonian Slave Body: Racial Enlightenment in the Atlantic World," *Atlantic Studies,* IX (2012), esp. 196–202.

13. William C. C. Claiborne to Thomas Jefferson, Sept. 29, 1803, in Carter, ed., *Territorial Papers of the United States,* IX, 60 ("sudden"), Jan. 16, 1804, 161–162; Claiborne to Commandants of Districts, Circular [1804], in Rowland, ed., *Official Letter Books of W.C.C. Claiborne,* II, 74 ("impress"), Claiborne to James Madison, Jan. 10, 1804, I, 329.

CONQUEST

Between the 1810s and 1830s, Anglo-Americans achieved ascendancy in the Gulf South. In ways both direct and indirect, the experience and success of U.S. expansion encouraged a sweeping, though never complete, shift in how Anglo-Americans perceived other American groups' intellectual capacity to perform science and citizenship.

Military victories were essential to imposing Anglo-American domination. These victories included defeating the rebellious slaves of Louisiana's German Coast in 1811, the Creeks in 1814, the British military in 1815, the free black soldiers of the so-called Negro Fort in 1816, and Seminole and Spanish fighters during the First Seminole War of 1818. The shared purpose of all these wars was to ensure that Britain and Spain did not develop sustained relationships with local natives, blacks, and creoles that would limit either the United States' control over the Gulf South as a strategic frontier or Anglo-Americans' authority over other peoples. The acquisition of Florida in 1821, which resulted in part from Andrew Jackson's 1818 invasion, marked the end of other imperial powers vying for the region and, indeed, helped inspire federal officials to declare in the 1823 Monroe Doctrine that European empires could no longer compete for political influence anywhere in the Americas. With the end of imperial competition, black, native, and creole residents in the Gulf South lost the actual and potential alliances that had given them some clout in regional power relations.[14]

White slaveowners rose to unprecedented dominance over Gulf South blacks during the 1810s and 1820s. Slavery had been in decline in the United States before the acquisition of the Mississippi Territory in 1798 and Louisiana in 1803 ensured that slavery would remain one of the nation's principal economic engines. This rebirth of slavery occurred primarily because U.S. expansion into the lower Mississippi Valley coincided with booms in the production of sugar and cotton, plantation commodities that had begun to transform the region during the Spanish era but would ensure an intense demand for enslaved laborers throughout the antebellum period. The United States government played a decisive role in ensuring white supremacy in the Gulf South: not only did it eliminate the presence

14. Walter Johnson, *River of Dark Dreams: Slavery and Empire in the Cotton Kingdom* (Cambridge, Mass., 2013), 18–36; Meinig, *Shaping Of America*, II, 23–32, 188–197; Claudio Saunt, *A New Order of Things: Property, Power, and the Transformation of the Creek Indians, 1733–1816* (Cambridge, 1999), 233–290; Rothman, *Slave Country*, 106–168; Eliga H. Gould, *Among the Powers of the Earth: The American Revolution and the Making of a New World Empire* (Cambridge, Mass., 2012), 178–209.

of European empires that could support and arm blacks but it also provided whites with military man power to squash slave uprisings, legislated tariffs that encouraged sugar production, and turned Indian ground into plantations through purchases, removals, and surveys. Although federal officials restricted the importation of foreign slaves into the Mississippi Valley because they feared a Haiti-esque rebellion, the U.S. government did allow a massive influx of enslaved blacks from the eastern states through the internal slave trade. The entrenchment of slavery in the lower Mississippi Valley would make the institution even more central in the southern states on the Atlantic seaboard and promote its spread into western North America.[15]

Choctaws and other Gulf South Indians went from being influential regional powers to evicted peoples in the early nineteenth century. The collapse of the ecosystems and markets that had made deer hunting economically viable combined with the end of Spanish rule in Louisiana and West Florida to debilitate the Choctaws' potential to resist U.S. expansion. U.S. officials, no longer obliged to tread lightly in Choctaw affairs for fear of driving them into the arms of Spain, exploited the Choctaws' dependence on Euro-Americans for manufactured goods and liquor to get them in debt and then leveraged that debt to force land sales. Many Anglo-Americans encouraged this state of dependence because it made Choctaws and other natives more willing to accept—and, seemingly, more in need of—philanthropy. Despite calls for the Choctaws' removal by southern planters and Mississippi officials, the 1819 Civilization Act made philanthropy into policy by ordering the War Department to pay missionaries to get Indians to act like Anglo-Americans in the hope that they might eventually be incorporated. As federally funded agents of civilization spread the gospels of Christianity, agriculture, and thrift, they also deepened divisions between traditionalistic Choctaw leaders and an emerging cadre that historian James Taylor Carson has termed "cosmopolitans," mostly mixed-race Choctaws who accrued personal wealth through cattle ranching and, in fewer cases, cultivating slave-picked cotton. These cosmopolitans endorsed a wide-ranging platform of reform, including creating a Choctaw national state, in an effort to resist removal. But the political and ideological rifts among the Choctaw leadership ultimately worked to the advantage of Anglo-Americans, whose power

15. Kastor, *Nation's Crucible*, 62–66; Rothman, *Slave Country*, 32–62, 177–183, 203–204, 219; Johnson, *River of Dark Dreams*, 32–34; John Craig Hammond, *Slavery, Freedom, and Expansion in the Early American West* (Charlottesville, Va., 2007), 31–40; Robert E. Bonner, *Mastering America: Southern Slaveholders and the Crisis of American Nationhood* (Cambridge, 2009), 4–5.

to effect removal became overwhelming after the 1828 election of Andrew Jackson as president. In January 1830, Mississippi officials claimed jurisdiction over the Choctaws' lands, and, in May, Congress passed the Indian Removal Act. The Choctaws became an early experiment in federal removal policies as U.S. troops enforced the emigration of most Mississippi Choctaws between 1831 and 1833. Some twenty-five hundred Choctaws died amid expulsion; most of those who chose to remain in Mississippi were cheated out of their government-promised land.[16]

The regional influence of Louisiana's French and Spanish creoles waned over the 1800s, but they did secure incorporation into the United States as the political equals of other white citizens. During the territorial era, which lasted from 1803 to 1812, Anglo officials debated whether creoles were intellectually prepared for citizenship and reserved the governorship and most other appointed positions for themselves. But creoles resisted both potential political disenfranchisement and cultural corrosion by allying with francophone immigrants from France and Haiti and emphasizing their commitment to the racial oppression of blacks. For a while, these strategies worked: French-speaking residents maintained numerical superiority in the area around New Orleans until the mid-1830s, voted in French-speaking mayors, and exercised influence as lawyers, merchants, and planters. Although many creoles resented U.S. rule, they nevertheless came to welcome it once it became clear that the U.S. military and Anglo-American officials would ensure white creoles' power over free and enslaved blacks. Louisiana's 1812 statehood inaugurated the full incorporation of its creoles into the United States' political community and marked the development of a definition of nationhood in which all white men shared power regardless of ethnicity.[17]

16. James Taylor Carson, *Searching for the Bright Path: The Mississippi Choctaws from Prehistory to Removal* (Lincoln, Neb., 1999), 87–97 (quote, 88), 113–126; Ronald N. Satz, *American Indian Policy in the Jacksonian Era* (Lincoln, Neb., 1974), 64–86; Sheehan, *Seeds of Extinction*, 151–155, 250; Richard White, *The Roots of Dependency: Subsistence, Environment, and Social Change among the Choctaws, Pawnees, and Navajos* (Lincoln, Neb., 1983), 95–143; Clara Sue Kidwell, *Choctaws and Missionaries in Mississippi, 1818–1918* (Norman, Okla., 1995), xv, 35–44, 135.

17. Kastor, *Nation's Crucible;* Joseph G. Tregle, Jr., "Political Reinforcement of Ethnic Dominance in Louisiana, 1812–1845," in Lucius F. Ellsworth, ed., *The Americanization of the Gulf Coast, 1803–1850* (Pensacola, Fla., 1972), 78–81; Domínguez, *White by Definition,* 101–115; Paul F. Lachance, "The Foreign French," in Hirsch and Logsdon, eds., *Creole New Orleans,* 101–130, Tregle, "Creoles and Americans," 131–188; Jay Gitlin, *The Bourgeois Frontier: French Towns, French Traders, and American Expansion* (New Haven, Conn., 2010), 50, 159.

EXCLUSION AND EXPERIMENTAL EXPANSION

The United States' conquest of the Gulf South coincided with, and did much to encourage, fundamental shifts in ethnographic discourse. Anglo-American students of man increasingly described intelligence as something fixed that varied at the ethnic and racial levels, and they considered the ability to be scientific—that is, as disciplined, experimental, and objective as they congratulated themselves on being—a key variable for ranking American peoples' minds. These rankings (which, put bluntly, went Anglo-Americans, European-descended creoles, Indians, blacks) had taken shape by the 1830s and were closely related to the power relationships that crystalized as Anglo-Americans won supremacy in the Gulf South. Anglo-Americans were, of course, also exchanging ideas with men of science in Europe, some of whom were becoming convinced that bodies, and particularly heads, governed potential intellectual achievement. It was by no means inevitable that this anatomy-based ethnography would take root in the United States: U.S. expansion played a critical role in making race and intelligence inseparable for many Anglo-Americans. Conquest, and the unequal relationships it engendered, meant that Anglos had a redoubled interest in justifying their dominance of other peoples. Conquest also put the region's other peoples in states of degradation that seemed to correspond with their supposed mental incapacity, ensured that their bodies would be available as objects of research, and undercut alternative networks of information and patronage through which non-Anglos might have been able to repudiate theories about their own inferiority.[18]

The oppression of blacks was at the heart of the Gulf South's new political order. Although enslaved men and women continued to produce natural knowledge and, sometimes, share it with whites, they had almost no chance of being accepted as legitimate scientific practitioners. Soon after the Louisiana Purchase, for example, a newly formed committee of health commissioners fired the French physician in charge of the New Orleans Charity Hospital because he employed a black surgeon; the physician's retort that this surgeon was more capable than the white commissioners fell

18. On perceptions of racial difference overlapping with U.S. expansion, see Horsman, *Race and Manifest Destiny*, esp. 4–6, 85–115; and Thomas R. Hietala, *Manifest Design: American Exceptionalism and Empire* (Ithaca, N.Y., 2003), 171. On shifts in how men of science in the United States and Europe viewed race and intellectual potential, see Carson, *Measure of Merit*, 76–95; Jordan, *White over Black*, 436–446, 537; and Ann Fabian, *The Skull Collectors: Race, Science, and America's Unburied Dead* (Chicago, 2010), 1–46.

on deaf ears. The hospital did continue to employ enslaved blacks throughout the antebellum era, but, with very few exceptions, their duties were confined to nonmedical labors. The medical commissioners also took issue with the Charity Hospital's patroness being a woman, the French creole Louise Castillon, and ousted her and her daughter from the hospital's management through a combination of legal attacks and extralegal strong-arming. The society of Spanish New Orleans had been comparatively open in allowing blacks and women to have property rights and positions of responsibility. In the medical vein, as in many other aspects of life, these opportunities became far more limited as the United States brought its own brand of civilization to the Gulf South.[19]

The living conditions that white Americans imposed on Gulf South blacks, including enforcing ignorance through violence, seemed to offer evidence that African Americans were as inherently weak minded as many Anglos were coming to believe. Massachusetts clergyman Timothy Flint led congregations of enslaved blacks in Louisiana and endeavored to educate them in the process. Writing in 1826, he claimed that blacks picked up "the rudiments of reading quicker than even the whites" but that "it would be difficult to teach them arithmetic, or combination of ideas or abstract thinking of any kind." Since Flint was reluctant to see Louisianan planters' treatment of slaves as inhumane, he was left wondering why the state's enslaved blacks seemed so incapable of scientific thought. "Whether their skull indicates this by the modern principles of craniology, I cannot say," he mused, "but I am persuaded, that this susceptible and affectionate race have heads poorly adapted to reasoning and algebra." Despite writing against the evils of slavery as an institution, it was easier for Flint to accept that blacks might be naturally inferior than to acknowledge that the objectification they endured might make it tough to pick up algebra during a short weekly lesson.[20]

19. Rudolph Matas, *The Rudolph Matas History of Medicine in Louisiana,* ed. John Duffy, 2 vols. (Baton Rouge, La., 1958–1962), I, 324, 421–422; John Salvaggio, *New Orleans' Charity Hospital: A Story of Physicians, Politics, and Poverty* (Baton Rouge, La., 1992), 26; John Castellanos, "The Early Charity Hospital," 1897, MSS 307, 29, HNOC; "Certified Copy of Act of Cession by Micaela Almonester to Her Hereditary Rights as Patroness of Charity Hospital to the City of New Orleans," Mar. 9, 1811, Charity Hospital Collection, 4145 b, LSM; Jennifer M. Spear, *Race, Sex, and Social Order in Early New Orleans* (Baltimore, 2008), 178–214.

20. Timothy Flint, *Recollections of the Last Ten Years, Passed in Occasional Residences and Journeyings in the Valley of the Mississippi, from Pittsburg and the Missouri to the Gulf of Mexico, and from Florida to the Spanish Frontier . . .* (Boston, 1826), 345.

Around this same time, Anglo-American physicians were becoming less inclined to recognize the medical expertise of natives as legitimately scientific. In 1821, nearly four hundred white and black inhabitants of Darlington, South Carolina, petitioned the state assembly to let one John McKenzie practice medicine without a license. McKenzie had trained with "the oldest, most skillful, and most experienced physicians of the Choctaw Nation of Indians" and learned "their mode of practice, which consist[s] in the proper administration of roots and herbs." The petitioners all attested that McKenzie had cured them of various ailments that European-trained doctors had failed to relieve, and they insisted that the state's quackery laws were not meant "to prevent those who had made that science their peculiar study and the chief object of their attention" from practicing medicine "merely because their researches have induced them to a course of practice unknown to most of the Physicians in our state." The physicians and legislators who formed the special committee to review this petition were impressed by McKenzie's record of cures and suggested that practitioners of "the science of medicine" still had much to learn about "active and highly useful vegetables." Yet these learned gentlemen, unlike the petitioners, were clear that McKenzie's "art" was not science and that practicing native healing methods was distinct from "a scientific course" of medical practice. The committee members agreed to let McKenzie practice his art in South Carolina on the condition that he shared "his remedies, to what diseases they are adapted, and his mode of administering them" with the medical society of Charleston. These officials and physicians reinforced the boundaries of legitimate science even as they sought to learn more about alternative healing techniques.[21]

More importantly, many ethnographic observers, including many philanthropists, were coming to doubt whether Choctaws or other Indians were capable of civilization and assimilation. Anglo-Americans increasingly identified natives' seeming inability to achieve progress as innate instead of a product of Anglos' own efforts to push them off their land and foster economic dependence. Writing from New Orleans in 1819, architect Benjamin Henry Latrobe held that "the rights of humanity" ought to be "common to Men of all races," but he was hesitant to accept his own kinship with Indians, "a nation whose anatomical, as well as mental structure, differs

21. Jacob Cook et al. to the President and Members of the Senate, Nov. 7, 1821, Petitions to the General Assembly, 1776–1866, Ser. S165015, no. 32, reel ST1460, SCDAH ("oldest"), William Anderson, Special committee report, Dec. 17, 1821, Petitions to the General Assembly, 1776–1866, Ser. S165005, no. 53 ("science of medicine").

so widely from my own." He doubted that "the efforts of honest agents and missionaries to civilize and convert them" could succeed before "the cruel encroachments of our frontier settlers," "the more specious ... purchases of their Lands by our government," and other *"natural events"* caused Indians' extinction. "The march of white Society," he noted, "tends inevitably to their extirpation ... [and this] is the will of a higher power, that power which we call Nature." Latrobe pointed to the impoverished Choctaws he saw "crawl about" New Orleans in an "intermediate existence between annihilation and savage vigor" as supporting evidence for this "philosophic view." Despite recognizing the many ways U.S. expansion had undermined the Choctaws' prosperity, Latrobe still averred that nature—Indians' inferior anatomy and whites' superiority over all earthly things—meant that "the sum of human happiness would be greater ... if the Indians did not exist, and the country were peopled by a civilized nation." Both the process and aims of U.S. expansion encouraged Latrobe to see natives as incapable of equality and destined for extinction.[22]

Anglo-Americans' conquest of the Gulf South made the bodies of blacks and Indians available to white experts who studied corporeal evidence to gauge mental prowess. A few Anglos had collected human remains from the lower Mississippi Valley during the early years of U.S. expansion, but skull collecting became far more common in the region after 1830, when phrenology came into vogue in the United States and, perhaps more importantly, the objectification of blacks and expulsion of Indians ensured that would-be collectors had plenty of specimens. These collectors included phrenologists working in the Gulf South like Joseph Rodes Buchanan (who had about 175 skulls from Gulf South native groups and some 80 or 90 skulls from the region's blacks) and William Byrd Powell (whose collection included about 200 Indian crania from the Gulf South) as well as the more famous Philadelphia craniologist Samuel George Morton (who accrued few Gulf South crania before the Second Seminole War). Buchanan, for his part, toured the Gulf South in the late 1830s and did public phrenological analyses of

22. Edward C. Carter, II, John C. Van Horne, and Lee W. Formwalt, eds., *The Journals of Benjamin Henry Latrobe, 1799–1820: From Philadelphia to New Orleans*, 3 vols. (New Haven, Conn., 1984–1988), III, 233–235 ("nation," "efforts," "cruel," "more specious," "*natural*," 233, "rights," "common," "march," "tends," "crawl," "intermediate," "philosophic," 234), 239 ("sum"), emphasis in original. On Anglo-Americans' growing doubts about natives' improvability, see Sheehan, *Seeds of Extinction*, 241, 249, 254; White, *Roots of Dependency*, 124; and Robert E. Bieder, *Science Encounters the Indian, 1820–1880: The Early Years of American Ethnology* (Norman, Okla., 1986), 12.

some of the black and native skulls that local whites seemingly had in abundance. Like many phrenologists, Buchanan espoused that there was "no fatalism" in phrenology because brains and thus skulls were malleable and improvable. But he still chose to confirm his white audience members' presumptions that black skulls evinced servility and reverence for their masters whereas Indian crania revealed a propensity for violence and an incapacity for more than "some mechanical skill in [a] rude way."[23]

Morton was eventually able to acquire Choctaw skulls because removal made these remains accessible. While the Choctaws resided on their ancestral homelands, would-be skull collectors had a difficult time: conchologist Timothy Abbott Conrad wrote to Morton in 1833 from southwestern Alabama, an area that Choctaws still occupied, that "it would scarce be possible to get Indian skulls; as the Indians will not allow their bury places to be ransacked." The enforcement of Indian removal caused southeastern Indians to abandon burial grounds while, at the same time, the disease and starvation that accompanied removal generated thousands of cadavers that at least some army surgeons were eager to snatch up. The skulls that conquest made available became evidence for Morton's arguments for polygenesis and the inherent inferiority of blacks and North American Indians. White supremacists deployed these arguments to help legitimate the world that U.S. imperialism was creating, one where slavery was a positive good and Indians would soon disappear.[24]

One sign of the bifurcation that had developed between how Anglo-Americans understood whites and nonwhites in the years following the conquest of the Gulf South was that no one was amassing creole skulls. For Anglo citizens, European-descended creoles were racially and politi-

23. Joseph Rodes Buchanan to Samuel George Morton, Oct. 29, 1839, Morton Papers, Ser. I, APS, William Byrd Powell to Morton, Aug. 6, 1838; "Dr. Buchanan in Florida," *American Phrenological Journal and Miscellany,* I (1839), 338–339 ("no fatalism," 339); "Dr. Buchanan in Alabama," *American Phrenological Journal and Miscellany,* I (1839), 242 ("some mechanical"); see also "Lectures of Dr. Buchanan in Louisiana," *American Phrenological Journal and Miscellany,* II, (1840), 370.

24. T. A. Conrad to Morton, Dec. 28, 1833, Morton Papers, Ser. 1 (quote), Zina Pitcher to Morton, Mar. 4, 1834. See also Fabian, *Skull Collectors,* 6; and Donna L. Akers, "Removing the Heart of the Choctaw People: Indian Removal from a Native Perspective," in Clifford E. Trafzer and Diane Weiner, eds., *Medicine Ways: Disease, Health, and Survival among Native Americans* (Walnut Creek, Calif., 2001), 11. The literature on Morton and so-called scientific racism is vast, but see Stanton, *Leopard's Spots,* 54–72; Michael O'Brien, *Conjectures of Order: Intellectual Life and the American South, 1810–1860,* 2 vols. (Chapel Hill, N.C., 2004), I, 215–250; and Fabian, *Skull Collectors.*

cally part of the national community, other white Americans with whom they shared authority over the lands and nonwhite peoples that the United States had subjugated. In the political arena, therefore, the imperative of racial domination decreased the importance of emphasizing ethnic distinctions among whites. Many Anglo ethnographers were nevertheless reluctant to accept that the intellectual abilities of French and Spanish creoles equaled their own. Examining the ways that Anglo-Americans wrote about creole ignorance does much to explain why generations of Anglo-American historians elaborated and entrenched these views. For nineteenth-century historians like Francis Parkman, portraying French and Spanish Americans as simpleminded and ill prepared for republican citizenship helped make Anglo-American expansion seem like a natural and positive development. This interpretation was not necessarily cynical nor one that simply stemmed from historians' biases. Rather, portraying creoles as ignorant probably seemed founded on good science, a perspective built from decades of ethnographic writings by seemingly credible observers.[25]

Anglo-Americans looked for ways to denigrate creoles and exclude them from leading roles in the scientific community without insinuating that a white people was innately inferior. Geographer William Darby, for example, blamed the lack of educational opportunities under the Spanish and territorial governments for the purported intellectual weakness of creole women. "If the women of Louisiana are found deficient in mental endowment," he explained, "the reason is obvious: want of the means of acquirement." He also held that creole women were more virtuous and "more under the guidance of reason" than creole men and could, as republican wives and mothers, help uplift their male counterparts. But, although Darby advocated education because "the minds of the Creole women, remarkably active and tenacious, are much less ignorant than is generally supposed," he made it clear that "their rank in the scale of intelligence" would never be "exalted," regardless of the extent of their schooling. Darby's observations on creole women had, in one convoluted argument, emphasized the capacity of creoles to have the same political role in the United States as Anglo-Americans

25. On the political and racial alliance of Anglo-Americans and creoles, see O'Brien, *Conjectures of Order,* I, 302; Kastor, *Nation's Crucible;* and Johnson, *River of Dark Dreams,* 34. For the historiography of creoles' ignorance, see Tregle, "Creoles and Americans," in Hirsch and Logsdon, eds., *Creole New Orleans,* 141–160, Lachance, "Foreign French," 122–123; Daniel Usner, "Between Creoles and Yankees: The Discursive Representation of Colonial Louisiana in American History," in Bradley G. Bond, ed., *French Colonial Louisiana and the Atlantic World* (Baton Rouge, La., 2005), 1–21; and Gitlin, *Bourgeois Frontier,* 2–10, 158–161.

while also hinting that they would be hard-pressed to ever achieve scientific greatness.[26]

Timothy Flint perpetuated environment-based explanations for human difference as a way to justify creoles' failings. Flint—who "amused [him]self with observing the countenance and gesture of this simple race of ignorant creoles" while in Louisiana—contended in 1826 that the state's climate gave creole minds different propensities from those of people born in the North. Not only did human bodies develop more quickly in Louisiana's hot climate, but "the apprehension is quicker, the imitative arts more easily acquired, and the faculties unfold earlier." "Children born at the North . . . attain greater combination of thought, and think more profoundly," he continued, "but the Creole learns more easily to write a fair hand, to sketch a drawing, or copy a rose." Anglo-American men often considered detail-oriented observational skills and steady hands to be particularly pronounced in women, and effeminacy was a common charge that Anglo-Americans leveled at French and Spanish creoles. Flint's creoles were precocious, but their gendered talents limited higher thinking.[27]

The way Anglo-Americans viewed French and Spanish creoles—as people who could illustrate, describe, or collect nature but were unqualified to theorize about it—was strikingly similar to the way European experts had viewed American-born naturalists during the colonial era. Even the famous ornithologist John J. Audubon, who was born in Saint Domingue but identified himself as a Louisiana creole, crafted a persona as a rugged frontier collector in deliberate contrast to eastern urban naturalists who merely interpreted specimens sent to them. The same kinds of hierarchies that had structured scientific networks in Europe's early modern empires remained central to organizing the scientific community of the imperial United States.[28]

26. William Darby, *A Geographical Description of the State of Louisiana . . .* (Philadelphia, 1816), 262, 268 ("more under"), 269 ("women," "reason," "minds," "rank," "exalted").

27. Flint, *Recollections,* 338 ("apprehension," "Children," "but the Creole"), 370 ("amused"). For a similar argument about precocity, lack of reason, and effeminacy, see Stoddard, *Sketches, Historical and Descriptive, of Louisiana,* 329. On females' artistic and observational talents, see Susan Scott Parrish, *American Curiosity: Cultures of Natural History in the Colonial British Atlantic World* (Chapel Hill, N.C., 2006), 189–200; and Elizabeth B. Keeney, *The Botanizers: Amateur Scientists in Nineteenth-Century America* (Chapel Hill, N.C., 1992), 111–114.

28. On Audubon as a masculine frontier naturalist, see Gregory Nobles, "John James Audubon, the American 'Hunter-Naturalist': A New Species of Scientist for the New

Agriculture was at the heart of Anglo-Americans' justifications for expanding into the lower Mississippi Valley, expelling its Indians, and entrenching the enslavement of its blacks. Agriculture was also a foil through which Anglo-Americans compared their own abilities to develop nature, and even develop themselves, with the region's creole planters. Surveyor John Landreth thought Louisiana's soils should be exploited by Anglo-American planters because, ironically, Louisianan creoles were too obsessed with plantation agriculture. He wrote in 1819 that creoles "have but little chance to get information, and of course care but little about it." "Money Negroes Sugar and Cotton and land seems to engross all their time and attention but I believe [that] once a majority of Real Americans [are] here it would soon be very different." Despite creoles' ostensibly un-American singleminded pursuit of mastery and wealth, Landreth suggested that "they have generally left intirely uncultivated their minds and their soil," a significant character flaw considering that "from the great fertility of the soil they are inabled to keep a continual feast for their minds were they only disposed to read." He was clear, however, that "this disposition is only referring to what I call the natives of the place the criols descendent of the French and Spanish and doesn't apply at all to the new American emigrants." For Landreth, being a real American planter meant taking advantage of the leisure and wealth afforded by slave-based agriculture to cultivate oneself as an enlightened gentleman.[29]

Anglo-Americans defined themselves as a people who understood the power of agronomic experimentation. Florida, for example, was a territory that many Anglo-Americans looked on as a vast agricultural laboratory just waiting for experimenters. They were not the first whites to anticipate that experimentation would make Florida sprout wealth: Pedro Menéndez de Avilés tested sugar and other commodities in the sixteenth century, as did British planters in the eighteenth. In 1821, the year that Spain transferred Florida to the United States, one eager geographer wrote that the acquisition of Florida "opens a wide field of enterprize and philosophical experiments to our citizens." Even authors who were pessimistic about the potential of Florida to live up to its ballooning expectations promoted it as an experimental space. U.S. Army engineer John Eatton Le Conte stated that "Florida has always

---

Nation," *Common-Place,* XII, no. 2 (January 2012), http://www.common-place-archives.org/vol-12/no-02/nobles/.

29. John Landreth, "Journal of John Landreth on an Expedition to the Gulf Coast" (1818–1819), 26 ("have but"), 123 ("they"), NARA, microfilm at PKY.

been over valued; it therefore becomes our duty to lay aside the expectations of an El Dorado or a fountain of immortality, and by a diligent scrutiny [and] practical experiments ... strive to discover the best uses to which our newly acquired territory can be applied." And many Anglo-Americans did just that: projectors formed companies to test the viability of growing coffee; the Florida Institute of Agriculture experimented with grapes and sent their results to would-be Swiss immigrants; individual planters sowed new tropical crops; and, with the help of a federal land grant, Henry Perrine established an experimental plantation on Indian Key to test the viability of introducing Mexican and Caribbean plants into U.S. territory.[30]

Florida's status as an agronomic laboratory even promised to bolster U.S. science on the whole. In "An Oration, Delivered ... to the Florida Institute for Agriculture, Antiquities and Science," Colonel James Gadsden heaped praise on the many benefits of "the application of science as guides to agricultural operations, a spirit of theorising and testing theory by experiment, of reasoning on premises and pursuing them to their practical results." He claimed that sugar, rice, tobacco, and many tropical fruits would thrive in Florida: "Am I extravagant! let experiment decide—will you condemn me for speculating?—Test the reality." Gadsden told his audience of white Floridians that "no class of men have greater opportunities for intellectual improvement and varied learning than the southern planters" and that experimental agriculture would enable them to cultivate both the wealth needed for, and the proclivity to perform, experiments in the basic sciences.

> Comparatively few persons in Europe and still fewer in this country, have the means or faculties of prosecuting any branch of experimental science

30. William Darby, *Memoir on the Geography, and Natural and Civil History of Florida* (Philadelphia, 1821), vii ("opens"), viii; Richard Adicks, ed., *Le Conte's Report on East Florida* (Gainesville, Fla., 1978), 34 ("Florida"); "Pedro Menéndez' Contract for the Settlement and Pacification of Florida," Mar. 20, 1565, trans. Eugene Lyon, in Lyon, ed., *Pedro Menéndez de Avilés* (New York, 1995), 82; Joyce E. Chaplin, *An Anxious Pursuit: Agricultural Innovation and Modernity in the Lower South, 1730–1815* (Chapel Hill, N.C., 1993), 154; David Hancock, *Citizens of the World: London Merchants and the Integration of the British Atlantic Community, 1735–1785* (Cambridge, 1995), 143–171; Peter Stephen Chazotte, *Facts and Observations on the Culture of Vines, Olives, Capers, Almonds, etc. in the Southern States, and of Coffee, Cocoa, and Cochineal in East Florida* (Philadelphia, 1821); David B. Macomb, secretary of the Florida Institute, near Tallahassee, July 10, 1827, in [Richard S. Hackley, ed.], *Documents in Proof of the Climate and Soil of Florida Particularly the Southern Section* (New York, 1832); John Hoxie, "Account Book and Diary, Florida, 1822–1825," PKY; Nelson Klose, "Dr. Henry Perrine, Tropical Plant Enthusiast," *FHQ*, XXVII (1948), 189–201.

> to the extent they desire. It becomes important therefore that those who have abundant means, should contract a taste for such arts and sciences, as are most conducive to the morality and well being of the people—chymistry and mechanical philosophy, when made objects of pursuit and interest to the less informed classes of mankind, cannot but powerfully conduce to benefit the morals and to enlighten and strengthen the understanding. In their elementary principles and still more in their advanced state, aided by the countenance and influence of the wealthy, the physical sciences, must always help in extending and improving the mental energies of the people and be made a source of amusement, happiness and profitable exertion.

Florida planters' wealth, free time, and taste for experiment promised to make them natural leaders of the United States' scientific community and cultivators of both good crops and good citizens.[31]

THE PERSISTENCE OF IMPROVABILITY

From the 1810s to the 1830s, Anglo-Americans' conquest of the southeast borderlands and the developing science of ethnography had combined to delineate hierarchies that ordered which ethnic and racial groups were capable of inclusion in the United States' political and intellectual life. But literate blacks and natives, creole historians of science, and even some Anglo-American skull collectors and planters spoke out against hardening notions that Anglo-Americans were innately better. Enlightenment-era theories that all humans were improvable, U.S. officials' own rhetoric about uplifting other peoples, and memories of Spanish colonial society all motivated and informed Gulf South inhabitants' efforts to resist the changes wrought by U.S. imperialism.[32]

Free people of color publicly challenged the imposition of the United States' racial order in the Gulf South. They were particularly opposed to new legislative measures that would make blackness synonymous with enslavement and, therefore, erase free blacks as a class. In an open letter to the *Working Man's Advocate* in 1831, "A Free Colored Floridian" employed the language of intellectual improvability to call out Anglo-Americans for being too uneducated to recognize that their views on race were merely prejudices

31. James Gadsden, *Oration, Delivered by Col. James Gadsden, to the Florida Institute of Agriculture, Antiquities and Science, at Its First Public Anniversary . . .* (Tallahassee, Fla., 1827), 9–15 ("application," 12, "Am I," 13, "no class," 15, "Comparatively," 14–15).

32. On historical memory in western borderlands, see Lisbeth Haas, *Conquests and Historical Identities in California, 1769–1936* (Berkeley, Calif., 1995).

spawning from ignorance. This anonymous author, probably the son of a white Florida planter and his black wife or mistress, cited "the history of the world" as proof that "jealousy and prejudice are the offspring of ignorance or want of liberal education." And, since laws offered "a standard measure by which the comparative state of civilization of [a] country may be fairly estimated," he considered the territorial laws working to more fully oppress free blacks in Florida evidence enough of Anglo-American legislators' ignorance. He blamed most of humanity's intellectual deficiencies on "neglect in cultivating the rational faculties," and the most glaring sign of "human imbecility" in the United States was "the general prejudice ... against complexion or difference of color." Perhaps playing to Anglo-Americans' self-conscious identity as experimenters, he claimed that "experience is the test of truth, and proves [racial difference] to be fictitious." He contrasted the prevalence of racial bias in the United States with "every other American nation or colony situated as we are," where "prejudice against complexion, if it does exist at all ..., is so modified and subjected to reason ... that no perceptible evil is produced from it." He pushed U.S. legislators to give the same "constitutional protection to [free blacks'] person and property" that was "granted to all free people in all civilized countries." Reason, experimentation, civilization: the author based his case for equality on the same concepts that Anglo-Americans had used to justify their ascendancy. He closed by recommending that free blacks in the United States "look towards Mexico as a place of safety and permanent refuge." Not only would darker-skinned people thrive in Mexico's hot climate, but its inhabitants were *"free from all prejudice against complexion,"* and its laws *"recognize no difference of merit on account of color."* This Floridian, who had probably grown up under Spanish rule, held that Spanish American nations were exemplars of enlightened views on human difference because they distinguished ability from race.[33]

Many Choctaw leaders championed the educational and institutional

33. A Free Colored Floridian, "Prejudice Against Color," *Working Man's Advocate* (New York), Oct. 1, 1831, reproduced in Daniel Stowell, ed., *Balancing Evils Judiciously: The Proslavery Writings of Zephaniah Kingsley* (Gainesville, Fla., 2000), 76–81 ("history," 77, "prejudice against complexion," 78, "look," *"free,"* 79, *"recognize,"* 79–80). Daniel Stowell suggested that the author of this text was Zephaniah Kingsley, Jr., or his mixed-race son, George. I see no reason to believe that the white planter Zephaniah would sign only this one of his publications as a person of color; it is much more likely that George or one of the other educated sons of Floridian planters and black women authored this piece. On black Americans' use of scientific discourse to argue against their oppression, see Britt Rusert, *Fugitive Science: Empiricism and Freedom in Early African American Culture* (New York, 2017).

reforms introduced by Anglo-American philanthropists as part of their efforts to forestall removal. These cosmopolitan Choctaws enthusiastically undertook religious, technical, and scientific education to prove that they were capable of civilization and to limit their nation's economic dependency. They invited missionaries to establish schools within their territory, designated funds for mechanics to host apprentices, and both supported and attended the Choctaw Academy, a secular school in Kentucky that housed scientific apparatus and taught reading, mathematics, geography, astronomy, and land surveying to future Choctaw leaders. Peter Perkins Pitchlynn, one of the most prominent Choctaw reformers, recorded (in the Choctaw language) how Choctaw statesmen passed laws to create blacksmith shops, promote cotton production, and regulate relationships among whites, natives, and enslaved blacks. These statesmen established that whites and Choctaws could legally marry but forbade Choctaws from marrying or having sex with black slaves. Like European-descended creoles, Choctaws sought to prove their equality with Anglo-Americans by enforcing legal distinctions between themselves and African Americans.[34]

In the years preceding their removal, Choctaw cosmopolitans were embracing the optimism intrinsic to the language of improvement and were looking forward to a bright future. George Washington Harkins, a young mixed-race man who had attended the Choctaw Academy, wrote to Pitchlynn in 1827 that the Choctaws were quickly advancing beyond "darkness and superstition" and would soon develop into "one of the most enlightened nations upon the face of the earth." Five years later, Harkins was on a steamboat bound for Indian Territory. He had refused to sign the Treaty of Dancing Rabbit Creek in 1830, but it had nevertheless authorized his removal from Mississippi. Writing a farewell letter "To the American People" from the departing boat, Harkins again evoked the language of mental improvement. But his optimism was gone. He began by apologizing to the "American people, knowing and feeling sensibly my incompetency; and believing that your highly and well improved minds would not be well entertained by

34. Carson, *Searching for the Bright Path,* 86–105; White, *Roots of Dependency,* 103–143; Kidwell, *Choctaws and Missionaries in Mississippi,* 35–44, 101–104; Christina Snyder, "The Rise and Fall and Rise of Civilizations: Indian Intellectual Culture during the Removal Era," *JAH,* CIII (2017), 386–409; Peter Perkins Pitchlynn, *A Gathering of Statesmen: Records of the Choctaw Council Meetings, 1826–1828,* ed. Marcia Haag and Henry Willis (Norman, Okla., 2013), 69–70, 100, 139–142. On racial slavery among nineteenth-century Choctaws, see Barbara Krauthamer, *Black Slaves, Indian Masters: Slavery, Emancipation, and Citizenship in the Native American South* (Chapel Hill, N.C., 2013).

the address of a Choctaw." Harkins asserted that the Choctaws would strive to build a nation based on the "example [of] the American government" and asked that "we may for the future be cared for as children." The rhetoric of improvement had morphed into a piteous lament.[35]

Some Anglo-American ethnographers, particularly those who actually interacted with natives and blacks, were eager to demonstrate that race did not determine intelligence. These included Dr. William Byrd Powell, a prolific skull collector who lived and traveled in the Gulf South throughout the 1830s. Powell, like most other skull collectors in the region, analyzed crania through the interpretive lens of phrenology, a science predicated on the theory that brains had distinct "organs," usually thirty-seven in number, that indicated one's intellectual capacity, moral quality, and behavioral traits. More importantly, Powell and most other phrenologists reasoned that experience, not inheritance, shaped an individual's brain, and reformers of various stripes embraced phrenology to promote human improvement. Powell wrote that "the crania of all healthy young people are thin, and have, comparatively, their several parts of very uniform thickness, showing an equal action of all parts of the brain," a circumstance that suggested "young persons are adapted to any system of education." It was the education and life experiences of blacks and Indians that usually prevented them from achieving Euro-American levels of advancement. Whereas the heads of "civilised men" varied within ethnic groups because they reflected an individual's experiences, the crania of natives and blacks were "of pretty uniform thickness through life" because "all savages live in the same manner—all hunt, fight, lounge, muse, revenge, and rely on the Great Spirit." "One is not a planter, another a banker, another a merchant," he explained, "hence the reason why all savages of the same clan or tribe have heads similarly formed." Powell also claimed to see evidence in his personal collection of crania from "free and slave Africans, and our southern slaves" that only the skulls of enslaved blacks developed a thickness indicating idiocy, whereas "those whose ancestors were always free, are as thin as the best Caucasian crania." Powell did not follow this line of reasoning to its logical conclusion and speak out explicitly against slavery.[36]

35. G. Harkins to Pitchlynn, Dec. 20, 1827, quoted in White, *Roots of Dependency,* 127 ("darkness"); George W. Harkins, "To the American People" (December 1831), from *Niles' Register,* no. 41 (Feb. 25, 1832), 480, Sequoyah National Research Center, University of Arkansas, Little Rock, https://ualrexhibits.org/trailoftears/letters/george-w-harkins-to-the-american-people-december-1831/ ("American people").

36. John D. Davies, *Phrenology: Fad and Science, a 19th-Century American Crusade* (New Haven, Conn.,1955), 1–88 ("organs," 6); William Byrd Powell, "Remarks on

Powell strongly rejected the idea that all Indians as a race shared the same traits, a theory he attributed to ignorant phrenologists, particularly northerners and Europeans, who tried describing Indian character without any experience interacting with natives. "From the manner in which northern and European phrenologists speak of our savages," Powell wrote, "I am sure they know little about them." Powell targeted Ephraim Langdon Frothingham, a Bostonian fond of making racialized blanket statements about Indians (such as they had "little desire or power to think") and who had even insisted on "the *great imperfectness* of phrenology, as a science" because the faculties evident in native skulls did not correspond "with the known character of this race of men." Powell drew on his encounters with natives and collection of skulls to trash Frothingham. In a challenge to Frothingham's "ventures to make a mass of ruins of the present phrenological edifice," Powell wrote, "I have about two hundred Indian crania, and have mixed with the living tribes, observed their manners and customs, and if, by such means, a correct notion can be had of *our* Indians, Mr. Frothingham has not made one correct remark about them." There was, at least for Powell, no inherent contradiction in being a skull collector and an opponent of racial stereotyping.[37]

Powell took on a more eminent figure than Frothingham when he disputed Philadelphia craniologist Samuel George Morton's scientific methods. Powell had offered Morton "the use of [his] cabinet"—a collection that included "about two hundred crania" of "Choctaw, Creek, Alabama, Uchee, Hitchatee, Cherokee, Chickasaw, Attackapas, Monumental and Natchez Indians"—in 1838 as Morton was preparing his *Crania Americana,* a book devoted to measurements and illustrations of Indian skulls. Powell let Morton know that *Crania Americana* was not such an original idea, as it had "partially anticipated" his own plans "to publish a phrenological view of savage crania with plates." More critically, Powell pushed Morton to incorporate phrenological methods into the supposedly more objective, quantification-based craniology that Morton made famous. After finishing *Crania Ameri-*

the Human Skull," *American Phrenological Journal and Miscellany,* I (1839), 431–432 ("crania," "young persons," "civilised," "pretty uniform," "free," "those whose," 431, "all savages," 431–432, "One is not," "hence," 432).

37. W[illiam] Byrd Powell, "Remarks on the Organ of Watchfulness," *American Phrenological Journal and Miscellany,* I (1839), 474 ("From the manner," "I am"), 476 ("ventures," "I have"), emphasis in original; E. L. Frothingham, "On the Character of the North American Indians," *Annals of Phrenology,* II (1835), 315 ("little desire"); Frothingham, "New Phrenological Theories . . . ," *Annals of Phrenology,* II (1836), 420 ("*great imperfectness,*" "with the known"), emphasis in original.

cana—in which Morton proclaimed, but did not elaborate on, the "singular harmony between the mental character of the Indian, and his cranial developments as explained by Phrenology"—Morton apparently told Powell that he "would like to have had the advantage of [Powell's] 'judgment and advice'" while completing the work. Powell pounced on this opportunity to chide Morton that "each tribe has a well marked national formation of the head, and between the several tribes there is frequently an immense difference of cranial configuration." Powell averred that skull shape, as determined by phrenology, was more significant in determining an individual's intellectual potential than the cranial capacity measurements practiced by Morton: he told Morton that "to obtain the *form* [and not the internal volume] for yours in every instance should have been the prime object." In this critique, Powell promoted a more nuanced approach to reading skulls, one that evaluated Indians as improvable individuals and nations instead of as a race with fixed traits. Although some racial theorizers in the Gulf South would loudly propound the inherent inferiority of Indians and blacks, this view was by no means shared by all ethnographers based in the region, especially not before the 1840s.[38]

Louisiana's French and Spanish creoles had long resented Anglo-Americans' denigrations of their supposed ignorance. So it must have been especially gratifying when a creole intellectual applied the history of science and technology toward critiquing and repurposing Anglo-Americans' associations between scientific ability and citizenship. Charles Gayarré, a proud creole and the grandson of agronomic experimenter Jean Étienne de Boré, is best known for his four-volume history of the dominations that France, Spain, and the United States imposed on Louisiana. Yet Gayarré also studied the history of science and technology, and he used this field to challenge the definition of science, boast about Louisianans' accomplishments, and, in a reversal of long-standing Anglo-American criticisms, declare that many northerners were too ignorant for republican citizenship.[39]

38. Powell to Morton, Aug. 6, 1838, Morton Papers, Ser. 1 ("use," "about two hundred," "Choctaw," "partially," "to publish"), emphasis in original, Powell to Morton, Aug. 12, 1839 ("each tribe," "to obtain," "would like to"); Samuel George Morton, *Crania Americana* ... (Philadelphia, 1839), i ("singular").

39. On creoles' rejection of accusations of intellectual inferiority and efforts to ensure their full incorporation into U.S. political life, see Matas, *Rudolph Matas History of Medicine in Louisiana*, ed. Duffy, I, 301–325; Tregle, "Political Reinforcement of Ethnic Dominance," in Ellsworth, ed., *Americanization of the Gulf Coast*, 80–82; Gitlin, *The Bourgeois Frontier*, 50–59; and Julien Vernet, *Strangers on Their Native Soil: Opposi-*

Gayarré wrote two lectures, collectively titled the "Influence of the Mechanic Arts on the Human Race," which he presented at the Mechanic's Institute of New Orleans and the Franklin Institute of Mobile in 1854. He began these talks by interrogating why science meant what it meant. He found that, even though "the mechanic [may] secretly feel that he is unfolding the wings of intellectual ascension" when practicing a craft, "his fellow beings" derided these tasks as unscientific because they shared "an accepted definition, which is not more accurate than most definitions," that mechanical arts were pursuits "in which the hand and the body are more concerned than the mind." Gayarré rejected this intellect-based definition of science because the mechanical and liberal arts were, at heart, quite similar. "Two men are stuffing capons," he imagined, one of whom was "a mechanic" while the other was "Francis Bacon . . . stuffing a fowl with snow, to make an experiment." Gayarré suggested the explanation for why "it is within the power of man, to ennoble even the stuffing of a capon" as scientific was historical and derived from class differences. The bifurcation that distinguished between mechanical arts and science began in the ancient world and was rooted in slavery: since masters gave slaves mechanical tasks, the master class came to associate these with drudgery while distinguishing their own intellectual activities as liberal arts, "the pursuits of the free." For Gayarré, the distinction between science and technology had been "nursed in the lap of slavery" in the ancient world but, now that liberty and civilization were spreading, the "scholastic distinctions established between the Arts ought to be abolished." Ennobling technological expertise would make doxological science—"the study of nature [which] leads to the knowledge of God"—available to all, not just an overeducated few.[40]

The disparagement of the mechanical arts had made it easier for Anglo-Americans to criticize creoles as unscientific, but Gayarré was proud that Louisianans were beginning to prove their capacity for all sorts of science. Gayarré shared his "Influence" lectures with the graduating class of the Centenary College of Louisiana, and he told them that Louisianans of

tion to United States' Governance in Louisiana's Orleans Territory, 1803–1809 (Jackson, Miss., 2013), esp. 58–59.

40. Charles Gayarré, *Influence of the Mechanic Arts on the Human Race: Two Lectures* (New York, 1854), 7, 8 ("mechanic"), 9 ("Two men"), 9–10 ("Francis"), 10 ("it is within"), 14 ("pursuits"), 32 ("scholastic"), 42 ("study of nature"). On Gayarré's histories of Louisiana, see O'Brien, *Conjectures of Order*, I, 292–302. On doxological science, see Theodore Dwight Bozeman, *Protestants in an Age of Science: The Baconian Ideal and Antebellum American Religious Thought* (Chapel Hill, N.C., 1977).

earlier generations were not to blame for their ignorance since they had repeatedly asked their various imperial dominators for schools. Although Louisiana's rulers had neglected these requests, the graduates of Centenary College and other institutions were finally presenting "a victorious answer to the sneers of detraction" with which observers had long characterized Louisianans: "No longer shall it be said, that our bright sun smiles only on agricultural and commercial wealth, but blesses not with fertility the fields of the intellect." The graduates' scientific education, which they received in "our lovely South," would "carry [them] up beyond the earth, on the wings of astronomy, to listen to the music of the planets and penetrate into the mysteries of the celestial spheres." "It may make you, as a mathematician, agriculturalist, mechanician, or physician," he exhorted the graduates, "the pride and blessing of your country, and even of the human race." In this short commencement address, Gayarré established that Louisianans had always treasured knowledge, had finally silenced their detractors, and had become fully integrated with the southern section of the United States.[41]

Gayarré posited that the mechanical arts and U.S. imperialism developed hand in hand. "The march of industry has been such," he claimed, "that the tool of the Mechanic may be said to be now the sceptre of the world, and that the superiority of a nation over all others would be surely the result of its ascertained superiority in the Mechanic Arts." He had no doubt that it was "particularly the United States which may be said to be the destined home of the Mechanic Arts, and the seat of that power which they will ultimately extend all over the earth." This was not just an accurate prediction; it was an observation based on the United States' North American conquests since 1803. "It is as much to our excellence in the Mechanic Arts as to the beauty of our political institutions which have secured the development of those Arts, that we are indebted for all our territorial acquisitions; among which California and New Mexico may be compared to two magnificent portals which have lately opened their wide gates to the introduction of American industry, American enterprise[,] and American institutions over our whole continent." Although he resented many aspects of the United States' seizure of Louisiana, Gayarré was nevertheless eager to count himself among the Americans who were conquering the continent and honing technological expertise amid encounters with western nature.[42]

Still, the progress of technology in the United States also had deleteri-

41. Gayarré, *Influence of the Mechanic Arts on the Human Race,* 76 ("victorious"), 78 ("our lovely"), 82 ("carry").

42. Ibid., 59–61 ("march," 59, "tool," 59–60, "particularly," 61).

ous consequences. Gayarré proclaimed that technology had become such a dominant factor in the northern states that it was making many white Americans too ignorant for republican citizenship. He was, in effect, throwing Anglo-Americans' well-worn argument that creoles' mental ignorance made them less capable of participating in national politics back in the Yankees' faces. He reminded his audience that "the Mechanic Arts had been contemptuously abandoned to the slaves" in ancient societies but that these skills had thrived over the last two centuries because widespread liberty had helped dissolve some of the artificial distinctions between thinkers and tinkers. But he added that the very success of industrialization in the nineteenth century "may lead back to slavery." He was not talking about southern black slavery; indeed, his lectures never mentioned it. Rather, Gayarré held that capitalism and technology were fostering a type of "bondage" in northern industrial centers "by which a man becomes the slave of another without becoming his property, and in whose preservation his master takes no interest." This reasoning was similar to the insistence of other proslavery southerners that the paternal bonds of slavery were more humane than the wage-based labor relations of industry. For Gayarré, the more pointed critique was that industrialization compelled white workers to focus all their mental powers on a single menial task that sucked "the last spark of human intellect" out of their brains. Gayarré, who liked to flavor his histories with bitter ironies, smugly took on the airs of a philanthropist and advocated for the education of his ignorant northeastern countrymen in the hope that they, too, might one day achieve equality in national political and intellectual life. "It ought therefore to be the wish of every one who has at heart the improvement of the human race in every respect, and in connexion with all the Arts—particularly the most important of all—the Art of government—to cause the stream of education to flow" to America's centers of industry. The United States might have defeated the other imperial powers vying for the Gulf South, but ongoing competitions inspired by expansion, including the growing tension between southern and northern whites, continued to shape diverse Americans' ethnographic discourses.[43]

Not every white planter in the Gulf South welcomed the United States' stricter racial hierarchies, and none were more outspoken than Zephaniah Kingsley, Jr. Born in England, Kingsley became a South Carolina loyalist, a Danish citizen, and a slave trader in Africa and the Caribbean before pledging allegiance to Spain and settling in East Florida in 1803. He

43. Ibid., 64 ("Mechanic Arts"), 66 ("bondage"), 68 ("last spark"), 71–72 ("ought"). On Gayarré's penchant for irony, see O'Brien, *Conjectures of Order*, I, 296–301.

had well over one hundred slaves by the time the United States incorporated Florida, but he also had a Wolof wife, enjoyed lasting relationships with other black women, and recognized and emancipated his sexual partners and mixed-race children. Like other planters who had lived in Spanish Florida, Kingsley seemed to love the social world that Spain's three-caste system of whites, free blacks, and slaves made possible, and he fought against new U.S. laws that made free black Floridians far less free, illegalized interracial marriage, and declared all offspring of such unions illegitimate. Since he was losing these legal struggles, Kingsley tried appealing to Anglo-Americans' reason.[44]

Kingsley's writings, most notably his *Treatise on the Patriarchal, or Co-operative System of Society,* first published in 1828, argued that enslaved blacks were both necessary to reaping profits from torrid climates and fully capable of equality with whites. Kingsley, who announced that he was "a slave owner, and has a right to express his opinion, having lived by planting in Florida for the last twenty-five years," reminded his readers that Florida was destined to be the center of agricultural experimentation in the United States. But, since the "want of health" that whites experienced in this semi-tropical environment made them unable to enrich themselves or the United States through their own labor, the enslavement of black workers was justified. Like other apologists for slavery, he affirmed that it was a legitimate and benevolent social relationship, claiming that "the condition of slaves may be equally happy and more independent of the ordinary evils of life, than that of the common class of whites denominated free." But unlike many slave-owners, Kingsley emphasized that "unprejudiced people" ought to agree that enslaved blacks were in no way innately inferior. "Few, I think, will deny that color and condition ... are two very separate qualities," he conjectured, "but the fact is, that ... our legislators, for want of due consideration, have mistaken the shadow for the substance, and confounded together two very different things." In 1842, the New York abolitionist L. Maria Child interviewed Kingsley about his efforts "to prove that *colour* ought not to be the badge of degradation; that the only distinction should be between *slave* and *free*," and she was shocked to learn that he also advocated "intermar-

44. Daniel L. Schafer, *Zephaniah Kingsley Jr. and the Atlantic World: Slave Trader, Plantation Owner, Emancipator* (Gainesville, Fla., 2013), esp. 1–6, 178–179. For another Florida planter who opposed the United States' racial order, see Cameron B. Strang, "Planters and Powerbrokers: George J. F. Clarke, Interracial Love, and Allegiance in the Revolutionary Circum-Caribbean," in Jorge Cañizares-Esguerra, ed., *Entangled Empires: The Anglo-Iberian Atlantic, 1500–1830* (Philadelphia, 2018), 142–158.

riage between the races" because it produced "a great improvement of the human race." Despite her antislavery position, Child considered Kingsley's ideas so baffling that she observed that his head would "prove as great a puzzle to phrenologists, as he himself is to moralists and philosophers."[45]

Kingsley's view that nonwhites were both capable of civilization and legitimate subjects of enslavement did not seem so odd to Florida planters who had shared his experiences with life in the Spanish circum-Caribbean. The Florida-born Anglo-Spanish planter George J. F. Clarke, who, like Kingsley, had marriage-like relationships with black women and recognized his mixed-race children as heirs, wrote that "the only difference in man, laying aside his color, is the difference of opinion; and that difference of opinion arises from the difference of education." "How illiberal the mind to the contrary," Clarke railed, "how little or how partially must they have studied human nature! Are we not all the children of habit, the mere reflections of education and manners?" He invited

> the sceptical *[sic]* in this part of the philosophy of human nature [to] turn his eyes to the city of Mexico, and see there the examples of talents natural and acquired, in the fine arts and belle lettres, manifest among Indians; let him look into the Havana and see the many finished workmen in the useful and elegant crafts, to be found there among the Africans ... and then let him declare if he is not always the master-piece of Nature's works and the only master of arts. And after thus seeing what he has acquired, will the caviller *[sic]* attempt to say what bounds have been prescribed to his acquirements by his country or his color?

Kingsley, who had "lived long in different slave holding countries," took an even broader view, citing examples from Spanish Florida, Brazil, and British, Dutch, French, and Spanish islands in the Caribbean to bolster his case that blacks were capable of intellectual excellence and that slavery thrived best in societies that maintained the three-caste system.[46]

45. Z[ephaniah] Kingsley, "A Treatise on the Patriarchal, or Co-Operative, System of Society, as It Exists in Some Governments, and Colonies in America, and in the United States, under the Name of Slavery, with Its Necessity and Advantages; by an Inhabitant of Florida," in Stowell, ed., *Balancing Evils Judiciously,* 40 ("condition"), 41 ("want of health"), 62 ("slave owner"), 66 ("unprejudiced," "Few," "but the fact"); L. Maria Child, Letter from New York, July 7, 1842, reproduced ibid., 107–109 ("to prove that *colour,*" 107–108, "intermarriage," "great improvement," 108, "prove as great a puzzle," 109).

46. George I. F. *[sic]* Clarke, quoted in Charles Vignoles, *Observations upon the Floridas* (1823) (Gainesville, Fla., 1977), 137 ("only difference"); Kingsley, "Treatise," in Stowell, ed., *Balancing Evils Judiciously,* 42–46, 55, 62 ("lived long").

But Kingsley's arguments in favor of free blacks' rights were not simply nostalgic yearning for the bygone Spanish era or a rose-tinted perspective on life in other slave societies. Rather, his experience with the fierce competitions for power between empires and among ethnic groups in Florida and the wider Caribbean inspired Kingsley to contend that it was in whites' best interest to perpetuate the three-caste system and win the attachment of free blacks. Kingsley had experienced the Haitian Revolution, the sacking of his own plantation by Seminoles in 1812, and Anglo-American invasions of Florida during the 1810s, and he remained convinced that the enslavement of blacks was a continuing power contest in which U.S. officials and white planters needed to negotiate authority with free blacks in order to protect their lives and property. He pronounced that Anglo-Americans' "prejudice" against blacks "neutralizes the physical strength of the country, by placing one portion of the inhabitants in hostile array against the other." The surest means for U.S. officials to make certain that all the territory's people fought together against a foreign power would be to win the allegiance of blacks by treating slaves well and rewarding free blacks with legal rights. This policy, he claimed, had proven successful in Brazil, where that empire's slaves and free blacks remained loyal in its war against the republic of Buenos Aires despite the latter's efforts to incite Brazil's blacks to rebel.[47]

Kingsley's greater fear was competition within Florida, and he warned that Anglo-Americans were importing overly violent and oppressive master-slave relations into the territory that they were too weak to enforce without the support of free blacks. The master class could rely on "fear and force" to oppress blacks in states like South Carolina where white power was firmly established and where large populations of "back country" whites were ready "to put down or exterminate all the colored people in case of insurrection." Floridians, on the other hand, needed to govern blacks with "wisdom and policy" to ensure that slaves remained contented because there were too few whites to resist them if they decided to challenge their masters. Whites had to negotiate authority with free blacks because "co-operation of the free colored people is absolutely necessary when the white population is scanty." If whites would only abandon their prejudice and grant free blacks legal protections, free blacks would "become identified with the whites [by interest] on one side, and with the slaves by descent on the other; a connexion which perfectly cements the three castes . . . each being perfectly con-

47. Kingsley, "Treatise," in Stowell, ed., *Balancing Evils Judiciously,* 43–44, 53 (quotes).

tented with its permanent, lawful privileges." For Kingsley, the weakness of Florida's planter class, its lack of backcountry whites, and the failure of its rulers to win the attachment of free blacks made the territory a powder keg that resembled pre-revolutionary Haiti more than it did the rest of the southern United States.[48]

Kingsley advocated a philanthropic campaign to educate whites and blacks alike as the best way to prevent competitions from erupting into violence. Along with repealing "the most oppressive parts [of] the laws now in force," Kingsley sought to implement "a more liberal education" for Florida whites because "their present conduct towards colored peoples seems to be graduated by the cultivation of their minds." He also worried that both Christian missionaries and black religious leaders—such as "Gullah Jack or Jack the Conjurer," who had brought "his conjuring implements with him in a bag" to Kingsley's plantation—inspired "ignorance and want of rationality" among slaves that made them more prone to reject their enslavement. He thus endorsed educating blacks in order to limit the intellectual and political influence of Christian and African religious leaders: "Let any slave owner reflect and say how much advantage the country would derive from preaching up industry, economy, and a local attachment to the slaves ... instead of preaching up terror and dismay, misery and discontent." Even planters' own mental development depended on implementing policies that would decrease blacks' incentives to challenge their authority. "The [slave] owner," Kingsley wrote, "no longer a kind of state prisoner, hovering over the movements of his negroes and overseers, could liberalize and improve his mind by travelling, and satisfy his thirst for knowledge wherever the advance of science offered the greatest field for acquirement." Florida planters would only achieve distinction in the sciences if they implemented humane slave laws and protected the rights of free blacks.[49]

Kingsley's was a losing battle. Anglo-American migrants to Florida were eager to perform mastery through violence and had not experienced the destruction that accompanied decades of geopolitical competition for Florida. They were not, therefore, interested in negotiating authority with anybody, especially blacks. So, in the mid-1830s, Kingsley freed his slaves and relocated them and his mixed-race children to Haiti, where, he trusted, they

48. Ibid., 45 ("become identified"), 51 ("fear and force," "back country," "put down," "wisdom"), 65 ("co-operation").

49. Ibid., 60 ("most oppressive," "more liberal," "their present"), 66 ("the [slave] owner," "no longer"), 68 ("Gullah," "his conjuring"), 70 ("ignorance"), 74–75 ("Let any slave owner," 74).

would thrive under that nation's more enlightened laws. Kingsley had given up hope that class, race, and perceived intellectual potential would be anything other than inextricable in the United States.[50]

• • •

Ethnographic knowledge and rhetoric in the United States, particularly discourses about mental abilities, developed amid territorial expansion. Many Anglo-Americans performed ethnographic observations during and, especially, after the United States' conquest of the southeast borderlands that seemed to reinforce the hypothesis that nonwhites had inherently inferior brains and could never achieve equality in the nation's political and intellectual communities. However, blacks, natives, creoles, and even a few Anglo-Americans used ethnography to challenge the new order that U.S. rule was making, and they tended to favor the increasingly old-fashioned perspective that all men were created equal. As had been the case during the Spanish, French, and British eras, imperialism—and the social and political relationships it engendered—conditioned how all peoples in the Gulf South understood nature and each other. The Anglo-American master class had achieved dominance throughout most of the region by the 1830s. But ongoing encounters with local inhabitants, combined with increasingly tense relations between northern and southern whites, continued to affect the ways Anglo-Americans pursued, circulated, and applied natural knowledge throughout the antebellum era. Indeed, violence-based relationships with enslaved blacks helped reveal the structure and history of the earth itself.

50. Baptist, *Creating an Old South*, 88–153.

{ 6 }

DEEP HISTORY, DEEP SOUTH

Slavery and Geology in the Antebellum Era

• • •

In the decades before Charles Darwin's 1859 *On the Origin of Species*, geology was the science that most challenged Euro-American understandings of the connection between God, living things, and the increasingly deep history of the planet. Yet contemporary power relationships could determine how individuals developed and interpreted geological knowledge, and, in the antebellum Gulf South, these relationships centered on slavery. Plantation slavery—and the labor, patronage, and networks it provided—enabled collections and observations that defined the Gulf South's geohistory while emerging geotheories inspired new means of justifying and furthering slavery.[1]

Anglo-American planters dominated Louisiana, Alabama, Mississippi, and northern Florida by the 1830s, and some planters chose to exploit their authority over lands and enslaved blacks to study geology. Planters' fascination with fossils and strata was by no means atypical: naturalists, entrepreneurs, and government officials alike developed a profound interest in geology during the 1820s and 1830s. Historians have even argued that the geological surveys that several states sponsored during this period represented a defining moment in the history of American science because they were among the earliest collaborations between naturalists and government in the United States. Research into the deep history of the Gulf South, however, did not develop through state patronage. Rather, planters filled the role that governments and learned societies played in other parts of the country by taking the lead in fostering geological investigations. Plantation slavery supplied resources, labor, and networks that supported sci-

1. On geology in the late eighteenth and early nineteenth centuries, see Charles Coulston Gillispie, *Genesis and Geology: A Study in the Relations of Scientific Thought, Natural Theology, and Social Opinion in Great Britain, 1790–1850* (Cambridge, Mass., 1951); Martin J. S. Rudwick, *Bursting the Limits of Time: The Reconstruction of Geohistory in the Age of Revolution* (Chicago, 2005); and Rudwick, *Worlds Before Adam: The Reconstruction of Geohistory in the Age of Reform* (Chicago, 2008).

entific work in ways that the federal and state governments would not for years to come.[2]

Slavery was the most significant social context that impinged on geology in the Gulf South, but relationships between southern and northern whites—including both warm friendships and bitter competitions—also affected the pursuit and patronage of geological research. On the one hand, Alabama planter Charles Tait, his wife Sarah, and their slaves collaborated with Philadelphian naturalists to generate new perspectives on the region's geohistory. The Taits' and other Alabamans' generous support for natural history highlights that networks of information and patronage in the imperial United States were polycentric and not organized by hierarchies of place in which northeastern cities were always central and southern and western spaces were always peripheral. Mississippi planter Dr. Rush Nutt, on the other hand, viewed northerners as enemies. Nutt looked to new geological theories, especially uniformitarianism, for weapons to deploy against religious northerners who, he believed, sought to destroy slavery, science, and liberty in order to pave the way for ignorance, theocracy, and disunion. Nutt incorporated geological perspectives into his broader natural philosophy to defend slavery as a positive good and also envisioned geo-engineering projects that would promote the expansion of plantation agriculture by transforming North America's environment. Slavery and science strengthened each other in the early United States, and the geological knowledge and perspectives they generated influenced how Americans understood the deep history, present, and future of their nation.[3]

2. On geology in early America, see Walter B. Hendrickson, "Nineteenth-Century State Geological Surveys: Early Government Support of Science," *Isis,* LII (1961), 357–371; George H. Daniels, *Science in American Society: A Social History* (New York, 1971), 174–205; James X. Corgan, ed., *The Geological Sciences in the Antebellum South* (University, Ala., 1982); Paul Lucier, "Commercial Interests and Scientific Disinterestedness: Consulting Geologists in Antebellum America," *Isis,* LXXXVI (1995), 245–267; Andrew J. Lewis, *A Democracy of Facts: Natural History in the Early Republic* (Philadelphia, 2011), 129–153; and Conevery Bolton Valenčius, *The Lost History of the New Madrid Earthquakes* (Chicago, 2013). Some southern states, especially the Carolinas, were among the first to enact official geological surveys in the 1820s, but state-sponsored geological surveys came late to the Gulf states of Alabama (1848), Mississippi (1850), Louisiana (1864), and Florida (1907).

3. Scholarship on slavery and science in the United States has been surprisingly limited, focusing primarily on how scientific practices influenced health and agriculture on plantations or southern whites' views of race. See Sharla M. Fett, *Working Cures: Healing, Health, and Power on Southern Slave Plantations* (Chapel Hill, N.C., 2002); and Lester D. Stephens, *Science, Race, and Religion in the American South: John Bachman

Charles Tait and Rush Nutt were both prominent figures whose shared interest in natural history was common enough among planters. More critically, their cases cast doubt on some of historians' central assumptions about natural knowledge in the antebellum era, particularly the notion that the idiosyncratic ideas and practices of northeastern naturalists somehow typified science in the vast United States. Slavery and imperialism (and not just liberty and democracy) conditioned studies of nature, plantations (and not just learned societies and the state) supported science, and hostility toward Christianity (and not just efforts to confirm the truth of scripture) permeated natural philosophy. Moreover, the relationships that mattered most to Tait's and Nutt's scientific work—violence against enslaved blacks and exchanges and competitions with other whites—were far from atypical; these same kinds of interactions had been shaping natural knowledge in America for centuries. Violence, competition, and exchange undergirded geological investigations and interpretations in the United States, and, considering the growing political and economic importance of slavery, brutality against nonwhites might have been even more central to intellectual life in the antebellum period than it had been in the colonial era.[4]

---

and the Charleston Circle of Naturalists, 1815–1895 (Chapel Hill, N.C., 2000). An exception is Britt Rusert, *Fugitive Science: Empiricism and Freedom in Early African American Culture* (New York, 2017). Far more than histories of the United States, studies of the Caribbean and Brazil have demonstrated that enslaved blacks and the institution of slavery could both support and challenge European science at the imperial level. See Londa Schiebinger, *Plants and Empire: Colonial Bioprospecting in the Atlantic World* (Cambridge, Mass., 2004); James H. Sweet, *Domingos Álvares, African Healing, and the Intellectual History of the Atlantic World* (Chapel Hill, N.C., 2011); and James Delbourgo, *Collecting the World: Hans Sloane and the Origins of the British Museum* (Cambridge, Mass., 2017).

4. For intellectual histories focused on planters and their families, see Michael O'Brien, *Conjectures of Order: Intellectual Life and the American South, 1810–1860*, 2 vols. (Chapel Hill, N.C., 2004); and Elizabeth Fox-Genovese and Eugene D. Genovese, *The Mind of the Master Class: History and Faith in the Southern Slaveholders' Worldview* (Cambridge, 2005). Historians have reified the idea that northeastern science was quintessentially American by focusing on the absolute or relative progress of science in the South instead of analyzing how natural knowledge in the region was part of—and might change how scholars think about—American science more broadly. See Thomas Cary Johnson, Jr., *Scientific Interests in the Old South* (New York, 1936); Robert V. Bruce, *The Launching of Modern American Science, 1846–1876* (Ithaca, N.Y., 1988), 54–63; and Ronald L. Numbers and Janet S. Numbers, "Science in the Old South: A Reappraisal," in Numbers and Todd L. Savitt, eds., *Science and Medicine in the Old South* (Baton Rouge, La., 1989), 9–35. On the centrality of violence to slavery and slavery's significance in the

PLANTATION PATRONAGE AND NORTH AMERICA'S TERTIARY PERIOD

Born in Virginia, Judge Charles Tait (1768–1835) moved to the Georgia piedmont and became a state senator, judge, and U.S. senator. He pushed his fellow congressmen to grant Louisiana, Mississippi, and Alabama statehood and, during the Missouri Crisis of 1819, argued that federal policies meant to restrict the expansion of slavery would push "the slaveholding states [to] form a separate Confederacy." Tait relocated again in 1819, this time to Claiborne, Alabama, where he established a plantation and took a position as a federal judge. Tait, his family, and the roughly 60 slaves they brought with them were among hundreds of thousands of migrants from the Atlantic states to the cotton lands of Alabama and Mississippi during the antebellum era. By the time of his death sixteen years later, Tait had almost doubled his slave holdings to 115, making him one of the wealthiest men in the state.[5]

Like several other Alabama planters, Tait developed a keen interest in geology, conchology, and paleontology. He pondered the fossil shells embedded in the bluff near the town of Claiborne and, while visiting Philadelphia in 1826, attended geological lectures and befriended some of the city's leading naturalists, including Isaac Lea and Samuel George Morton. Tait was elected to the American Philosophical Society before returning to Alabama, where he set to collecting fossil shells in earnest and, in 1829, began sending them to Lea at the Academy of Natural Sciences of Philadelphia (ANSP).[6]

Lea recognized that the collection and circulation of these specimens relied on enslaved blacks. In February 1829, Lea asked Tait for "land and river shells, bivalves and univalves" of both living and extinct species, writing that "your rivers and brooks must produce many which must be new and these would be a *treasure* to me." He added that Tait's "servants" could easily make these collections "in a few leisure hours" and advised that these slaves ought to pack the shells in boxes stuffed with Tait's cotton to ensure

---

antebellum era, see Edward E. Baptist, *The Half Has Never Been Told: Slavery and the Making of American Capitalism* (New York, 2014).

5. Charles H. Moffat, "Charles Tait, Planter, Politician, and Scientist of the Old South," *Journal of Southern History,* XIV (1948), 206–233 (quote, 220); Moffat, "The Life of Charles Tait" (Ph.D. diss., Vanderbilt University, 1946), 225; James David Miller, *South by Southwest: Planter Emigration and Identity in the Slave South* (Charlottesville, Va., 2002), esp. 4–10.

6. On the extent of geological interests in Alabama and Mississippi, see Johnson, *Scientific Interests in the Old South,* 57–83.

safe shipment to Philadelphia. It seems that Tait, who had only one leg and probably would not have braved the shell-rich bluff alone, employed at least one of his slaves as a conchological specialist, though Lea had a low opinion of this individual's abilities. After thanking Tait for a "valuable present" of shells and requesting another shipment, Lea noted that "it will be very difficult for your servant to select the rarer species" and suggested that this slave collect a wide variety of potentially valuable specimens on the chance that some of these would prove new. Although the identity of this enslaved shell collector is not clear, several of Tait's slaves had received some education, and a few—including Hezekiah, a carpenter in charge of Tait's cotton gins—were technological specialists.[7]

Lea welcomed Tait's shells as valuable additions to the ANSP's collections, but he did not initially realize their importance as Tertiary fossils. The Tertiary period—now also known as the Cenozoic—began after the extinction of the dinosaurs around sixty-five million years ago. The French naturalist Georges Cuvier recognized that a fundamental change in the history of life on earth had occurred when strata preserving evidence of an ancient age of giant saurians (the Secondary) gave way to the age of mammals (the Tertiary) in which mankind continued to live. Cuvier and others examined the shift between the Secondary and Tertiary for minerals and fossils that might better explain this awe-inspiring revolution. Moreover, as Charles Lyell argued in the early 1830s, investigating the Tertiary could reveal mysteries about why and how the earth changed over time. For Lyell, the Tertiary was the test case for uniformitarianism, the theory that presently observable causes—erosion or the slow accumulation of sediment, for example—could account for the earth's past and present features. In other words, since Tertiary strata were still being formed, they presented a chance to test if actual causes, by which Lyell meant the observable forces that transformed the earth, could in fact be applied to the Secondary and Primary formations. Tertiary strata also facilitated geological comparisons across the globe. Naturalists since Cuvier had compared fossil species from

7. Isaac Lea to Charles Tait, Feb. 19, 1829, Tait Papers, ANSP ("land and river"), emphasis in original, Lea to Tait, June 22, 1830 ("valuable present"); Moffat, "Charles Tait," *Journal of Southern History*, XIV (1948), 229–231. Although it is not apparent that any distinct African American knowledge tradition influenced the information produced by Tait's slaves, blacks did inform Europeans' knowledge of North America's deep history during the colonial era, such as when "native *Africans*" enslaved in South Carolina assured Mark Catesby that some of the mammoth fossils they unearthed were "the Grinders of an Elephant." See Catesby, *The Natural History of Carolina, Florida, and the Bahama Islands* . . . , 2 vols. (London, 1754), I, vii, emphasis in original.

sites across Europe to provide relative dates for strata; after Lyell, they began comparing the ratio of extant to extinct species in a given stratum in order to make temporal comparisons without discounting regional differences in species.[8]

North America's Tertiary formations presented an opportunity to study the beginning of life as we know it in the New World. Americans' interest in geology was peaking in the 1820s and 1830s, and the expertise needed to order strata chronologically contributed to the increased specialization of geologists, paleontologists, and conchologists. William Maclure, one of the first geological authorities in the United States, recognized only the Primary and Secondary formations in his 1817 overview of U.S. geology; analyses of the North American Tertiary as such only began in the 1820s. By 1828, Morton, who specialized in geology before becoming more exclusively devoted to craniology, had identified Tertiary deposits near the Atlantic coast from New Jersey to Virginia but could only gesture to vague reports about analogous formations in Georgia, Florida, and Alabama. He did stress, however, that Tertiary fossils were "totally different" from those of the Secondary, serving as signposts with which to gauge the relative age of the major New World strata. Morton also supposed that since about 150 new shell species had been found in one small Tertiary deposit in Maryland, an astounding number of unknown Tertiary species probably remained to be identified throughout the Deep South.[9]

Lea would not get around to inspecting the Alabaman shells that Tait had sent him until 1833, and this brief period of conchological neglect allowed the young Timothy Abbott Conrad to emerge as the nation's leading expert on the Tertiary. Although Conrad had no college education, the New Jersey native became close to the community of naturalists that cohered around Morton at the ANSP. The publication of the first two numbers of Conrad's *Fossil Shells of the Tertiary Formations of North America* in 1832 marked the beginning of research dedicated principally to the nation's Tertiary. Although the 1832 numbers dealt largely with the Atlantic states, Conrad described some specimens he received from the Gulf South,

8. See Rudwick, *Bursting the Limits of Time,* esp. 543; and Rudwick, *Worlds Before Adam,* esp. 135, 379–389.

9. Samuel George Morton, "Geological Observations on the Secondary, Tertiary, and Alluvial Formations of the Atlantic Coast of the United States of America," *Journal of the Academy of Natural Sciences,* VI, part 1 (1829), 67 (quote); William Maclure, *Observations on the Geology of the United States of America . . .* (Philadelphia, 1817); George H. Daniels, *American Science in the Age of Jackson* (New York, 1968), 30, 38; George P. Merrill, *The First One Hundred Years of American Geology* (New Haven, Conn., 1924), 93, 117.

including shells that a New York physician had collected near Claiborne and fossils sent by the Louisiana planter Henry Bry.[10]

Conrad's most important work would not appear until after his 1833–1834 expedition to Alabama, but the 1832 numbers of his *Fossil Shells* demonstrate the explanatory and nationalistic significances that Anglo-American naturalists attached to shells and Tertiary deposits. For one, the "vast abundance of fossil shells which characterize the Tertiary formations throughout the world" made them ideal specimens with which to compare geographically distant strata. Conrad was also enthusiastic about the patriotic aspects of conchology, for it could "awaken in this country [a] laudable zeal for extending the boundaries of scientific inquiry." Though Conrad believed the "beauty" of "our Tertiary fossils" would be enough to "recommend them to the notice of the mere Conchologist," he stressed that their "connexion with Geological phenomena" and their geographically vast beds made them "even more important than the most celebrated contemporaneous deposits in Europe." In his estimation, American Tertiary shell deposits were "scientific treasures." The garish shells of extant species living in foreign seas might "adorn the cabinets of the curious," but, Conrad argued, "the science of Geology has given to the more homely fossils" a more meaningful role as "mute interpreters of those strange revolutions, of which the memory of man has preserved not a solitary trace." Tertiary conchology was a key to both the future rise of American science and the past revolutions that long anteceded that which created the nation.[11]

10. Gilbert Dennison Harris, "Introduction," in Harris, ed., *Republication of Conrad's Fossil Shells of the Tertiary Formations of North America* (Washington, D.C., 1893), 5; Harry Edgar Wheeler, "Timothy Abbott Conrad, with Particular Reference to His Work in Alabama One Hundred Years Ago," *Bulletins of American Paleontology,* XXIII, no. 77 (Sept. 2, 1935), 9–10; William Healey Dall, "Determination of the Dates of Publication of Conrad's 'Fossils of the Tertiary Formation' and 'Medial Tertiary,'" *Bulletin of the Philosophical Society of Washington,* XII (1895), 216; T. A. Conrad, "Fossil Shells of the Tertiary Formations of North America," no. 2 (1832), in Harris, ed., *Republication of Conrad's Fossil Shells,* 42. For other examples of Henry Bry's geological contributions, see Richard Harlan, "Notice of Fossil Bones Found in the Tertiary Formation of the State of Louisiana," American Philosophical Society, *Transactions,* New Ser., IV, (1834), 397–408; and "Donations Received by the American Philosophical Society, Since the Publication of Vol. III–New Series," American Philosophical Society, *Transactions,* New Ser. IV, (1834), 494.

11. T. A. Conrad, "Fossil Shells of the Tertiary Formations of North America," no. 1 (1832), in Harris, ed., *Republication of Conrad's Fossil Shells,* 15–17 ("vast abundance," 15, "beauty," 16, "adorn," 17). Conrad was as yet unaware of Lyell's quantitative method of comparison; in 1832, he advocated Cuvier's method of comparing types of fossil species as well as Cuvier's catastrophist theory of geological change over time.

Conrad was not the only American to promote conchology as a patriotic science in 1832. One anonymous reviewer in the *American Journal of Science* grudgingly acknowledged that "even the conchologist is forced to summon both his philosophy and patriotism, ere he can admit the pale Purpura, the homely Venus, and the uncolored Pecten to take their respective places in the cabinet by the side of their gaudy congeners from foreign seas." If the shells of America's oceans were unimpressive, he observed, "it is far otherwise with the shelly inhabitants of our inland seas, and fresh water rivers, where the family of Naiades revel in a profusion and beauty unsurpassed in the known world." Inland shells were not just pretty: they inspired the ruder sorts of people settling the West to develop scientific interests and contribute to national knowledge and collections. "These shells," according to the reviewer, "attract the curiosity of the uneducated" in the western states, and, "what is more important, they have in numerous instances been the occasion of awakening a taste for conchology, and have become the basis of scientific collections in natural history." Although Tait was well educated, he was among the "friends, resident at the west," who supplied Philadelphian experts like Lea and Morton with their specimens. As this reviewer would have hoped, Tait's residence near remarkable inland shells did lead him to devote his retirement to science.[12]

The specimens, observations, and patronage that Tait offered from his Alabama plantation motivated Conrad's expedition to the Gulf South. Along with the 250 or so specimens that his slaves had collected, Tait had sent Lea a description of the mineral deposit that "passes through the whole extent of South Alabama, called the Shell Limestone Country." Tait, who had recently been named a corresponding member of the ANSP, suggested this was part of a single massive formation that stretched from western Louisiana "through South Carolina, Georgia and Florida, parallel to the sea coasts." More importantly, he encouraged Lea and Morton to send "an agent of the Academy . . . to this region for the purposes of Geological investigation," adding that it would be "a duty, as well as a pleasure" to help fund the expedition and give the agent "a home in my house." Morton was eager to comply since, as he told Tait, he expected that a fuller understanding of the Claiborne deposits would prove essential to "developing the *Geology* of this country." Lea, however, warned Tait to be "exceedingly careful to whom [you] extend your very liberal offer, for there are often pretenders

12. "Conchology—Mr. Lea on the Naiades, in the Transactions of the American Philosophical Society," *American Journal of Science and Arts,* XXII, no. 1 (1832), 169–170 ("even the conchologist," 169, "friends," 170).

and quacks in natural sciences as well as in the med. science." But Morton put such doubts to rest by assuring Tait that "T. A. Conrad, a member of the Academy, a zealous cultivator of Science and a gentleman of the most estimable Character," had agreed to undertake the expedition. Unlike earlier inhabitants of the Gulf South (like William Dunbar and Barthélémy Lafon) who had bent over backwards to convince eastern patrons of their trustworthiness, the leaders of Philadelphian scientific institutions now aimed to convince an Alabama planter that their representative was credible enough to merit support.[13]

In December 1832, Conrad became the latest in a long line of Philadelphian naturalists, including William Bartram and William Maclure, to embark on an expedition to the Gulf South. The first few months of Conrad's journey took him through the Carolinas and Georgia, but he would spend most of the year between February 1833 and February 1834 examining the bluff near Claiborne, Alabama, and enjoying the hospitality of Charles and Sarah Tait. Soon after arriving in Alabama, Conrad found he had "stumbled on the land of promise, a perfect El dorado of fossils." Not only were these fossil deposits rich and spread out over a "vast tract" of the state, they were also older, more unique, and more significant than Conrad had imagined. After a month residing with Tait, Conrad realized that the shell species he had collected from a stratum halfway down the bluff were not, as he had suspected, from the Middle Tertiary, but Eocene deposits that, as the oldest of the Tertiary strata, hearkened back to the dawn of the age of mammals. The Eocene stratum exposed on the bluff was directly above a recognizably Secondary formation, and the strata above the Eocene represented nearly all of the more recent Tertiary eras. The Claiborne bluff was nothing less than a timeline of North American geohistory, and Conrad, who had a pronounced romantic streak, eulogized the site in the lyrics of a song, "Claiborne," which he shared with Morton in April 1833:

Thou great magician Science, who cannot tell
The love of other worlds, thy wand advance
And lo! in earth what mute historians dwell,
That give to Truth the halo of Romance.

13. [Charles] Tait to Isaac Lea, July 20, 1831, in Lea, *Contributions to Geology* (Philadelphia, 1833), 22 ("passes"); Tait to Samuel George Morton, Oct. 16, 1832, ANSP Official Correspondence, ANSP ("agent"); Morton to Tait, Apr. 16, 1832, Tait Papers ("developing"), emphasis in original; Lea to Tait, Dec. 11, 1831, ANSP Official Correspondence ("exceedingly"); Morton to Tait, Dec. 5, 1832, Tait Papers ("T.A. Conrad"); Moffat, "Life of Charles Tait," 215.

Cuvier, like Prospero, hath peopled earth
With forms restored from ages which have been
A populous world, ere human joys had birth,
Ere lovely woman came to light the scene!

Not only would these fossils redefine America's ancient past, they would bolster national pride. As Conrad told Morton, "I shall collect such a mass of facts as render your Synopsis [Morton's geological text-in-progress] a match for any in Europe, in copiousness of detail and geological interest." The research Tait was funding was not just southern; it promised specimens and observations that would define North America for national and international audiences.[14]

Like other geologists of the period, Conrad studied fossils as part of a wider interest in natural history. He wrote at length about Alabama's martens, frog songs, and a "Talleyrand lizard"—so named for its changeability—that "possesses the most impertinent curiosity of any animal I know, womankind excepted." Along with shells, Conrad was keen to provide Morton and the ANSP with larger remains. He sent Morton some "saurien *[sic]*" bones recently unearthed at a nearby plantation; this deposit of huge Secondary fossils would, Conrad hoped, "constitute a new *Golgotha,* for geological devotees." Conrad knew that Morton, a specialist in the Secondary, would value these giant fossils because they could demonstrate that the Secondary deposits of Alabama were "synonymous" with those of western Louisiana, from which a local planter had previously supplied Morton with dinosaur bones. Conrad also tried, and failed, to procure Morton some Choctaw skulls.[15]

14. T. A. Conrad to S. G. Morton, Apr. 20, 1833, Morton Papers, Ser. 1, APS ("stumbled"), Conrad to Morton, Apr. 3, 1833 ("I shall collect"). On the strata at Claiborne Bluff, see Conrad to Morton, Sept. 25, 1833, ibid.; and Conrad, "Observations on the Tertiary and More Recent Formations of a Portion of the Southern States," *Journal of the Academy of Natural Sciences,* VII, part 1 (1834), 116. For an overview of Conrad's travels across the South and within Alabama, see Wheeler, "Timothy Abbott Conrad," *Bulletins of American Paleontology,* XXIII, no. 77 (Sept. 2, 1935), 24–56. In a prose description of the same sentiments he expressed in the song "Claiborne," Conrad observed that "Claiborne is a most interesting spot to the geologist." "No where," he pronounced, "will he find in greater profusion those mute historians of the ancient revolutions of our globe, who seem as if by the wand of a magician to come forth from the eternity of the past." See Conrad, "Claiborne, Alabama: From the Note Book of a Traveller," *Advocate of Science, and Annals of Natural History,* I (1834), 29.

15. Conrad to Morton, Apr. 20, 1833, Morton Papers ("Talleyrand"), Conrad to Morton, Dec. 6, 1833 ("saurien"), Conrad to Morton, Dec. 28, 1833.

Conrad's encounters with the nature and peoples of the Gulf South left him ambivalent about southerners and racial slavery. After a month in Claiborne, he described the "delightful Climate . . . of the South, but what a barbarous race of people, and yet better than I expected." "In illustrating the beauties of Slavery," he continued sarcastically, "I will observe that within 14 months, four women have been murdered by their slaves in the vicinity and a lady at present is ill in Claiborne, having been poisoned by one of her servants." The many ambiguities in this description are telling: it is unclear whether he means whites, blacks, or all southerners as the "barbarous race," and, though his observations about the murderous "beauties of slavery" were probably tongue in cheek, he certainly had a low opinion of the "little black devils" he met throughout his travels.[16]

Conrad approached Alabama as a foreign land, but he relished the chance to live like a cotton planter. When Charles and Sarah were away visiting their son's plantation, Conrad remained at their home with "an uncommonly interesting girl, not 12 years of age." As he told Morton, "We are master and mistress pro temps; I doubt whether any lady could do the honours of the table more gracefully than this little niece of Mrs. Tait's." Conrad also began to eat and drink like Tait, dining "every day on Butter (but have as ever a religious abhorrence of quadrupeds) and the Judge and I drink to each others' healths every day in a bumper of the best Madeira." He confessed to Morton that "I had seriously debated my inclinations whether I should ever return to Philadelphia." Far from inspiring animosity or pro-northern sentiments, Conrad's encounters with southern planters made him want to be one.[17]

Neither the lifestyle nor discoveries that Conrad enjoyed in Alabama would have been possible without the wealth, laborers, and networks that whites' power over enslaved blacks put at the command of cotton planters and cotton traders. Conrad had intended to fund much of his expedition with subscriptions from his friends at the ANSP that he would repay with specimens and by selling sets of Claiborne fossils. However, the moneys at his disposal proved inadequate to cover his needs in Alabama, so Conrad decided to "spunge upon the planters" for cash, food, and board. And rich southerners gave generously. Tait palmed Conrad money with no expectation of reimbursement as "a contribution to science," and the Mobile cotton

16. Conrad to Morton, Apr. 3, 1833, ibid. ("delightful"), Conrad to Morton, Jan. 30, 1833 ("little").

17. Conrad to Morton, Aug. 25, 1833, ibid. ("uncommonly"), Conrad to Morton, Apr. 3, 1833 ("every day").

merchant J. B. Toulmin gave Conrad "25 dollars, towards collecting fossils for the Academy." Toulmin also offered "to make his house my home whenever I visit Mobile," largesse that Conrad interpreted as Toulmin's "uncommon view of the concessions due to Science." Despite Conrad's diet of "oranges and champagne," the seeming eagerness of the planter class to feed, house, and fund him made his expenses "trifling." Moreover, Tait's patronage allowed Conrad to remain in Alabama far longer than he had expected. "[Tait] is convinced that no part of the United States offers so interesting a field for geological discovery as South Alabama," he reported, "and that to *do it justice* I must remain at least one year." Like other researchers, Conrad needed money to live and time to work, and cotton planters and merchants provided him with both.[18]

Conrad's new planter friends made it possible for his geological investigations to span a large swath of the state by providing him with free transportation. "Judge Tait and Mr. Toulmin," he told Morton, "have procured me the freedom of nearly all the steam boats on the Alabama and Tombechbee *[sic]* rivers, so that I can now stop anywhere between Mobile and Tuscaloosa, and Mobile and Montgomery, a great part of which extensive country is covered with ... fossils." In the acknowledgements of an 1834 publication, in which he recognized Tait, Toulmin, and other Alabamans whose wealth stemmed from cotton, Conrad thanked a steamship owner for "offering me, as a missionary in the cause of science, a free passage in his steamboats on the waters of Alabama, a privilege which has never been extended for a similar purpose to an individual in any other state in the Union." Conrad also needed to roam the countryside, and he relied on planters like Dr. Robert Walker Withers for horses "to traverse the prairie countries which are fairly covered with organic remains." Amazed that Alabama's elites gave "so much for science," Conrad dreamed about "its [science's] future honours when its importance is *duly* appreciated by the mass." In sum, Conrad was surprised that "in this new country Science meets with more indulgence by far than I anticipated," and, perhaps, he saw this interest in the sciences as a nationally progressive byproduct of westerners' fascination with America's shells.[19]

18. Conrad to Morton, Apr. 20, 1833, ibid. ("spunge"), Conrad to Morton, Sept. 11, 1833 ("contribution"), Conrad to Morton, June 1833 ("25 dollars"); Conrad to C. Tait, May 18, 1833, Tait Papers ("to make"); Conrad to Morton, May 8, 1833, Morton Papers ("oranges"), Conrad to Morton, Apr. 3, 1833 ("[Tait] is convinced"), emphasis in original. On how Conrad planned to sell or trade Claiborne fossils, see Wheeler, "Timothy Abbott Conrad," *Bulletins of American Paleontology*, XXIII, no. 77 (Sept. 2, 1935), 24; and Conrad to Morton, Apr. 3, 1833, Morton Papers.

19. Conrad to Morton, May 8, 1833, Morton Papers ("Judge Tait"); T. A. Conrad, *New*

Conrad's geological efforts in Alabama relied on slave labor, but, like many other white naturalists, Conrad effaced blacks' assistance from most of his writings. The only instance in which he mentioned benefiting from local blacks was a September 1833 letter that noted how, "In company with a servant of Judge Tait's, I paddled a canoe 50 miles down the Alabama River" in search of shell deposits. This "servant" might have been the same individual to whom Lea was indebted for his boxes of Claiborne fossils, but Conrad made no further mention of him. More ambiguously, Conrad described how some fossils "were collected under my supervision." Instead of praising the contributions of these (probably black) excavators in his publications, Conrad thanked Tait and other whites.[20]

Planters' networks were also essential to the success of Conrad's expedition. For one, Tait's personal connections acquainted Conrad with military officers, physicians, and clergy living throughout the state who, in turn, directed Conrad to new fossil deposits. More importantly, the success of plantation slavery itself relied on the flow of information and commodities across global markets, and these networks enabled the circulation of geological data from Alabama. Conrad and Tait sent boxes of specimens to Morton on the same ships that carried cotton and plantation goods between Mobile, Philadelphia, and New York. The merchant Toulmin even drew on his prosperity and transatlantic connections to offer Conrad "free passage in a fine ship to France" so that he might sell Claiborne fossils in Europe and "get French fossils in exchange." Naturalists in Europe were well aware that slavery made Gulf South fossil specimens available and that the cotton trade circulated them. British geologist John Finch, an expert on the Tertiary, wrote to Tait in 1834 that "I have seen Mr. Lea's description of the shells which you sent him from Alabama, and I wish to possess the *shells themselves*." "As you are on the spot," Finch continued, "and have laborers to command, I presume you could easily have a small barrel collected." Moreover, since "there are always ships for Liverpool, and London" sailing from

---

Fresh Water Shells of the United States ... (Philadelphia, 1834), 23 ("offering"); Conrad to Morton, May 8, 1833, Morton Papers ("traverse"), emphasis in original.

20. Conrad to Morton, Sept. 11, 1833, Morton Papers (quotes). For men of science effacing laborers and informants, see Neil Safier, *Measuring the New World: Enlightenment Science and South America* (Chicago, 2008). On how enslaved naturalists were also essential to Benjamin Wailes's state geological survey of Mississippi in the early 1850s, see Christopher Morris, *The Big Muddy: An Environmental History of the Mississippi and Its Peoples from Hernando de Soto to Hurricane Katrina* (New York, 2012), 206–208. For Conrad crediting his white supporters, see Conrad, *New Fresh Water Shells of the United States*, 22–23.

New Orleans and Mobile, Finch figured "it would not be much trouble" for Tait to forward specimens that "would be extremely useful" for the geological lectures Finch gave in London.[21]

Conrad's and Lea's work on the Tertiary fossils from Claiborne reveals that Gulf South planters and Philadelphian naturalists experienced something of a role reversal, or at least balancing out, of the patron / client and center / periphery relationships that metropolitan men of science promoted as key to disciplined knowledge production. Historians have argued that the geological surveys sponsored by state governments in the 1820s and 1830s were key to increasing the authority of naturalists in the United States. Yet Philadelphian experts could also derive authority from earning the patronage of a Gulf South planter. The value of Tait's support was apparent during the so-called Conrad-Lea controversy, when both geologists deferentially played up their reliance on him. The controversy began in 1833 when Lea, after learning that Conrad had identified the Claiborne shells as new Tertiary specimens, rushed to classify the Claiborne shells himself in his *Contributions to Geology*. Conrad was outraged: he cried to Morton that "Mr. Lea talks about his labours for 4 years past in collecting the fossils and geological information in relation to them from Claiborne—Good God! *his labours indeed!!!*" Morton sympathized and helped him fight back, dashing off two new numbers of Conrad's *Tertiary Fossils* in which Morton offered his own names for Claiborne species. The two coexisting nomenclatures that emerged from these 1833 publications remained a source of confusion for conchologists for decades. Throughout this controversy, one way that Lea and Conrad competed for Tait's approval was by naming new species after him. Lea identified one shell as "Taitianus" while Conrad presented *Cassis Taitti:* "I dedicate this beautiful species to my kind friend Judge Tait, of Claiborne, whose love of science first brought into notice the rich deposits of fossils near the town in which he resides." Naming specimens to honor benefactors—and, hopefully, secure their ongoing support—was an old practice among naturalists, and both Conrad and Lea recognized that an Alabama planter, and not just officials and natural philosophers in eastern cities, could be pivotal to patronage networks that reached eastward across the United States.[22]

21. Conrad to Morton, Apr. 20, 1833, Morton Papers, Conrad to Morton, Apr. 3, 1833, Conrad to Morton, May 8, 1833 ("free passage"); John Finch to Charles Tait, May 14, 1834, Tait Papers ("I have seen"), emphasis in original; Moffat, "Life of Charles Tait," 218. Conrad did not go on this trip to Europe. On Finch's work on the Tertiary in New Jersey during the 1820s, see Merrill, *First One Hundred Years of American Geology,* 93.

22. Conrad to Morton, Sept. 25, 1833, Morton Papers ("Mr. Lea"), emphasis in origi-

Conrad and Lea addressed Tait with the kind of deference that learned Philadelphians like themselves usually expected to receive from curious men in the West. Conrad flattered Tait by stressing the Alabaman's identity as a legitimate geologist: "I fear some new trouble on the plantation may have served to disquiet you," he wrote Tait in 1834, "and to divert your mind from the pursuit of the Geologist, and his pleasing dreams of remote antiquity; Daily do I recall your kind and benevolent manners, your love of knowledge, and your indefatigable industry in pursuit of it." Tait was no lazy planter but an industrious naturalist whose contributions were known "to the Geological Society in London" and who had made Claiborne "a classic spot to the Geologists." Lea, too, made every effort to spell out his deference in two separate dedications to Tait in his *Contributions.* He emphasized that his own work on "the Tertiary Formation of Alabama [was] owing to [Tait's] kindness" and that Tait had "a strong claim upon me." It was not just Lea who was in Tait's debt: "To him science owes the great obligation of having first brought [these fossils] to light." Conrad's publications outdid Lea's, however, by emphasizing his personal bond with the Taits, "whom I must always remember with feelings akin to filial attachment." Being a poet helped as well, and Lea offered no dedication quite as attuned to the sublime as Conrad's wish that "May the evening of [Tait's] life be calm and serene, and, as the meridian was passed in exertions honourable to himself and useful to his country, my friend needs no eulogium from a humble votary of Natural Science." Tait, when portrayed by Conrad's pen, was the ideal patriot-naturalist.[23]

Whereas Lea, a rich man living in Philadelphia, could bring his own resources and those of a major city to bear in the controversy, Conrad—who was still in Alabama when the dispute began—struggled to make himself relevant to America's newfound interest in Alabaman fossils. Still, being in Alabama did present Conrad with one key advantage: he could solicit the

---

nal; Moffat, "Life of Charles Tait," 212 ("Taitianus"); Conrad, "Observations on the Tertiary and More Recent Formations of a Portion of the Southern States," *Journal of the Academy of Natural Sciences,* VII, part 1 (1834), 146 ("I dedicate"). For the kinds of scientific networks that naturalists in eastern cities envisioned, see Lewis, *Democracy of Facts,* 129–153. Tait offered to finance further expeditions from the ANSP to Alabama and hoped that Morton could recommend an intrepid botanist to study the "rare or new [plant] species" near Claiborne. See Conrad to Morton, Sept. 11, 1833, Morton Papers.

23. Conrad to Tait, Nov. 14, 1834, Tait Papers ("I fear"); Lea, *Contributions to Geology,* iii ("Tertiary Formation"), 27–28 ("To him," 28); Conrad, *New Fresh Water Shells of the United States,* 23 ("whom I must").

support of Charles and, perhaps more importantly, Sarah from the comfort of their own living room. The Taits retained significant control over the fossils that enslaved blacks and Anglo geologists unearthed near Claiborne, and Conrad convinced Sarah that he, and not Lea, should receive these specimens. Soon after learning about "the nefarious conduct of Mr. Lea" in a letter from Morton, Conrad replied that "I read your letter to Mrs. T and it did my heart good to hear her express so much admiration at your conduct, whilst at the same time, she lost all interest in Mr. Lea, and insisted that I should select from among the shells *she had ordered to be collected for him,* all such as I had any wish to have, and I did as; therefore *our friend* will in future get no more unios [river mussels] from here." This letter not only reveals that Sarah controlled access to Claiborne fossils but also that she had ordered their collection in the first place. Charles Tait has received the historical credit for making Claiborne's fossils known to science, but Sarah and the slaves she dominated were, at the very least, also deeply involved in this process. Conrad was well aware of the value of Sarah's support, and he was just as florid in expressing deference to her as to her husband. After his return to Philadelphia, Conrad wrote that "I need not assure Mrs. Tait of my gratitude to her; her kindness nothing but Death can obliterate from my daily remembrance."[24]

The information and specimens that the Taits and their slaves made available was integral to how Lea and Conrad elaborated the geohistory of North America. Since Lea never visited Claiborne, his efforts to situate the Gulf South's Tertiary relied on Tait's depictions of the visible strata of the Claiborne bluff and on the fossil shells that Tait and his wife had ordered their slaves to collect. In his *Contributions,* Lea wrote that "we look to the Tertiary Formation with peculiar interest, as its lower deposit or period contains the incipient state, or as Mr Lyell says, 'the *dawn* of the existing state of the animate creation.'" Lea followed Lyell's method of ordering strata chronologically based on the percentage of extant species among their fossils; consequently, even though he was "not perfectly satisfied that a single species is strictly analogous to those from the Eocene period of Europe," he "had no hesitation in referring [the Claiborne shells] to the same period." The quantitative analysis of fossils needed for Lyell's system

24. Conrad to Morton, Sept. 11, 1833, Morton Papers ("nefarious," "I read"), emphasis in original; Conrad to Tait, Apr. 14, 1834, in Wheeler, "Timothy Abbott Conrad," *Bulletins of American Paleontology,* XXIII, no. 77 (Sept. 2, 1935), 117 ("I need not"). For other examples of Conrad expressing his gratitude to Sarah Tait, see Conrad's letters to Tait from June 7, 1833, Dec. 16, 1833, Jan. 14, 1834, all ibid., 117.

of ordering required a large number of specimens, the kind of mass collection that slave labor facilitated enormously.[25]

Although Lea was relatively modest in his interpretations—he mostly settled for applying Lyell's *Principles* to the Gulf South—Conrad was more eager to speculate about the causes that made the region an almost geologically unique part of North America. This could be a highly imaginative exercise, for, as Conrad wrote to Morton, "Nature will not conform to our theories [so] we must contrive one to suit her sportive and eccentric ladyship." In a series of published essays, he noted that "I am acquainted with no theory which satisfactorily accounts for the manner in which strata of organic remains like those of Claiborne have been deposited; but it seems evident that they have been forced up from the bottom of the sea." Following the tenets of catastrophism, Conrad initially forwarded the explanation that "this line of coast was upheaved by volcanic agency." Sometime in 1834, however, Conrad changed his tune. As he told Tait, "Geology has been completely revolutionized of late, by Lyell, and all our poetic dreams of catastrophes, and violent revolutions, so far at least, as relates to newer secondary and tertiary formations, are like the passing visions of Slumber." "Lyell," he stated, reversing course, "proves that all these changes were referable to causes similar to those now in operation." Conrad then adopted Lyell's mode of determining the relative age of strata, Lyell's nomenclature for the Eocene, and Lyell's uniformitarian theory that actual causes like the slow deposit of mud from the Mississippi River accounted for the region's stratigraphy.[26]

Conrad drew on onsite observations, planters' knowledge, and his perceptions of the region's history to construct the Gulf South as a geologi-

25. Lea, *Contributions to Geology,* 14 ("we look"), 18 ("had no hesitation"), 19 ("not perfectly"), 27. The rarity of Eocene deposits increased their value all the more since, as in Europe, America's Tertiary groups "have detached and isolated positions, while the Secondary period extends over great areas." See ibid., 13.

26. Conrad to Morton, Dec. 6, 1833, Morton Papers ("Nature"); Conrad, "Claiborne, Alabama: From the Note Book of a Traveller," *Advocate of Science, and Annals of Natural History,* I (1834), 29–30 ("I am acquainted," 29); Conrad, "Mobile, Alabama: From the Note Book of a Traveller," *Advocate of Science, and Annals of Natural History,* I (1834), 59 ("this line"); Conrad to Tait, [n.d.] 1834, in Wheeler, "Timothy Abbott Conrad," *Bulletins of American Paleontology,* XXIII, no. 77 (Sept. 2, 1935), 84 ("Geology"); Conrad, "Sketches from the Note Book of a Traveller," *Advocate of Science, and Annals of Natural History,* I (1834), 160; Conrad, "Observations on the Tertiary and More Recent Formations of a Portion of the Southern States," *Journal of the Academy of Natural Sciences,* VII, part 1 (1834), 116.

cally coherent part of the planet. His fossil hunting revealed that, unlike in Europe, there was greater difference between the shell species of the Eocene and Pliocene (the youngest strata of the Tertiary) than between the Eocene and the Secondary, an observation that suggested a different periodization for mass extinctions in the New and Old Worlds. He asserted that the fossils he found in Claiborne "prove doubtless that our *Eocene* [in America] is older than that of Europe," a patriotic claim that challenged the Neptunian perspective, made famous by Georges-Louis Leclerc, comte de Buffon, that the New World was younger than the Old. Conrad also relied on data derived from the daily workings of plantations across the region to argue that much of the land from Georgia to Louisiana was one extended Tertiary bed. He learned about southern Georgia's Eocene oyster shells because they were "extensively quarried by the Indigo planters, who convert them into lime." Conrad compared late Secondary and early Tertiary limestone deposits in Florida with those of Alabama based on the descriptions planters sent him of local building materials. One H. B. Croom informed Conrad about two different kinds of limestone near his Tallahassee plantation, a hard siliceous variety used to make millstones and "a rock so soft as to be easily *sawed* or *cut* into blocks for building houses and chimneys, to which purpose it is in some places applied." Conrad argued that, "with regard to the siliceous rock which Mr. Croom alludes to, I cannot doubt its identity with the Eocene of Alabama, where it is likewise used for millstones." Without visiting Tallahassee, Conrad used the practices of the area's planters to reconstruct its physical past. Thanks largely to Conrad's inquiries and interpretations, American naturalists were coming to see the Deep South as a geologically, and not just climatically and politically, distinct section of North America.[27]

For Conrad, the Gulf South's geohistory was fully interwoven with its environmental and human histories. In "Mobile, Alabama," an 1834 essay, Conrad imagined the deep history of the city from the most ancient past to the present. His narrative of Mobile was steeped in a vision of American—not biblical—history, specifically a whiggish version of Anglo-America's ostensible destiny to expand across and improve a continent that was, according to his own account, itself steadily expanding. Mobile was once beneath

27. Conrad, "Observations on the Tertiary and More Recent Formations of a Portion of the Southern States," *Journal of the Academy of Natural Sciences,* VII, part 1 (1834), 123 ("prove"); Conrad, "Fossil Shells of the Tertiary Formations of North America," no. 3 (1835), in Harris, ed., *Republication of Conrad's Fossil Shells,* 79 ("extensively"), 80–84; Conrad, "Notices of the Geology of West Florida," *Advocate of Science, and Annals of Natural History,* I (1835), 351–352 ("rock," 351, "with regard," 352).

"the waves of a fathomless ocean," but, as time marched on and the seafloor rose, "succeeding ages witnessed, in place of the dark waste of interminable waters, the calm surface of a shallow bay." The Gulf Coast became terra firma "after the elevation of the land had forced the waters to retire, and the flower, the bird, and the insect, soon added beauty and animation to the scene." "Then," as millennia passed, "came the giant inhabitants of woods and savannahs, the Mastodon, Megalonyx, Megatherium ... whose giant forms ... furnish the mind with ... many a strange vision of the mysterious and mutable past." After another "change came over the earth," the next being to enter the scene was "the savage, in all the wildest features of uncultivated nature." But, even as the environment around Indians continued to change, they remained untouched by the passage of time. "Over the savage supremacy the arts of civilization exalted the European," Conrad boasted, "and the Indian was subdued; he became more degraded, and although he concealed the stubborn energy of his unconquerable passions, his nature remained the same." Since Anglo-Americans, in step with the progressive history of the Mobilian earth itself, had come to dominate the region, the unchanging Indian had no other option than to vanish. "And now," Conrad concluded, "a city is built which is rapidly rising in wealth and population, and the products of a vast extent of country are by the agency of steam propelled along the bosoms of noble rivers; the native is gradually retiring from a scene so uncongenial to his spirit, and ere long he will disappear altogether from this region, like the bison, and at last even from the cognizance of civilized man." The disappearance of the Indians and ascendance of a higher stratum of civilization was, for Conrad, as inevitable as the extinction of Eocene mussels.[28]

The support that Conrad received from planters and cotton merchants reveals that learned men and officials in eastern cities were not always at the top of the United States' patronage networks. Plantation masters and mistresses in the Deep South funded scientific work, facilitated access to specimens, and raised the status of northeastern Anglo naturalists. Historians have become increasingly aware of just how influential plantation slavery was in the United States' economy and politics. The example of the

28. Conrad, "Mobile, Alabama: From the Note Book of a Traveler," *Advocate of Science, and Annals of Natural History*, I (1834), 59–60 ("waves," 59, "Then came," 59–60, "change," 60). For how Conrad claimed that the Gulf coast was steadily expanding into the Gulf of Mexico as the continent's rivers deposited sediments that slowly lifted the seabed, see Conrad, *Fossils of the Tertiary Formations of the United States* (Philadelphia, 1838), xii.

Taits—a family of planters who enabled geologists to rewrite the history of the continent itself—suggests that slavery had a similarly vital role in U.S. science.[29]

RUSH NUTT'S THEORY OF THE EARTH

Dr. Rush Nutt (1781–1837) used geotheory to praise slavery and heap scorn on religion, particularly that of northern Presbyterians who, he claimed, schemed to destroy slavery, science, and democracy in the United States. Several southern planters were interested in geology because it yielded data about soils and mineral fertilizers that promised to make agriculture more profitable. Nutt went much farther: he expounded a geology-based natural philosophy to argue that plantation slavery was an enlightened and patriotic institution, and he proposed geo-engineering projects that would help expand plantation agriculture across the continent. Historians of the earth sciences in the early United States have focused on the writings of a handful of northern Anglo-Americans to argue that natural knowledge and religion not only coexisted peacefully in the antebellum United States but that the harmony between them was crucial to the development of American science on the whole. According to Nutt, however, the way that northern Christians studied nature was neither American nor scientific. Plantation slavery, he insisted, was at the center of American, and not just southern, modernity and science.[30]

Like Charles Tait, Rush Nutt was among the many Virginians who migrated to the Gulf South and established cotton plantations. He had received his M.D. at the University of Pennsylvania working under Benjamin Rush, in whose honor he shortened his given name of Rushworth. Nutt moved to southwestern Mississippi in 1805 and briefly ran a medical practice before marrying Eliza Ker, a scion of one of the territory's leading fami-

29. For the importance of slavery to economics and politics, see Walter Johnson, *River of Dark Dreams: Slavery and Empire in the Cotton Kingdom* (Cambridge, Mass., 2013); and Matthew Karp, *This Vast Southern Empire: Slaveholders at the Helm of American Foreign Policy* (Cambridge, Mass., 2016).

30. On applied geology in the South, see William K. Scarborough, "Science on the Plantation," in Numbers and Savitt, eds., *Science and Medicine in the Old South,* 79–102; and Richard C. Sheridan, "Mineral Fertilizers in Southern Agriculture," in Corgan, ed., *Geological Sciences in the Antebellum South,* 73–82. For studies of science and religion in the early United States (studies that disproportionately conflate the experience of northeastern Anglos with Americans on the whole), see Daniels, *Science in American Society,* 206–222; Herbert Hovenkamp, *Science and Religion in America, 1800–1860* (Philadelphia, 1978), 119–146; and Lewis, *Democracy of Facts,* 107–128.

lies, and establishing a plantation near Rodney, Mississippi. Over the next three decades, Nutt served as a militia surgeon and justice of the peace and, in 1830, was among the cofounders of Oakland College in Lorman, Mississippi. In 1833, Rush and his younger son, Rittenhouse (named after the astronomer), set off on a grand tour of Europe, Egypt, and the Holy Land "on purposes connected with science." Rush's older son, Haller (named after physician Albrecht von Haller), has become the most well known member of the Nutt clan for undertaking the construction of Longwood plantation, an architecturally remarkable mansion and tourist destination in Natchez. Rush Nutt was also involved with the United States' internal slave trade and, in 1817, travelled back to Virginia to purchase slaves for himself, his wife, and his fellow Mississippi planters. Nutt's role as a slave trader, like that of Barthélémy Lafon around this same time, serves as a reminder that some of the most learned white men in the lower Mississippi Valley not only owned slaves but dealt them for profit.[31]

As one of the earlier Virginian migrants to the Gulf South, Nutt moved to the region when it was still a contested borderland. He conducted business deals and scientific observations in Spanish Florida and wrote a detailed ethnography of the Chickasaws (part of a longer but mostly lost travel diary from 1805) that prefigured many of the theories of human nature that would prove central to his natural philosophy during the 1830s. For one, he belittled the Chickasaw's religious practices as "superstitions" in much the same way that he would harangue the rituals of Christian priestcraft. He was glad that the Chickasaws accepted divorce and polygamy, both of which he later endorsed, and noted how racial mixing—in this case among Indians and whites—created biologically improved offspring, a theme he reiterated in his later writings on race. Moreover, Nutt was among the first Anglos to disinter human remains in search of evidence about the Mississippi Valley's native past. After taking up "at least 35 or more human sculls and a vast number of bones" in Tennessee, Nutt concluded that "the [former] inhabitants were in the habit of interring the dead in the same manner as the Chickasaw do at present, which is in their houses under the Cabins." Nutt's perspectives on the region's nature and peoples emerged amid encounters

31. On the Nutts' grand tour, see George Robbins to Allen Robbins, May 7, 1833, Nutt Papers, box 3, nu 262 (quote), and Rittenhouse Nutt to Haller Nutt, Aug. 30, 1834, box 3, nu 256. On Rush Nutt's slave trading, see Rush Nutt to Eliza Nutt, Nov. 7, 1817, ibid., box 1, nu 261, and Rush Nutt to Eliza Nutt, Nov. 28, 1817, box 1, nu 261. By the 1820s, Nutt, like most other planters in the lower Mississippi Valley, was purchasing Virginian slaves in New Orleans. See John Ker to Rush Nutt, Jan. 17, 1824, ibid., box 2, nu 172.

with Spanish officials, living and dead natives, and, of course, the blacks he bought, sold, and forced to labor on his plantation.[32]

Nutt achieved lasting renown as one of the Gulf South's most noteworthy experimental planters, but his writings on natural philosophy have been forgotten. He designed an improved gin that ran on steam power and, more importantly, played a critical role in developing the hybridized Petit Gulf strain of cotton, a staple that made Mississippi's planters some of the most wealthy agriculturalists in human history. Nutt's role in generating this staple was certainly important, but scholars have neither explored his views on history, geology, religion, and slavery nor incorporated them into larger arguments about America's intellectual milieu. This neglect has stemmed most obviously from the fact that the majority of Nutt's writings were never published. However, southerners active in the decades after Nutt's death also worked assiduously to distort his legacy and, perhaps, those of other southern skeptics. Nutt's arguments about race and religion embarrassed southern thinkers who, during the 1840s and 1850s, made Christianity, romanticized traditionalism, and whites' innate racial superiority essential to their proslavery arguments. The only aspect of Nutt's science acceptable to this generation of southerners was his agronomy, so they ignored or effaced his less orthodox ideas. Nutt's own descendants took the lead in this erasure. In 1859, Rittenhouse Nutt sent a reminiscence of his father to historian J. F. H. Claiborne that stressed Rush's medical expertise and role in conquering the Gulf South by directing "his zeal and firmness in assisting to . . . curb the unruly spirit so prevalent in border life." But Rittenhouse focused primarily on listing the agricultural innovations that made Rush "distinguished as a cotton planter" and disproving slanderous claims that his "father was an unbellever in the entire Divine Revelation." Claiborne's 1880 *Mississippi as Province, Territory, and State* republished much of Rittenhouse's letter and elaborated how "the Petit Gulf seed were introduced, it was commonly said from *Mexico,* by Dr. Rush Nutt, a distinguished planter and scientist." By conflating Nutt's wide-ranging studies with agronomy and Christianity, these southerners seem to have deliberately misremembered Nutt's intellectual life.[33]

32. Rush Nutt, "A View of the Chickasaw Nation," in "Diary of a Tour through the Western and Southern Parts of the United States," (1805), 12 ("superstitions"), 15, 29 ("at least"), MDAH; Enrique De Grand-Pre Letter, May 4, 1821, Kislak Collection, MS 218, LOC; James Cowden to Rush Nutt, May 19, 1821, Nutt Papers, box 1, nu 70, Rush Nutt, "Geological Remarks upon Caves and Sink-holes," box 8, nu 363.

33. Rittenhouse Nutt to J. F. H. Claiborne, May 20, 1859, in *Semi-Weekly Mississippian* (Jackson, Miss.), May 27, 1859 ("his zeal"), Rittenhouse Nutt, "The Late Dr.

Nutt and his family had wide-ranging interests and were well integrated into regional and national knowledge networks. Nearby planters looked to Nutt to vet their designs for new cotton technologies, and a Massachusetts gentleman sought the "instruction and patronage" that Nutt could give his son, who had moved to Mississippi and hoped to become a physician. Nutt's wives sent and received specimens independently of their husband. One male correspondent told Eliza, Rush's second wife, that "I have sent you such slips and Roots as I thought it Possible you might wish to have" and acknowledged that the "Fig slips or roots you sent me I prize very highly." Eliza also took charge of administering physic on the plantation, including dosing her children with the dangerous vermifuge pinkroot, while the doctor was off leading slave coffles back from Virginia. After Eliza's death, Rush and his third wife exchanged specimens with French geologist Joseph Nicollet, who had befriended the Nutts during his expedition to Mississippi and Arkansas. Nicollet sent Rush some minerals from "Sioux Country" and enclosed some *"curiosités sauvages* pour Madame Nutt" in the parcel. The Nutts were a minor node in the webs of information, patronage, and specimens that reached into both the Northeast and the West.[34]

Rush Nutt was more than a scientific planter, collector, or ethnographic observer. He aspired to be a natural philosopher and, starting in the early 1830s, began drafting his theory of the earth, a text in which he formulated a grand vision of natural and human history. Nutt began this project around the same time that Charles Lyell published his *Principles,* a work that explicitly sought to develop a new geotheory that explained the earth's history

Rush Nutt," ("father"); J[ohn] F. H. Claiborne, *Mississippi, as a Province, Territory, and State ...,* I (Jackson, Miss., 1880), 141 ("Petit Gulf"). On Petit Gulf cotton, see Johnson, *River of Dark Dreams,* 151. For later works that reinforced the notion that Nutt's research was limited to agronomy, see John Hebron Moore, *Agriculture in Ante-Bellum Mississippi* (New York, 1958), 49–50; and Scarborough, "Science on the Plantation," Numbers and Savitt, eds., *Science and Medicine in the Old South,* 80. For an essay that notes how Nutt's geological work has been ignored but does not actually explore this topic, see Martha Coleman Bray, "Southern Influences on the Career of Joseph Nicollet," in Corgan, ed., *Geological Sciences in the Antebellum South,* 116. On the intellectual interests of southerners during the 1840s and 1850s, see Fox-Genovese and Genovese, *Mind of the Master Class,* esp. 473.

34. Ezekial Savage to Rush Nutt, Mar. 27, 1822, Nutt Papers, box 2, nu 303 ("instruction"), James Cowdon to Eliza Nutt, Mar. 10, 1823, box 2, nu 71 ("I have sent you"); J. N. Nicollet to Rush Nutt, Nov. 15, 1836, J. N. Nicollet Papers, box 1, fol. 2, LOC ("Sioux Country"), emphasis in original; J. R. Bedford to Rush Nutt, Feb. 22, 1822, Nutt Papers, box 2, nu 23, John Ker to Rush Nutt, Feb. 26, 1822, box 2, nu 170, Eliza Nutt to Rush Nutt, July 25, 1817, box 1, nu 226.

as a whole. Although geotheory was an old genre in Europe, serious naturalists like Georges Cuvier had given up the seemingly impossible quest of constructing an all-encompassing theory of the earth and, instead, focused on investigating discrete aspects of the deep past. But Lyell's volumes revived interest in geotheory, and it seems that his *Principles* and the theory he espoused in it—uniformitarianism—structured Nutt's approach. Nutt never completed his magnum opus, though he did publish two articles "extracted and abridged from unpublished MSS. on a theory of the Earth" in the *American Journal of Science* in 1833. Nutt divided his manuscript into several chapters on a vast array of topics, but uniformitarianism was the primary lens through which he interpreted the world and everything in it.[35]

Nutt's hostility to religion was apparent throughout his theory of the earth, and his skepticism stood in stark contrast to the writings of leading northern geologists. Benjamin Silliman and Edward Hitchcock, northerners who historians have considered exemplars of American geology in the 1820s and 1830s, contemplated the earth's formations to better appreciate God's creation. Historian Theodore Dwight Bozeman called this a doxological approach to nature, one that lauded empiricism but considered man's knowledge too limited to develop any geotheory that contradicted the Word. Nutt, on the other hand, was a deist and freethinker who intended his theory of the earth as a heroic assault on ignorance and priestcraft, both of which he associated with the North. There was a popular resurgence of freethought in America during the second quarter of the nineteenth century, but Nutt's skepticism was part of an older tradition he shared with Thomas Jefferson and other Virginian deists; indeed, the South was a stronghold of freethought in America until the late 1830s, when Christianity became a more pronounced mark of gentleness for planters. One point on which Nutt did agree with northern Christian geologists was that empiricism—as opposed to a "fire-side geological survey of the globe"—ought to be the basis of American epistemology. According to Nutt, "Our distinguished and engenious *[sic]* country-men, [David] Rittenhouse and [Eli] Whitney, were philosophers of the very first order and were indebted to their father's plantation and workshops for their greatness." He added that he had "never met

35. Rush Nutt, "On the Origin, Extension, and Continuance of Prairies; Extracted and Abridged from Unpublished MSS. on a Theory of the Earth," *American Journal of Science,* XXIII (1833), 40–45; Nutt, "Miscellaneous Geological Topics Relating to the Lower Part of the Vale of the Mississippi [. . .] from Unpublished MSS. on the Theory of the Earth," *American Journal of Science,* XXIII (1833), 49–65. The manuscript of Nutt's book-length theory of the earth is at the Huntington Library, Nutt Papers, box 8, nu 363. On Lyell and uniformitarianism, see Rudwick, *Worlds Before Adam,* 297–314.

with more than five or six men who were scholars, that had more than ten or twelve well arranged ideas." "Farmers, mechanics, and merchants have more," he contended, before concluding "No man can be a philosopher unless he acquires knowledge from the storehouse of nature herself." The book of nature, read without religious bias, fostered genius.[36]

Nutt based his theory of the earth on his own observations in North America and throughout the world, but he also drew on the findings of French, Anglo, and Spanish residents of the Gulf South. He substantiated his claim that the deposition of soils from America's rivers drove North America's terrestrial expansion into the Gulf of Mexico with reports from "the pilots of the Balise," William Dunbar, "Spanish and French tradition," Alabaman steamboat operators, and Antoine-Simon Le Page du Pratz. He also looked to "the records of French settlers on the Red River" and the writings of early Natchitoches pioneer John Sibley for details about the trans-Mississippi West. Nutt rarely hesitated to formulate grand theories, but he did ground them on his own observations and those of credible witnesses.[37]

Uniformitarianism informed how Nutt understood geohistory, environmental history, and human history. For example, he applied a uniformitarian point of view to local geological research to reveal "that the land of the [Natchez] district has been gradually formed by the slow and uniform action of the rains, in bringing down the earth from the higher places." Nutt

36. Nutt, "Miscellaneous Geological Topics," *American Journal of Science,* XXIII, no. 1 (1833), 60, 61 ("fire-side"); Rush Nutt, Incomplete manuscripts, Nutt Papers, box 8, nu 363, 2 ("Our distinguished"), 4 ("never met"); Hovenkamp, *Science and Religion in America,* 119–146; John C. Greene, "Science and Religion," in Edwin S. Gaustad, ed., *The Rise of Adventism; Religion and Society in Mid-Nineteenth-Century America* (New York, 1974), 57–60; Daniels, *Science in American Society,* 206–222; Gillispie, *Genesis and Geology;* Theodore Dwight Bozeman, *Protestants in an Age of Science: The Baconian Ideal and Antebellum American Religious Thought* (Chapel Hill, N.C., 1977), 75–79. On freethought, see Henry F[arnham] May, *The Enlightenment in America* (New York, 1976), 42–65, 88–101; O'Brien, *Conjectures of Order,* I, 11; and Albert Post, *Popular Freethought in America, 1825–1850* (New York, 1943), 226–232.

37. Rush Nutt, "Further Inquiries into the Geological History of the River Mississippi," Nutt Papers, box 8, nu 363, 10 ("Spanish"), 12 ("the pilots"), Rush Nutt, "Geological Remarks upon the Valley of the Red River," 4 ("records"), Rush Nutt, "Geological Remarks upon the Appearance, Position, and Decomposition of Granite and Quartz, with Their Many Varieties, as Observed on a Tour through a Part of Georgia, Alabama, Arkansaw, Louisiana, and Mississippi," 13, 27, Rush Nutt, "Remarks on the Hibernation of the Bear, with Objections to the Opinion of a Similar Habit in the Raccoon," 1; Nutt, "Miscellaneous Geological Topics," *American Journal of Science,* XXIII, no. 1 (1833), 60.

was most keen on uniformitarian geotheory because it offered powerful evidence against Biblical miracles like Noah's flood. He wrote that "all the attempts to account for Noah's flood, upon natural principles have proved abortive," a jab at religious men like Hitchcock and Silliman who looked to Neptunist geotheory to confirm the reality of the Deluge. Instead, "The mountains have obtained their present altitude by slow and imperceptible means," he averred," and "to suppose [they] have been covered with water since they have obtained their present elevation, is contrary to all the known laws of nature, and the assertion stands in opposition to all geological phenomena." Nutt also mustered the seemingly armor-clad evidence of empirical observation to ridicule the notion that Noah's flood could explain the ethnic varieties of human beings. "We have no knowledge that is not the result of our experience and if we compare the past with the present condition of mankind, it will be found that the reasons assigned for the dispersion of Noah's descendents are very unsatisfactory." A uniformitarian perspective, in Nutt's view, demolished both geological and ethnological arguments based on miracles in Genesis.[38]

As for the history of life, Nutt struggled to identify the actual cause that explained the succession of species evident in the fossil record. He argued that "no time will or can change" the organic structure of particular species. "If species were subjected to change," he proposed, "we should be able to detect the individuals of the transforming tribes among the innumerable relics of animal remains that now constitute a considerable part of the sedimentary rocks." "But," he continued, "the truth is that no such transforming species appear and consequently the fact may be considered as established, that species are governed by fixed laws and never change." His own observations on Mississippi's fossilized snail shells made it clear that the size and pattern of species varied across strata, yet he also marshaled his own "geological investigation" to belittle the theory that spontaneous generation could explain these changes. This theory, he held, "not only exposes the naturalist to the chances of error, but betrays inattention and a want of extensive research in the science." In its place, Nutt offered only the cau-

38. Nutt, "Miscellaneous General Topics," *American Journal of Science,* XXIII, no. 1 (1833), 53 ("that the land"); Rush Nutt, "Remarks upon Some of the Prophecies and Miracles," Nutt Papers, box 8, nu 363, 9 ("all the attempts"), Rush Nutt, "Criticisms and Explanations of Many Passages of the Bible," 9 ("We have no"). On how eighteenth-century European historians incorporated catastrophist geotheory into arguments about the origin and character of various peoples, see Jorge Cañizares-Esguerra, *How to Write the History of the New World: Histories, Epistemologies, and Identities in the Eighteenth-Century Atlantic World* (Stanford, Calif., 2001), 44–48.

tious hypothesis that "life is an indivisible something, common to all organized bodies, and if it is anything apart from matter, it is an imponderable substance,—possibly electricity." All forms of life—"elephants, horses, fish, worms, and trees"—developed out of other organized bodies, such as seeds and embryos, which also had organic origins. Although Nutt reached no conclusion about the succession of species, he believed that his line of reasoning did offer evidence for deism. "Life appears to be the same substance in all living bodies, and the organized form is due to the cause of causes," Nutt observed. "He ordered it, and it was done." Unlike doxological natural history, which paid homage to God's presence in nature, the only place for God in Nutt's world was as First Mover.[39]

Although a uniformitarian approach to geohistory was typical of the 1830s, Nutt used this same principle—that presently observable processes could help make sense of bygone eras—to interpret human history. Historians have credited late eighteenth- and early nineteenth-century geologists, particularly Cuvier, with historicizing geology, using the language and perspective of human history to analyze the deep past. Nutt, however, geologized history. "The laws by which we are governed are universal and immutable," Nutt argued deterministically, "and are as invariable to men and animals, as to the vegetable and mineral kingdoms." "Man is the same now as in the beginning." Nutt, of course, was eager to apply this method to demonstrate the absurdity of divine interventions in nature and mock the doctrine of priests. He even maintained that a uniformitarian approach to history was the only method for getting at the truth of past events because priests had, in their wickedness, obliterated the historical record. "Deprived of all knowledge of the first inhabitants of the earth," Nutt proclaimed, "from that wretched policy of ancient conjurors in destroying the records upon subduing a country, we are left to judge the past by the present." This was not merely an issue of acquiring a clearer image of the past: uniformitarianism would allow humans to identify and recover social practices like polygamy that, according to Nutt, were benevolent and "stamped upon the nature of man" but had been suppressed by the clergy.[40]

39. Nutt, "Remarks upon Some of the Prophecies and Miracles," Nutt Papers, 8 ("no time"), Rush Nutt, "Remarks on the Preservation of Seeds, with Objections to the Theory of Spontaneous Generation," box 8, nu 363, 6 ("geological"), 12 ("life is"). For Nutt's observations on fossilized snails, see Nutt, "Further Inquiries into the Geological History of the River Mississippi," ibid., 20–21.

40. Rush Nutt, "The Consideration of Man and His Destiny," Nutt Papers, box 8, nu 363, 6 ("The laws"), Rush Nutt, "Polygamy Supposed to be a Primitive Practice and Natural to the Human Species," box 8, nu 363, 1 ("Deprived"), 10 ("stamped"), Rush Nutt,

Nutt applied this uniformitarian lens to his favorite subjects—the absurdities of religion and the evils of priestcraft—to show that just as most religious beliefs were false and most clergy wicked in his own day, so, too, had they always been. Religion, especially "the belief in a future state of existence and of future rewards and punishments is common to all people, and at all times, and originates in our selfishness and vanity." Although propensity for religious belief was part of human nature, he blamed priests for exacerbating this problem by promoting self-serving superstitions. "The multitude of absurdities palmed upon the world as supernatural gifts or revelations from God are the efforts of ignorant and designing men, whose minds are most depraved, and whose object is to impose upon the credulous, to live by the fruits of their craft." The processes through which dogma itself changed over time were also uniform: "The religious theories of Asia and Africa, like the many revolutions of their empires, have risen and fallen successively and the new doctrines are ever incorporated with the wreck of the fallen." Nutt believed in fixed laws of nature that characterized entire species, and, during a discussion of the power of rattlesnakes to fascinate their prey, he could not resist comparing serpents to priests. The "treachery" of clergy enabled them to "prey upon ninety nine in a hundred, of all the men, women, and children who have ever come into the world, or may ever exist in the future." "Thus the priest ever pregnant in treachery ... will continue to practice their *[sic]* damnable tricks, so long as the snake shall retain his organization." Whether looking at "the Eastern world," "the brotherhood in Europe," or North America, Nutt's natural history of the clergy expounded that "man in the capacity of a priest, aided by power, is the most detestable being on earth." For Nutt, a uniformitarian approach to history made culture-specific explanations for human behavior unnecessary.[41]

Nutt insisted that legitimate natural knowledge had to be divorced from spiritual contexts. "Science and priest-craft never shake hands," he wrote,

---

"Origin of Priestcraft," box 8, nu 363, 40. For historicizing geology, see Rudwick, *Bursting the Limits of Time.* On how other southern intellectuals condemned polygamy as a barbaric practice that made slaves out of white women and, by negating their positive influence over males, inhibited social progress, see Fox-Genovese and Genovese, *Mind of the Master Class,* 513–514; and O'Brien, *Conjectures of Order,* II, 608.

41. Nutt, "Consideration of Man and His Destiny," Nutt Papers, 6 ("belief"), Rush Nutt, "An Attempt to Show That the Bible Was Not Writtin in Palestine: Nor by Persons Acquainted with the Country in Which the Transactions Are Said to Have Taken Place," box 8, nu 363, 16 ("religious"), Rush Nutt, "Remarks upon the Fascinating Power of the Snake," box 8, nu 363, 11–12 ("treachery," "prey"), 12 ("Thus the priest"), Nutt, "Origin of Priestcraft," 23 ("Eastern").

because the clergy have "ever [been] opposed to the cultivation of the sciences and aim particularly to deceive and to keep the people in ignorance." In an optimistic definition, Nutt added that "science is but a correct notion of things," and, therefore, "science and priest-craft occupy extreme points; the one claims your reason, the other your faith." "Science elicits truth," he proclaimed, "while priest-craft is nourished by falsehood." Historians have suggested that the notion of a great conflict between science and religion in the nineteenth century is anachronistic. This conflict was, however, very real to Nutt.[42]

Much of the reason why priests had such enduring power over human beings was that predisposition to religion was a disease. Nutt argued that all people were more or less prone to this "malady," but susceptibility could "differ in degree and in proportion to the peculiarities of organic development." "Religious ideas then," he claimed, "proceed from a want of equilibrium or loss of balance in the power of the mind." And no religion offered "clearer evidence of diseased manifestations of the mind" than Christianity, "a money making and power making machine which has spilt more blood and made more money than half the religions in the world." Mankind had not initially been weak enough to succumb to this disease; it was not until "the first nations of the earth had greatly degenerated . . . that the people were ready to hear, to surmise and to believe the pretended revelations of the author of nature." Humans had slowly changed over time to become more susceptible to religion, and degenerative diseases and priestcraft were the actual causes that drove this change in human nature.[43]

Nutt was particularly worried that the clergy were "now with a deadly hand marching westward," bringing the malady of Christianity to the United States. This was, in part, beyond anyone's control, for more people in America were "constitutionally disposed to religion than in Europe." He desired that "some event" would lead to the expulsion of all priests from the United States as a matter of public health in the "hope that many thousand valuable citizens who annually fall victims to disease from priest-craft,

42. Nutt, "Origin of Priestcraft," Nutt Papers, 42–43 ("ever"), 43 ("Science"), Nutt, "Attempt to Show That the Bible Was Not Writtin in Palestine," 17 ("science is but," "science and priest-craft," "Science elicits," "while priest-craft"); Ronald L. Numbers, "Science and Religion," *Osiris*, I (1985), 65–68; Rudwick, *Worlds Before Adam*, 564–565; Bozeman, *Protestants in an Age of Science*, 72–95.

43. Rush Nutt, "Remarks upon the Religious Manifestations of the Mind and the Causes Which Give Rise to Them," Nutt Papers, box 8, nu 363, 1 ("malady"), 8 ("clearer"), Nutt, "Attempt to Show That the Bible Was Not Writtin in Palestine," 16 ("money"), Nutt, "Origin of Priestcraft," 1 ("first nations").

will be preserved to usefulness, and their off-spring rescued from idiocy." Destroying Americans' capacity to reason was the clergy's goal, and, Nutt reckoned, they waged a campaign to subdue all the institutions of learning and science in America and supplant the empiricism that bolstered American liberty with the credulous ignorance that paved the way for theocracy. As Nutt asked his readers: "Are [priests] not the most indolent part of the community, and the most ignorant in practical and useful knowledge? . . . Are they not employed in pulling down every temple of science in America? Would they not still have been employed in hanging witches and Quakers, but for the preventive laws of the land, which are based upon humanity and the little science we have? . . . Is there not less knowledge in the United States than in any part of civilized Europe?" Priests, particularly those who educated America's youth, intentionally perpetuated the relative backwardness of the nation, and "the renouned temples of science, so long the pride of America are almost silenced and may soon be found without their ornaments and without their glory." Nutt's glowing portrayals of the former grandeur of American scientific institutions were nearly as hyperbolic as his blasts against priests.[44]

Worst of all, priests were spreading the disease of Christianity from its epicenter in New England into the South, infiltrating the last remnant of Enlightenment in America. Nutt warned that northern priests attacked slavery and fomented sectionalism in order to destroy the Union and eliminate the democratic government that restricted the power of church over state. "The sovereignty of the priests over the northern states, must be acknowledged by all, and it is to this, we owe the hatred which exists between the north and the south." "It is the slave states that have preserved the representative form of government," he declared, "and prevented the union of church and state." Nutt thought that northern priests and abolitionists actually cared little about slavery per se but had made it an issue with the purpose of instigating disunion and dissolving democracy. "Twenty years ago," according to Nutt, northerners "considered slavery perfectly in accordance with all their books of craft and a condition of human nature which grew out of circumstances emanating from the law of nature." Yet, once they realized that "the slave states were likely to prove an everlasting bar to the blessed union, when the fathers of the church could make laws for the

44. Nutt, "Remarks upon the Religious Manifestations of the Mind," Nutt Papers, 9 ("constitutionally," "some event," "hope"), 10 ("now with a deadly"), Rush Nutt, "Remarks upon Education," box 8, nu 363, 6 ("Are [priests]"), Nutt, "Origin of Priestcraft," 38 ("renouned").

children of the country [and] dispense curses on the heads of the men and blessings on the lovely persons of the women," Nutt pronounced, "they suddenly denounced slavery as a sin and have become very solicitous for emancipation." "Their efforts have been so far successful as to distract the union, which must very soon terminate in a division of the northern and southern states." For Nutt, it was a northern conspiracy of Priest Power, not the Slave Power plotting of elite southerners like himself, that had made slavery the central issue threatening the Union's integrity.[45]

Nutt's disdain for northern priests, especially Presbyterians, was visceral: "I say of the priests, what David said of his enemies: 'Oh Lord I do hate them with perfect hatred.'" Presbyterians had launched an attack against freethought and deism during the 1820s and 1830s, and they achieved remarkable success in taking over many educational institutions in the United States. Nutt deeply resented this, but his anti-Presbyterian views also related to contemporary debates in geology. He was familiar with the writings of Thomas Cooper, another natural philosopher living in the South who had attacked Presbyterians for seeking political power and savaged Silliman's geological writings as mere apologies for Christianity. The clergy had tried to oust Cooper from his position as president of South Carolina College because his lectures on geology excluded or mocked the Old Testament. Although Nutt did not publish his rants and never achieved Cooper's notoriety, he did use Cooper's career as a case study to illustrate how Presbyterians were squashing American science in its southern strongholds. "Thus," through the clergy's campaign against Cooper, "have we been deprived of the virtues and talents of the most learned and scientific man of our country." The priests had not stopped with South Carolina College, however; they wanted to infiltrate all of the South's temples of learning. "These enemies of science" were currently plotting against "that splendid establishment of Virginia University" while "a priest ridden monster was lately employed to bring a false accusation against the College of Tuscolusa *[sic]*." In Nutt's mind, the disputes over the discrepancies between geology and Genesis overlapped with the sectional struggle for the soul of the nation.[46]

45. Nutt, "Origin of Priestcraft," Nutt Papers, 34 ("the sovereignty"), 34–35 ("slave states"), 35 ("Their efforts"). Historians have suggested a similar timeline for northerners' shift toward an antislavery position. See David Brion Davis, *Slavery and Human Progress* (New York, 1984), 81–82. On how other southern intellectuals equated slavery with true Americanness, see Robert E. Bonner, *Mastering America: Southern Slaveholders and the Crisis of American Nationhood* (Cambridge, 2009).

46. Nutt, "Remarks upon Education," Nutt Papers, 6 ("I say"), Nutt, "Origin of

As a materialist, however, Nutt found himself agreeing with the Presbyterians in one key doctrine: predestination. He admitted that Calvinists who espoused predestination "have on their side the present improved state of science; inasmuch as we most clearly detect in the human species a striking difference in the organic development of the brain, by which it may be seen that one man is endowed with a greater degree of mental capacity, than another." In an argument that included several long quotes from phrenological pioneer Franz Joseph Gall, Nutt suggested that people were born with specific material organizations of the brain that limited their actions during life.[47]

Although Nutt did think that particular individuals and populations were less capable of improvement than others, he argued that multigenerational exposure to particular environments, and not inherited biological traits, accounted for contemporary variations in human stature, complexion, and intellectual capacity. He did, in this sense, have more in common with Enlightenment-era thinkers like his mentor Benjamin Rush than the harder racial theorists emerging in the 1820s and 1830s. "Like the animal and vegetable creation," Nutt reasoned, "the perfection of [man's] organs depend upon circumstances beyond his control." "His stature and muscular powers are influenced by climate, nay more, his mental capacity is contracted or expanded in proportion as temperature of heat, moisture or dryness of the atmosphere with vicissitudes of climate and constrained habits, influence the development of his brain." Despite all their supposed corporal and mental differences, Nutt declared that "there is nevertheless an intermediate shade of sameness, which would appear to blend us and favor the idea that all the races are but varieties proceeding from causes long continued." Uniformitarianism seemed to explain ethnic difference: humans were all essentially the same, but climate, disease, and social context were the actual causes that had made some groups more fit than others over deep time.[48]

Priestcraft," 37 ("Thus have we been"); May, *Enlightenment in America,* 88–101; Thomas Cooper, *On the Connection Between Geology and the Pentateuch, in a Letter to Professor Silliman …* (Boston, 1837); Post, *Popular Freethought in America,* 233; Dumas Malone, *The Public Life of Thomas Cooper, 1783–1839* (New Haven, Conn., 1926), 337–352; Hovenkamp, *Science and Religion in America,* 125–136.

47. Nutt, "Origin of Priestcraft," Nutt Papers, 38 (quote), Nutt, "Consideration of Man and His Destiny," 5.

48. Rush Nutt, "Remarks upon the Diffusion of the Human Species, and Causes of Their Many Complexions and Other Peculiarities of Organization," Nutt Papers, box 8, nu 363, 1 ("Like the animal"), 2 ("there is nevertheless"). On how Benjamin Rush made

Still, Nutt held that there was, at his particular moment in history, a clear hierarchy of races. Based largely on his observations in North Africa, Nutt contrasted the "fine organic development" of Europeans with "the defects in the bony structure, the want of muscular power, and the deficiency in the mental organs of the brain of some of the inhabitants of Africa." Over the long term, however, the hierarchies that differentiated ascendant from degenerate nations shifted. The mental and physical traits of a given group improved or declined based on the environment in which they lived and the populations with which they intermarried. Europeans were not inherently or permanently superior to Africans; they only happened to be so at the moment.[49]

Unlike southerners who looked to Christianity or polygenesis to legitimize the enslavement of blacks, Nutt considered the cycle of human degenerations and the political revolutions that accompanied them to be the root cause of, and justification for, slavery. "Nations degenerate from natural causes, though this law of our nature is hastened or protracted by physical means and during such a state of degeneracy and depopulation of countries, a train of evil propensities accumulate, which in numerous instances, give rise to the necessity of enslaving or extirpating them." Just as in Nutt's present, people throughout history who looked different had prejudices against each other that, when combined with the advantages that some groups enjoyed, led supposedly superior peoples to conquer and enslave lesser ones. "A difference of complexion as well as a superior or inferior development of the organs of the body, would give rise to a difference of feelings towards each other and those who were of superior mental and bodily endowment would quickly claim prerogative, and subject the inferior to bondage or extermination." In Nutt's worldview, bondage was a humane alternative to genocide.[50]

Blacks were not, for Nutt, the only contemporary group whose current state of degeneration warranted enslavement. He put Jews in this category, too. Nutt argued that Jews were "a nuisance in every country and both humanity and the best interests of nations require that they should be made slaves to the cultivation of the soil." "Such a measure," he reasoned, "is sanctioned by usage, as it has been a dernier resort with all eastern nations from

---

a similar argument about human difference, see David Freeman Hawke, *Benjamin Rush: Revolutionary Gadfly* (Indianapolis, Ind., 1971), 107.

49. Nutt, "Consideration of Man and His Destiny," Nutt Papers, 1.

50. Nutt, "Origin of Priestcraft," Nutt Papers, 1 ("Nations"), Nutt, "Remarks upon the Diffusion of the Human Species," 5 ("difference").

time immemorial and will necessarily occur to all countries so long as the human species shall exist." For Nutt, Jews seemed just as fit to pick his Petit Gulf cotton as blacks; that several other societies throughout European, Asian, and African history agreed that Jews deserved enslavement seemed to support both Nutt's specific argument about Jews and the uniformitarian theory undergirding it.[51]

Like his mentor Benjamin Rush, Nutt found that disease caused the degeneration that produced blackness. Unlike Rush, who used this theory to promote abolitionism, Nutt argued that these degenerative diseases had made Africans' enslavement an historical inevitability. This was particularly true of blacks in tropical Africa, for its harsh disease environment gave its inhabitants "marks of the greatest degeneracy, we shall consider them as a race whose black skin, crisped hair and deficient development of many of the organs of the body,—to proceed principally from disease and this train of morbid associations to originate in their country alone." Blackness was one manifestation of diseases peculiar to Africa that reduced Africans into historically inevitable, though not necessarily natural, slaves.[52]

According to Nutt, Africa's environment infected Africans with a degenerating disease, and slavery in America was its cure. This was apparent in the case of sodomy, which Nutt considered to be one of the manifestations of Africa syndrome. Sodomy, like blackness and a propensity for religion, was "a constitutional disease." "I say disease," he averred, "because it is a deviation from the natural fitness of things, and contrary to nature." "It consists of a morbid train of associated motions, and is generally accompanied with all other evil and unnatural propensities, such as lying, stealing, murder, and religious notions." Nutt thought there was only one "remedy for this disease": a "change of climate, and change of national habits and pursuits." "The negroes when brought to the U. States were addicted to this vice until they passed the 3rd or 4th generation," he explained, "when it began to decline, and [they became] as free from such morbid feelings in the 8th generation as nations least inclined to degeneracy." More generally, Nutt found evidence that blacks' bodies and minds improved by their "eighth

51. Rush Nutt, "Remarks upon the Original Complexion of the Jews and an Attempt to Show That They Are Not, at This Day, the Lineal Descendants of the Israelites," Nutt Papers, box 8, nu 363, 13.

52. Nutt, "Remarks upon the Diffusion of the Human Species," Nutt Papers, 7 (quote), 10, Rush Nutt, "The Influence of Climate in Modifying Disease and with Reflections on the Theory of Specific Contagion of the Plague," box 8, nu 363, 1. On Benjamin Rush and blackness as a disease, see Bruce Dain, *A Hideous Monster of the Mind: American Race Theory in the Early Republic* (Cambridge, Mass., 2002), 19–26.

generation in the United States." Their skins became lighter "independent of a cross with the white or red man." Their skulls changed as well, as "the forehead is raised and the cerebral hemispheres enlarged, the cheek-bone lower and less prominent, [and] the nose elevated and thinner," all of which craniologists associated with superior (white) heads. It took several generations, but Nutt found that enslavement in America did ameliorate the worst effects of Africans' endemic illness.[53]

If blacks seemed to be undergoing improvements in the United States, the opposite was true for whites. Siding with Buffon against Thomas Jefferson, Nutt observed that "we might also refer to the changes that have taken place in the European settlers for the same space of time [eight generations], whose organic development has materially diminished and deterioration of the species follows in every succeeding generation." White supremacy was not a given, and, although Nutt did not spell it out, he seemed to predict a trajectory in which the light-skinned descendants of Africans enslaved the degenerated offspring of their former masters. Even during the 1830s, when Anglo-American domination finally seemed secure in the Gulf South, Nutt recognized that white power could be a passing phase.[54]

Racialized slavery was not only a natural part of the rise and fall of nations; it was, from Nutt's deep historical perspective, a positive good because it saved blacks—and humankind on the whole—from degeneration and depopulation. "Nation rises after nation," Nutt wrote, "and in succession fill their period, when they degenerate, and are absorbed into other nations, or become extinct." "No matter what country—what climate—this fatality attends us, over which we can exercise no permanent control, as it is a law of our nature." Agricultural and pastoral nations alike lost their vigor if left to themselves, "but we will produce an instance of a people who

53. Nutt, "Attempt to Show That the Bible Was Not Writtin in Palestine," Nutt Papers, 5 ("constitutional"), 6 (" remedy"), Nutt, "Remarks upon the Diffusion of the Human Species," 11 ("eighth").

54. Nutt, "Remarks upon the Diffusion of the Human Species," Nutt Papers, 12 (quote). Nutt compared humans with sheep to help his readers make sense of the apparent degeneracy of people of European decent in America. Based supposedly on actual observations, he found that northern white sheep raised in the southern states degenerated until "black lambs are brought forth and the number increases from year to year. In some flocks we have noticed more than a third of the number to be black, whereas ten years before there was not one on the plantation or in the neighborhood. Suppose a flock is retained on a farm until all become black and then taken to other regions where they would improve in size and condition, would they return to their original color? We believe they would for we witness such a tendency in the human species." See ibid., 13.

from compulsory habits have become more prolific and populous," Nutt revealed triumphantly. "It is in the case of slavery in the United States of America." The climate, severe work regimen, and supposedly hearty food that accompanied enslavement improved blacks' fecundity so much that, according to Nutt's calculation, the enslaved population grew faster than that of whites. "It would appear that the multiplication of the species of depopulating countries ... can only be effected by coersion and this too in the way we have described as common with the Americans, whose slave population increases." This conclusion—that slavery was a humane institution that prevented blacks' extinction—was no doubt a convenient, though not necessarily disingenuous, perspective.[55]

Enslaving degenerated peoples also helped reinvigorate the human species as a whole because it encouraged the production of mixed-race children. Although all "nations degenerate and their countries depopulate from natural causes," Nutt proposed that this eventuality could be forestalled by "occasional removals from country to country and frequent marriages of the people of one country with another." Racial mixing was essential to the survival of human beings. "Will a drop of blood betwixt nations contribute to perpetuate or prolong the period of a nation's energy? We think it does, but it seems to depend on the people with whom the cross takes place." The offspring of Spaniards and American Indians, for example, "appears not to have renovated either," while intermarriage among European ethnic groups had more positive results. Nutt admitted though that there was a "serious difficulty attending this happy procession of nature in the improvement of our organization, and the only means by which we can arrest degeneracy:—it is avarice and pride." Like Zephaniah Kingsley, Jr., Nutt lamented that the prejudices of whites often prevented racial intermixing. But, unlike Kingsley, Nutt sympathized with prejudiced Anglo-Americans: "Who of us is willing to inter-marry with the degenerated African or Asiatic?" Although he did not go so far as to promote interracial sex, Nutt might well have considered it a way to improve the declining white population of America.[56]

Just as North America's environment altered humans' organic structure, Nutt found that humans had had an enormous effect on the continent in

55. Nutt, "Consideration of Man and His Destiny," Nutt Papers, 1 ("Nation"), Rush Nutt, "Depopulation of Countries," box 8, nu 363, 14 ("but we will"), 15 ("It would appear"). On how Enlightenment-era apologists for slavery similarly argued that slavery was fundamental to societies throughout history and, thus, natural, see Fox-Genovese and Genovese, *Mind of the Master Class*, 152, 201–202.

56. Nutt, "Depopulation of Countries," Nutt Papers, 15 ("Will a drop," "appears"), 20 ("nations," "occasional"), Nutt, "Remarks upon Education," 13 ("serious," "Who").

the not too distant past. Like Timothy Abbott Conrad, Nutt produced a history of the Gulf South in which humanity and the environment were mutually influential. He suggested, for example, that the combined agency of man and nature created and destroyed America's prairies through ongoing and observable processes, stating: "There is no fact that can be better established than that prairies are formed, and are now forming, by the operation of wind and fire. Very abundant proof was exhibited to the writer, more than twenty-eight years ago, when making a pedestrian journey through the distant and extensive regions of the west. He has seen the prairie in all its stages." Nature produced the requisite wind; fire was brought by "the tyranny of man." "He destroys vegetables and animals, to produce others in their stead, and thus maintains a constant warfare with animated nature." For Nutt, the battles of this prairie-producing war coincided with the main phases of Anglo-American expansion across the continent. The first stage, he figured, was led by cattle, whose weight compressed the earth and prevented fires from destroying the shoots of prairie grass. More importantly, "such circumstances favor the introduction of trees, which then immediately appear, and as they obtain sufficient size to shade the land, the grass itself is driven out; the cattle are thus driven to a greater distance from the plantations in some new direction, where they soon crop the grass, and place it beyond the reach of fire." The productive struggles between plants, animals, and humans pulled white men westward. "As the cattle recede," Nutt elaborated, "they are followed by the forest; and so soon as a farm can be enclosed by the young trees, the farmer, for the convenience of his stock, moves nearer to the prairie." "In this way," he concluded, "the prairies of Kentucky have disappeared; and those to the west of the Ohio and Mississippi, retreat as the settlements approach them." Perhaps Nutt's thirty years in the Gulf South had shaped his approach to environmental history by teaching him just how productive competition—in this case, among plants, animals, and people—could be.[57]

White men were not the only humans who changed North America: Indians had created the Mississippi River. "When the Indians arrived," Nutt wrote, "they must have found no small difficulty in penetrating the canebrakes." "To make their journeys the more readily, they doubtless adopted the plan of burning the cane every autumn, which ... would sometimes destroy, for the moment, the vegetation of immense regions." As the Indian population increased so did the extent of burning, and the cane, trees, and

57. Nutt, "On the Origin, Extension, and Continuance of Prairies," *American Journal of Science*, XXIII (1833), 41 ("There is no"), 43 ("tyranny").

grass—much like the nations that had vied for the Gulf South—struggled against each other "to take possession [of the land] and contend for the mastery." Animal, human, and vegetable life combined to cause waters that had hitherto been contained in nearby lakes to flow together and form a primitive stream. "The water began to show itself, and make its way slowly, through the thick canebrake, until taking advantage of the paths, made along the ravine by the beasts of the forest, it at length excavated a distinct channel." Nutt thought that "the Indians passed out of Asia into America" between the second and fifth centuries C.E. The Mississippi River was no older than that.[58]

Indians had helped create the Mississippi's channel, but Anglo-Americans could manipulate the river to make the Gulf South's environment even more exploitable. In his published articles of 1833, Nutt promised that the Mississippi River would soon "be perfectly subdued, and its vale brought within the control of man." Instead of vain efforts to prevent flooding with levees, he recommended cutting canals that would carry the river's water and, more importantly, sediment to the cypress swamps on the outskirts of plantations. The sediment would eventually settle in the swamps, and "all the lakes will be filled up." "With the accumulation of earthy matter," he surmised, "all the immense region of cypress forest, the most valuable timber in the world, will be reclaimed and brought within the reach of commerce and the arts of life." If planters followed Nutt's plan, he assured them, they would finally be able to reap wealth from swampland timber and realize the dream of being both lords and lumber cutters that Francis Philip Fatio had envisioned under the British and Spanish Empires some fifty years earlier.[59]

Nutt pushed his vision of reengineering Gulf South topography further in the unfinished manuscript of his theory of the earth. He began by emphasizing that the "high lands of Louisiana and Mississippi" consisted of clay that had "recently emerged from the gulf of Mexico" and was "deficient in the oxide of iron and destitute of vegetable matter, and is consequently sterile and devoid of nutritive properties." Between these loamless clay tracts and the low-lying swamplands, Nutt estimated that "not one fiftieth of the land [in the lower Mississippi Valley] can be cultivated, either on account of its liability to inundation, or from natural sterility." Luckily, both of these problems shared a common solution: direct the fertile alluvial soils carried

58. Nutt, "Miscellaneous Geological Topics," *American Journal of Science*, XXIII, no. 1 (1833), 57–59 ("When the Indians," 57, "To make," 57–58, "to take possession," 58).

59. Ibid., 63–64 ("be perfectly," 63, "all the lakes," 64).

by the Mississippi River onto the land. Based in part on his observation of irrigation systems along the Nile, Nutt suggested creating canals that would ensure that "while the river is up, the object will be to keep a steady current passing over the low lands, by which to leave a greater deposit and more speedily to elevate the plains." As things currently stood, Nutt regretted, "The very material, of which the few plantations are composed is suffered to pass the doors of the planter on the Red River and Mississippi and to be conveyed to the Gulf of Mexico, when at the same time, two thirds of almost every plantation consists of swamp land." He sought to "impose a stratum upon this extensive region of swamp land" by having every planter cut sluices from the river across their lands and, through this process, "a quantity of arable land would be raised upon the sterile clay [and be] most productive in quality." In time, Nutt believed this practice would allow the plantation society of the lower Mississippi Valley to expand far beyond the banks of the river. "Thus the work, of creating a superincumbent stratum, will go on, one planter settling behind another [further from the river], and continuing to extend the canal ... until, lakes and swamps, reptiles and carnivorous beasts, shall disappear, and this vast region now so frightful and deleterious in its effects, will become one continuous cultivated field, possessing a greatly improved climate and affording the most valuable and productive cotton land, in all the known world." Instead of alligators, swamps, forests, and disease, Nutt's environmental engineering program would convert the lands stretching both east and west from the Mississippi into one big cotton field.[60]

The process of effecting this terrestrial transmutation would also have the beneficial side effect of warming the region and making it more fit for sugar production. Nutt lamented that frosts often destroyed sugar crops in even the warmest parts of lower Louisiana and, worse, feared "that our climate becomes more and more unfriendly to the growth of sugar" as the average temperature steadily fell. Happily, directing river-borne sediment into the "plains would produce a wonderful effect upon the climate of the region." "The atmosphere would be much drier, and not only would it exert a friendly influence upon the health of the inhabitants ... but it would ob-

60. Rush Nutt, "Remarks upon the Swamp Land of the Mississippi and the Practicability of Improvement and Profitable Cultivation," Nutt Papers, box 8, nu 363, 1 ("high," "recently," "deficient," "not one fiftieth"), 4 ("while," "The very," "quantity"), 5 ("impose"), 6 ("Thus the work"). On other efforts to modify landscapes in the antebellum Gulf South in hopes of making the environment more salubrious, see Elaine LaFay, "Atmospheric Bodies: Medicine, Meteorology, and the Cultivation of Place in the Antebellum Gulf South" (Ph.D. diss., University of Pennsylvania [in progress]).

viate, in great part, the effects of early frosts, so that, instead of making one full crop of sugar, in every five years, (as appears to be a common calculation), it would be almost certain to reverse the rate of calculation." Geo-engineering would finally allow masters in the Gulf South to realize their old dream of living like Caribbean sugar planters.[61]

The expansion of the lower Mississippi Valley's environment and slave society was essential to perpetuating a virtuous America—one grounded in democracy, reason, slavery, and science—and forestalling corruption. Nutt saw cycles of corruption and regeneration everywhere in earth and human history: quartz was "forever decomposing and forever in a state of formation," the accumulation of sediment under the Gulf of Mexico and consequent enlargement of the continent "restore[d] to us the land that has been lost" through erosion, and nations degenerated through disease and the corrupting power of priests could be reinvigorated through migration, ethnic mixing, and enslavement. Nutt, like other Anglo-Americans in the early United States, argued that the degeneration of republican virtue and democratic institutions was only a matter of time, and he was convinced that priestcraft and industrial capitalism were the actual causes of corruption. The "Priests of America" sought to "influence the politics of the country and destroy the public institutions of science." Industrial capitalism, particularly "the precious tenures, attached to banks, rail-roads and gold mines," were a "canker and a curse in all countries." Gold, banks, and trains—which Nutt associated with the North—were "poetry in place of philosophy, priest-craft in place of common sense and the revelations of nature." Unlike these corrupting sources of wealth, plantation agriculture was patriotic, a social good, and grounded in natural law.[62]

Nutt's plan for extending the territory fit for cotton and sugar planting would have facilitated contemporary, and seemingly antithetical, visions of U.S. imperialism. On the one hand, many southern political thinkers of Nutt's generation agreed that expansion into new agricultural lands was essential to evading the corrupting forces of industrialization and market capitalism, and Nutt's method of terraforming promised white farmers

61. Nutt, "Further Inquiries into the Geological History of the River Mississippi," Nutt Papers, 9 ("that our climate"), Nutt, "Remarks upon the Swamp Land of the Mississippi and the Practicability of Improvement and Profitable Cultivation," 8 ("plains").

62. Nutt, "Geological Remarks upon the Appearance, Position, and Decomposition of Granite and Quartz," Nutt Papers, 2 ("forever"), 30 ("restore[d]"), Nutt, "Depopulation of Countries," Nutt, "Origin of Priestcraft," 38 ("Priests"), Nutt, "Remarks upon the Swamp Land of the Mississippi and the Practicability of Improvement and Profitable Cultivation," 8 ("precious").

boundless fertile fields for enlarging an empire for liberty. On the other hand, Nutt formulated his vision to reshape North America's environment at a moment when the lower Mississippi Valley was emerging as the hub of a cotton kingdom stretching deep into the continent, and the planters who drove this empire for slavery were thoroughgoing capitalists convinced that expanding the domain of plantation agriculture was key to transforming investment into greater wealth. Nutt, like many other planters, simultaneously found capitalism distasteful and used it to get rich, so he was happy enough to rationalize his plan for creating new lands—which would stimulate investment and economic growth—as a way to delay the corrupting effects of market capitalism.[63]

The often-contradictory motivations for Anglo-American imperialism and Nutt's own experiences with the nature and peoples of the Gulf South inspired his expansion-promoting geo-engineering scheme. Uniformitarian geotheory—which held that gradual processes could change the world—seemed to explain why this project could work. Uniformitarianism was not simply a means of understanding geological formations or human variations throughout deep history. For Nutt, uniformitarianism proved the viability of a massive experiment in which several small adaptations to the strata around the Mississippi River would transform the continent's environment and societies. By making North American nature more fit for plantation agriculture, Nutt hoped to encourage the ongoing expansion of a nation in which slavery, science, and democracy would thrive despite threats from northern Christians.

• • •

Plantation slavery supported geological research and affected its application. Slavery enabled Charles and Sarah Tait to offer patronage and recognition to northeastern naturalists, excavate and package the mass of fossils needed to characterize the Gulf South's geohistory, and circulate specimens and data through the networks built around the cotton trade. Rush Nutt drew on uniformitarian geotheory to legitimate African American slavery and proposed new geo-engineering techniques that would encourage the expansion of plantation agriculture. Slavery did not merely influence investigations and interpretations of nature in the South; it unearthed fossils that (once compared with those from other parts of the planet) helped develop a more complete history of the earth, and it inspired geologists to weigh in on national and international debates over the truth of scripture

63. See Drew R. McCoy, *The Elusive Republic: Political Economy in Jeffersonian America* (New York, 1982); Johnson, *River of Dark Dreams.*

and the source of racial difference. Interactions between white masters and enslaved blacks were, moreover, not the only violence-based relationships that generated knowledge in the antebellum era. Much of Florida remained Indian homeland, and, as had been the case since the 1500s, imperial expansion and the warfare resulting from it marked how whites and Indians alike understood the peninsula's nature and peoples.[64]

64. African American laborers in the nineteenth century—including freed slaves who dug trenches for the U.S. Army during the Civil War—unearthed evidence that has continued to inform new interpretations of deep history in the twenty-first century. See Nathan A. Jud and Leo J. Hickey, "Potomacapnos Apeleutheron Gen. Et Sp. Nov., a New Early Cretaceous Angiosperm from the Potomac Group and Its Implications for the Evolution of Eudicot Leaf Architecture," *American Journal of Botany*, C (2013), 2437–2449.

{ 7 }

SKULLS, SCALPS, AND SEMINOLES

• • •

The Second Seminole War (1835–1842) turned the last borderland space in southeastern America—the Florida peninsula—into part of the sectional South. U.S. officials and southern planters wanted all of Florida open to white settlement and could not tolerate that Indian lands in the peninsula provided a refuge for runaway slaves. They provoked the war to force the removal of the approximately five thousand natives who had resisted the federal government's efforts to relocate them to Indian Territory (now Oklahoma) and chose to stay in Florida to defend their lands and communities. The fighting began in earnest in December 1835 after the war leader Osceola killed Wiley Thompson, a U.S. Indian agent, and Charley Emathla, a native leader who promoted acquiescence to U.S. demands. There were pitched battles early in the war, but the Seminoles gradually retreated to southern Florida to fight with guerilla tactics. The war lasted until August 1842 and was the longest and most expensive of all U.S.-Indian wars. Despite the army's search-and-destroy campaign, about three hundred Seminoles remained in the peninsula when U.S. officials finally decided that the war had cost too many lives and too much money. Neither side won, though parts of southern Florida did stay native ground.[1]

The Second Seminole War was, to a great extent, fought over and against the bodies of the dead: spiritual, social, and political associations with human remains were inextricable from the war's cause, course, and outcomes. On the one hand, white Americans disputed the Seminoles' right to inhabit Florida as they dug up and decapitated the Indian dead. U.S. officials argued that the bones of pre-Seminole natives found in ancient

1. On the military and political history of the war, see J. T. Sprague, *The Origin, Progress, and Conclusion of the Florida War* ... (New York, 1848); John K. Mahon, *History of the Second Seminole War, 1835–1842* (Gainesville, Fla., 1967); and John Missall and Mary Lou Missall, *The Seminole Wars: America's Longest Indian Conflict* (Gainesville, Fla., 2004).

burial mounds undercut the Seminoles' assertions of ancestral roots in the peninsula while phrenologists found evidence in the skulls of recently killed natives that the Seminoles were an inherently violent people incapable of coexisting with Anglos. On the other hand, many Florida natives opted to wage war because removal from Florida and the defilement of their ancestors' graves threatened the bonds that connected living Indians to their dead kin. They also disinterred and mutilated the corpses of white Americans during the war, attacks that limited whites' own transgenerational ties to Florida. Although the war caused deep suffering for Florida natives, one of its productive results—the development of a Seminole ethnic identity with enduring ties to Florida—involved collecting the remains of their white foes.[2]

Incorporating the dead enriches current scholarship on Seminole ethnogenesis and illuminates how violence and revenge could engender knowledge and identities. In the mid-eighteenth century, diverse Creeks migrated from what is now southwestern Georgia to Florida and established towns that became politically independent of the Creek Confederacy. By

2. On how Indians' and whites' uses and understandings of violence against both the living and the dead were intertwined, see Erik R. Seeman, *Death in the New World: Cross-Cultural Encounters, 1492–1800* (Philadelphia, 2010); and Matthew Jennings, *New Worlds of Violence: Cultures and Conquests in the Early American Southeast* (Knoxville, Tenn., 2011). On nineteenth-century skull collecting and its role in shaping perspectives on ethnic and racial difference, see William Ragan Stanton, *The Leopard's Spots: Scientific Attitudes toward Race in America, 1815–59* (Chicago, 1966); and Ann Fabian, *The Skull Collectors: Race, Science, and America's Unburied Dead* (Chicago, 2010). On skull collecting amid wars of imperial expansion, see Simon J. Harrison, "Skulls and Scientific Collecting in the Victorian Military: Keeping the Enemy Dead in British Frontier Warfare," *Comparative Studies in Society and History,* L (2008), 285–303; Elise Juzda, "Skulls, Science, and the Spoils of War: Craniological Studies at the United States Army Medical Museum, 1868–1900," *Studies in History and Philosophy of Biological and Biomedical Sciences,* XL (2009), 156–167; and Samuel J. Redman, *Bone Rooms: From Scientific Racism to Human Prehistory in Museums* (Cambridge, Mass., 2016), 16–68. I am focusing on native Seminoles in Florida, and I use the term "Seminoles" in this chapter to refer to this group. Native Seminoles in Oklahoma, of course, also developed a Seminole identity, but this identity was distinct from that of Florida Seminoles because it was shaped by a different set of borderland encounters, particularly peaceful and violent interactions with Creeks, Anglos, plains Indians, and Mexicans in the West. Black Seminoles in Florida and Oklahoma developed unique ethnic identities as well. See Richard A. Sattler, "Seminoli Italwa: Socio-Political Change among the Oklahoma Seminoles between Removal and Allotment, 1836–1905" (Ph.D. Diss., University of Oklahoma, 1987); and Kevin Mulroy, "Ethnogenesis and Ethnohistory of the Seminole Maroons," *Journal of World History,* IV (1993), 287–305.

the 1760s, Euro-Americans were calling these Indians Seminoles. Florida natives had become increasingly unified politically in the early 1800s, but they had not yet developed a collective cultural identity. The Second Seminole War was the defining moment of the Florida Seminoles' ethnogenesis, the period when individuals within the political entity known as Seminoles came to see themselves as part of a coherent ethnicity with its own culture, historical consciousness, and communal attachments to Floridian land. As with many other groups in America's borderlands and the wider world, the pain and death that imperialism unleashed were driving forces underlying the Seminoles' ethnogenesis. However, natives also shaped their identities through their own harmful deeds, including scalping the enemy dead. The removal, circulation, and ritual uses of white scalps seems to have encouraged the Seminoles' emerging sense of being a distinct people by enabling the coherence of new communities that included both living and dead Florida Indians.[3]

The United States' conquest of Florida instigated destructive encounters with the dead that, for whites and natives alike, encouraged novel ideas about human difference. "In that place of skulls," as one disenchanted correspondent described wartime Florida, surgeons and officers in the U.S. Army collected Indian crania in the names of science and revenge, interpreted

3. The politically allied Seminoles of the early nineteenth century consisted of several ethnic groups: for example, contemporary writers often referred to the "Mikasukis," a large group of Florida Indians that might be the historical ancestors of the Miccosukees living in Florida today. See William C. Sturtevant, "Creek into Seminole," in Eleanor Burke Leacock and Nancy Oestreich Lurie, eds., *North American Indians in Historical Perspective* (New York, 1971), 92–108; and Richard A. Sattler, "Remnants, Renegades, and Runaways: Seminole Ethnogenesis Reconsidered," in Jonathan David Hill, ed., *History, Power, and Identity: Ethnogenesis in the Americas, 1492–1992* (Iowa City, 1996), 36–69. On how the Second Seminole War drove Florida Seminoles' ethnogenesis, see especially Brent R. Weisman, "Nativism, Resistance, and Ethnogenesis of the Florida Seminole Indian Identity," *Historical Archaeology*, XLI, no. 4 (2007), 198–212. On violence and ethnogenesis, see R. Brian Ferguson and Neil L. Whitehead, "The Violent Edge of Empire," in Ferguson and Whitehead, eds., *War in the Tribal Zone: Expanding States and Indigenous Warfare* (Santa Fe, N.Mex., 1992), 1–30. An overview of ethnogenesis is James Sidbury and Jorge Cañizares-Esguerra, "Mapping Ethnogenesis in the Early Modern Atlantic," *WMQ*, 3d Ser., LXVIII (2011), 181–208. Responses to Sidbury and Cañizares-Esguerra's article emphasized that historians must pay greater attention to the role of violence in Indians' and blacks' ethnogenesis; see Pekka Hämäläinen, "Lost in Transitions: Suffering, Survival, and Belonging in the Early Modern Atlantic World," *WMQ*, LXVIII (2011), 219–223; and James H. Sweet, "The Quiet Violence of Ethnogenesis," *WMQ*, LXVIII (2011), 209–214.

them through phrenology and craniology, and sent them off to nationally renowned collectors. Florida's Seminoles also cut off a great number of white scalps during this conflict and, like Euro-American skull collectors, delivered these trophies to elite medicine men in their own centers of ritual and calculation. Seminoles and whites decapitated, scalped, and disinterred each other's bodies, and both groups saw such desecrations as profane offenses that warranted equally brutal reprisals. Still, violence against the enemy dead was not simply destructive. By collecting, circulating, and analyzing the remains of each other's dead, Seminoles and whites both developed new knowledge about the Seminoles as a unique ethnic group.[4]

VIOLENCE IN THE SOUTHEAST

During the decades leading up to the Second Seminole War, the capacity to hurt people and to revenge injuries to oneself and one's kin had become an increasingly important component of white and native societies in the Gulf South. For white men, slights of honor led to duels and gouging matches; an obsession with all things military pervaded elite society; and brutality against blacks, Indians, and other whites were paths to social advancement. Fighting was a particularly significant aspect of the plantation

4. "From the *Philadelphia Public Ledger;* The Florida War," *Army and Navy Chronicle,* VIII (1839), 222 (quote). My approach in this chapter draws on anthropologist Ricardo Roque's argument that the collection of human remains in colonial settings needs to be studied in the contexts of entanglements between imperial and indigenous cultures, the material and social meanings of human remains within both cultures, and the violence that produced these remains and gave them meaning. See Roque, *Headhunting and Colonialism: Anthropology and the Circulation of Human Skulls in the Portuguese Empire, 1870–1930* (Basingstoke, Eng., 2010). On violence as an "interpretive concept" in borderlands history, see Ned Blackhawk, *Violence over the Land: Indians and Empires in the Early American West* (Cambridge, Mass., 2006), 5. On how violence and death were culturally and intellectually productive in zones of colonialism, see Joyce E. Chaplin, *Subject Matter: Technology, the Body, and Science on the Anglo-American Frontier, 1500–1676* (Cambridge, Mass., 2001); Jorge Cañizares-Esguerra, "Crusading and Chivalric Epistemologies: Iberian Influences on Early Modern Imperial Science," in Víctor Navarro Brotóns and William Eamon, eds., *Más allá de la leyenda negra España y la revolución científica/Beyond the Black Legend: Spain and the Scientific Revolution* (Valencia, 2007), 213–224; and Vincent Brown, *The Reaper's Garden: Death and Power in the World of Atlantic Slavery* (Cambridge, Mass., 2008). On how frontier violence contributed to the formation of an Anglo-American national identity, see Peter Silver, *Our Savage Neighbors: How Indian War Transformed Early America* (New York, 2008); and Carroll Smith-Rosenberg, *This Violent Empire: The Birth of an American National Identity* (Chapel Hill, N.C., 2010).

societies in northern Florida, where even bouts between white men could reinforce social hierarchies and the capacity of white males to dominate slaves and women. Collecting Indian body parts was, in part, an extension of competitions for status among white males. For instance, a lieutenant who flaunted two native scalps he had collected during the Second Seminole War explained that "I like to carry home some tangible *ividence [sic]* of my valor, so ... I can support my reputation in the neighborhood of being of *shere meat axe disposition!*" Moreover, as historians have argued, Anglo-Americans increasingly identified individual acts of vengeance—as opposed to state-sponsored punishment—as a legitimate means of responding to perceived offenses, particularly those of outsiders like Indians.[5]

The military in Florida also had its own institutional culture of violence. Some forty thousand whites served in Florida during the war, and, while many soldiers in the U.S. Army—which included a large number of foreign immigrants and, following the Panic of 1837, white urban poor—had little experience with Indian war, the volunteer companies from southern states included Indian fighters and slaveowners. These soldiers drank heavily, brawled with each other, and clashed with the white settlers. As in other wars throughout U.S. history, groups of soldiers in Florida developed their own standards of acceptable brutality that often went beyond those of white American society on the whole.[6]

Like Anglo-Americans, Florida natives incorporated warfare into their social and spiritual lives, and socially prescribed martial prowess—much of which derived from their Creek forbearers—might have been essential to the Seminoles' ability to endure their ongoing conflicts with the United

5. "For the N. H. Gazette; Seminole War—First Campaign; Extracts from the Journal of a Private; Battle of the Ouitilacoochee," *New-Hampshire Gazette* (Portsmouth, N.H.), May 16, 1837 (quote), emphasis in original; Elliott J. Gorn, "'Gouge and Bite, Pull Hair and Scratch': The Social Significance of Fighting in the Southern Backcountry," *AHR*, XC (1985), 18–43; Dickson D. Bruce, Jr., *Violence and Culture in the Antebellum South* (Austin, Tex., 1979); John Hope Franklin, *The Militant South, 1800–1861* (Cambridge, Mass., 1956); Edward E. Baptist, *Creating an Old South: Middle Florida's Plantation Frontier Before the Civil War* (Chapel Hill, N.C., 2002), 88–132; J. M. Opal, "Vengeance and Civility: A New Look at Early American Statecraft," *Journal of the Historical Society*, VIII (2008), 76–77.

6. Mahon, *History of the Second Seminole War*, 119, 325; James M. Denham, "'Some Prefer the Seminoles': Violence and Disorder among Soldiers and Settlers in the Second Seminole War, 1835–1842," *FHQ*, LXX (1991), 39–47. On troop-specific cultures of violence, see Clifton D. Bryant, *Khaki-Collar Crime: Deviant Behavior in the Military Context* (New York, 1979), 290–294.

States. Writing in 1791, Caleb Swan described how acts of war were vital to social standing for male Creeks, for "Young men remain in a kind of disgrace ... until they have performed some warlike exploit that may procure them a war name." Without a war name, young men were *"old women"* or simply "Este dogo, *you are nobody.*" Collecting a scalp was the usual proof needed to earn prestige as a fighter. As with white Americans, social pressures from within natives' own communities did much to promote mutilating the enemy dead.[7]

The understandings and uses of violence that developed among Florida Indians were based largely on Redstick nativism, a movement that emerged among Creeks and other southeastern native groups after the Shawnee warrior Tecumseh brought his call for a pan-Indian revival to Creek Country in 1811. Redsticks rejected white American society and practiced rituals, especially dances, that reinvigorated a self-consciously native identity and spirituality. During the Creek War (1813–1814), Creek Redsticks forwarded this vision of renewal through acts of destruction against Anglo-Americans and those Creeks who supported U.S. officials' vision of civilization. The Redsticks lost this war, and roughly two thousand Redstick refugees migrated to Florida, including the future war leader Osceola. Many Florida natives were sympathetic to Redstick views, and the peninsula became the nucleus of nativism in the Southeast.[8]

The rituals and sacred objects that native Seminoles developed during the total war of 1835 to 1842 suggest that combat had become increasingly central to their culture. Some warriors considered the capacity to kill essential to a happy afterlife. One army officer who asked a group of "Mickasukies" about their religious beliefs learned that, "when death lays them

7. Caleb Swan, "Report to Henry Knox on Creek Indians," May 2, 1791, Violetta Delafield-Benjamin Smith Barton Collection, 1783–1817, Ser. III: Bound Volumes, XXXII, APS, 35 ("Young men"), 36 ("*old*"), emphasis in original; William S[tephen] Belko, ed., *America's Hundred Years' War: U.S. Expansion to the Gulf Coast and the Fate of the Seminole, 1763–1858* (Gainesville, Fla., 2011). The wars of the early nineteenth century included the Patriot War (1812), the First Seminole War (1817–1818), and a tense period of "cold war"—which included skirmishes with settlers and slave raiders—that lasted from 1818 to the outbreak of total war in 1835. See Virginia Bergman Peters, *The Florida Wars* (Hamden, Conn., 1979), 61 ("cold war").

8. Joel W. Martin, *Sacred Revolt: The Muskogees' Struggle for a New World* (Boston, 1991), 133–149; Sturtevant, "Creek into Seminole," in Leacock and Lurie, eds., *North American Indians in Historical Perspective,* 106. On eighteenth- and early nineteenth-century nativism more broadly, see Gregory Evans Dowd, *A Spirited Resistance: The North American Indian Struggle for Unity, 1745–1815* (Baltimore, 1992).

below the ground, they say a sort of corporeal being like their former selves, (they cannot conceive of a spirit,) goes to that part of the world where the sun goes down; there he hunts, fishes, and enjoys himself, if he has been a good warrior, he finds plenty of game, if not, none." Centuries of violent encounters had molded how Florida natives thought about death and conceived of the cosmos.[9]

Moreover, both nativist and Afro-Atlantic approaches to nature influenced Florida Indians as they waged spiritually charged campaigns against Anglo-American expansion during the First and Second Seminole Wars. Early nineteenth-century Seminoles and Creeks employed force-endowed objects to empower their military activities, especially the "small leather pouch[es] [with] a few roots, bones, and other trifles" that Florida-born naturalist George J. F. Clarke called *"physic bag[s]."* Writing for a scientific audience in 1807, Indian agent Benjamin Hawkins described how Creek warriors would "carry in their shot-bags a charm, like Obi, a protection against all ills, called the 'war-physic'" that included pieces of horn and bone from extranatural animals. It seems that "Obi" was Hawkins's way of referring to the Obeah-associated conjure bags that Low Country blacks—many of whom encountered native societies as slaves, spouses, or allies—employed to heal the sick, harm enemies, defend against witchcraft, and endure and resist their enslavement. These Obi not only suggest that Gulf South blacks, like their predecessors during the French and Spanish eras, continued to employ force-endowed objects like *grisgris* but also that physic bags, as items whose agency was widely acknowledged, facilitated the convergence of black and Indian war powers against Euro-American domination. Native Seminoles and blacks were often close allies in the wars of the 1800s, and Florida Redsticks sometimes heeded the visions of both Indian and black prophets. As Captain Hugh Young noted in 1818, "[Josiah] Francis (Hillishija) and a negro girl were the only prophets among the lower tribes." "The girl always commences the process of divination by wrapping herself in a blanket, in which she made a singular whistling sound for several minutes. She said an invisible being then communicated a view of

9. "Correspondence of the *Savannah Georgian;* Florida; Copy of a Letter from a Young Officer of the Army to His Friend in This City, Dated Fort Lauderdale, E. F., June 20, 1839," *Army and Navy Chronicle,* IX (1839), 93 (quotes); Weisman, "Nativism, Resistance, and Ethnogenesis," *Historical Archaeology,* XLI, no. 4 (2007), 198, 202–206; Brent Richards Weisman, *Like Beads on a String: A Culture History of the Seminole Indians in Northern Peninsular Florida* (Tuscaloosa, Ala., 1989), 110; Louis Capron, "The Medicine Bundles of the Florida Seminole and the Green Corn Dance," in William C. Sturtevant, ed., *A Seminole Source Book* (New York, 1987), 162.

future events." The ways Afro-Atlantic prophets and healers used esoteric knowledge—much of which centered on mortuary rites—to defy the destruction of their social groups and form new communities probably had deep resonance with nineteenth-century Florida Indians.[10]

DOING SCIENCE IN A BORDERLAND WAR

In 1835, despite fourteen years of U.S. rule in the Territory of Florida and several scientific expeditions, the natural history and geography of central and southern Florida remained remarkably unknown to Euro-Americans. State-sanctioned violence during the Second Seminole War enabled surgeons and amateur naturalists serving in the U.S. Army and state militias to generate a vast amount of new information about Florida's plants, animals, and land. Florida was, however, hardly a safe place to conduct fieldwork. The context of war marked how naturalists wrote about the territory while the threat and reality of Seminole attack often dictated where natural knowledge could be produced.

10. "From the MS of the Late George I. F. Clarke, July 1st 1822, no. 6," *Florida Herald and Southern Democrat* (Saint Augustine, Fla.), Nov. 15, 1838 ("small leather"), emphasis in original; Benjamin Hawkins, "A Concise Description of the Creek Country, with Some Remarkable Customs Practised among the Native Inhabitants ... Communicated to Dr. Mitchill, December 14, 1805," *Medical Repository*, 2d Hexade, IV (1807), 41–43 ("carry," 42); Hugh Young, "A Topographical Memoir on East and West Florida with Itineraries of General Jackson's Army, 1818," *FHSQ*, XIII (1934), 94 ("[Josiah] Francis"). On Obeah, see Brown, *Reaper's Garden*, 145; and Laura de Mello e Souza, *The Devil and the Land of the Holy Cross: Witchcraft, Slavery, and Popular Religion in Colonial Brazil*, trans. Diane Grosklaus Whitty (Austin, Tex., 2003), 141. On spiritual life and resistance in Georgia and South Carolina, see Jason R. Young, *Rituals of Resistance: African Atlantic Religion in Kongo and the Lowcountry South in the Era of Slavery* (Baton Rouge, La., 2007), esp. 119, 184. On the social interactions between Creeks—including Redsticks—and southern blacks in the eighteenth and early nineteenth centuries, see Gary Zellar, *African Creeks: Estelvste and the Creek Nation* (Norman, Okla., 2007), 7–20. On the Redstick prophet Josiah Francis (also called Hillis Hadjo or Hillis Haya), see J. Leitch Wright, Jr., *Creeks and Seminoles: The Destruction and Regeneration of the Muscogulge People* (Lincoln, Neb., 1986), 161–207. On the alliance of black and Indian Seminoles during the Second Seminole War, see Jane G. Landers, *Atlantic Creoles in the Age of Revolutions* (Cambridge, Mass, 2010), 175–203. On the role of whistling in the divination practices of enslaved black women, see Javier Villa-Flores, "Talking through the Chest: Divination and Ventriloquism among African Slave Women in Seventeenth-Century Mexico," *Colonial Latin American Review*, XIV (2005), esp. 306–307. On enslaved ritual experts building communities, see Brown, *Reaper's Garden*, 60–91; and James H. Sweet, *Domingos Álvares, African Healing, and the Intellectual History of the Atlantic World* (Chapel Hill, N.C., 2011).

The Second Seminole War, and the men, money, and media attention it directed toward Florida, sparked a wide-ranging effort to explore the peninsula and fully incorporate this borderland into the United States. The military expedition to remove the Seminoles was also a scientific expedition: the army's topographical engineers, surgeons, and officer-naturalists explored, mapped, and described Florida and sent their discoveries to officials, scholars, and scientific publications. In an *American Journal of Science* article, Major Henry Whiting argued that Anglo-Americans knew so little about Florida because "much local information was . . . withdrawn" when many of the territory's learned Spanish residents departed in 1821. However, Whiting reassured his audience that "the war which has lately been carried on with the Florida Indians has opened the country generally to observation, and its character will hereafter be better, if not well understood." In the 1837 first edition of his *Territory of Florida,* John Lee Williams complained that Florida's interior was "wholly unexplored by white men," and he "hoped that our government will cause this part of the Territory to be carefully explored." By the time Williams penned the unpublished second edition (1844–1853), a work replete with geographic details gleaned from army personnel, he acknowledged that "this war, not withstanding the evils which have resulted, has occasioned a general exploration of the interior and southern portion of our peninsula, and in a great measure, developed the resources of the State." And Major General Thomas Jesup, then the commanding officer in Florida, told the secretary of war that, when the war began, "The greater portion of [Florida] was an unexplored wilderness, of the interior of which we were as ignorant as of the interior of China." "We exhibit in our present contest, the first instance . . . of a nation employing an army to explore a country." Jesup had apparently forgotten that Hernando de Soto had done the same thing in the same place three hundred years earlier.[11]

11. Henry Whiting, "Cursory Remarks upon East Florida, in 1838," *American Journal of Science and Arts,* XXXV (1838), 47–48 ("much local," 48); John Lee Williams, *The Territory of Florida* (New York, 1837), iv ("wholly"); "The Territory of Florida," unpublished MS, 2d ed., John Lee Williams Papers, 1820–1844, box 1, 1, PKY ("this war"); Th[omas] S. Jesup to J. R. Poinsett, Feb. 11, 1838, in Sprague, *Origin,* 200 ("greater"). On Anglo-Americans' overall ignorance of Florida's interior, see Mahon, *History of the Second Seminole War,* 129. On newspaper publishers' interest in the war, see E. Ashby Hammond, ed., "Dr. Strobel Reports on Southeast Florida, 1836," *Tequesta,* XXI (1961), 65. On the army's medical practices, see C. S. Monaco, "Shadows and Pestilence: Health and Medicine during the Second Seminole War," *Journal of Social History,* XLVIII (2015), 565–588.

An army of naturalists overran Florida between 1835 and 1842. Some of these were West Point-educated officers who pursued natural history as a refined hobby that set them apart from the rank and file while distracting them from hunger, disease, and boredom. One officer described the "small pocket herbarium that I had brought with me, expecting to relieve the tedium of the march ... in collecting a few treasures from this chosen region of Flora." A surgeon with a unit of Creek auxiliaries observed that "a very few of the officers devoted themselves to the acquisition of Botany, Ontology and Geology." "Every animal, plant or mineral," he noted, "became a subject of close scrutiny." State militia officers like the Charleston lawyer M. M. Cohen also collected plants and animals during their campaigns. Cohen planned to have his notes on Floridian "Topography, Philology, Geology, Zoology, and other *ologies*" read at Charleston's Literary and Philosophical Society. The most ubiquitous naturalists in Florida were army surgeons who conducted research to satisfy their own curiosity and advance American science. The medical staff was so prolific in its scientific work that the secretary of war reported that "the disposition they manifest to avail themselves of every opportunity to collect facts which may afford useful contributions to science, is justly appreciated by the Department, and will be encouraged."[12]

Army officers and surgeons had diverse scientific interests, including botany, zoology, geology, meteorology, and geography. Some pursued these interests in an official or semiofficial capacity, such as the topographical engineers who mapped much of the interior and surgeons like Dr. Samuel Forry who submitted a meteorological diary and observations on Florida's medical topography to the Surgeon General's Office. Most naturalists in the military worked for their own amusement, and some planned to publish their findings. Cohen explained that he would not spare "any pains to collect all the rare and beauteous products of this country; and then let the world tremble, as it anticipates a botanic disquisition, stamens, pistils, classes, orders, sex, habits, history, and all that Linnaeus or Jussieu could desire." Some naturalists complained that their military duties kept them from scientific work, such as an officer who was annoyed that rather than

12. [William W. Smith], *Sketch of the Seminole War, and Sketches during a Campaign, by a Lieutenant of the Left Wing* (Charleston, 1836), 142 ("small pocket"); W. P. Rowles, "Incidents and Observations in Florida in 1836," *Southron* (1841), 161, PKY ("a very few"); M[yer] M. Cohen, *Notices of Florida and the Campaigns* (Charleston, 1836), 5 ("Topography"), emphasis in original; J. R. Poinsett, "Report of the Secretary of War," Nov. 30, 1839, *Army and Navy Chronicle*, IX (1839), 404 ("disposition").

"finding opportunities ... to pick a flower, shoot a rare bird, or notice any thing uncommon," he was forced "to attend to the duties of my command as advance guard," which required him "to look for Indians and not for flowers." Even those soldiers more interested in hunting Indians than birds made noteworthy contributions. Colonel William Harney, who led brutal search and destroy campaigns in the everglades, "made some geographical discoveries" and "added something to our stock of knowledge in natural history" by shooting two manatees, "animals heretofore considered fabulous." Such discoveries prompted more learned men to speculate about valuable new resources, including a plan to harvest manatees for their oil and ivory-like bones.[13]

Hatred, fear, and both the performance and threat of violence cast a long shadow over how individuals interpreted the world during the war. One soldier mused that "a life spent in the wilderness ... with scarcely any subject of moving interest on which to employ the faculties, but this single one of pursuing or avoiding the Indian, superinduces a train of influences which cannot but operate unfavorably on the mind and character of the individual." Hunting and hiding from Seminoles stimulated such "rancorous hate," he argued, that "even the range of intellectual vision becomes circumscribed." "The mental powers are 'cribbed, cabined, confined;' and the moral faculties," he worried, "are in danger of becoming impaired and debased by the absorbing prevalence of the one feeling and the single aim of hating and killing the Seminole." War, in short, both enabled and affected soldiers' observations.[14]

Doing science in wartime Florida could be deadly work, and the threat of getting scalped circumscribed army naturalists' research. One Charlestonian militia officer noted that "the pleasure of [botanical] excursions was very much damped, by the thought that an Indian might be lurking in every bush," and he worried that while "I was stooping to pick a flower, one of the sneaking villains might pick off my scalp." He added that "I saw little

13. Cohen, *Notices of Florida,* 141 ("any pains"); [Smith], *Sketch of the Seminole War,* 142 ("finding"); "Florida," *New London Gazette* (New London), Apr. 7, 1841 ("made some geographical"); Thomas Lawson to Dr. S. Forry, Aug. 31, 1837, Letters and Endorsements Sent, 1818–1889, Office of the Surgeon General, RG 112, VIII, NARA; "Florida," *Army and Navy Chronicle,* IX (1839), 88. See also John Pickell, "The Journals of Lieutenant John Pickell, 1836–1837," ed. Frank L. White Jr., *FHQ,* XXXVIII (1959), 144, 150, 152–155; Rowles, "Incidents and Observations," *Southron* (1841), 55; and "Territory of Florida," unpublished MS, 2d ed., John Lee Williams Papers, 141.

14. R., "Scenes in Florida; A Leaf from My Journal, No. II," *Army and Navy Chronicle,* X (1840), 122.

therefore of the surrounding country, and the observations I have to make are consequently limited." The specter of scalping even dissuaded some of the United States' most intrepid men of science from exploring Florida. In 1836, Charleston naturalist John Bachman told ornithologist John J. Audubon that "with regard to Florida, nothing will be done by Naturalists for at least two years." "Your Indian friends, the cut-throats, have scalped almost every woman and child south of St. Augustine." He warned Audubon that even after the main force of Seminoles was eventually defeated "there will, undoubtedly, remain many small predatory bands that would make no bones of scalping an Ornithologist *secundum artem* [according to the art]; and would ask no questions whether he was the friend or enemy of William Penn."[15]

Some experts who did explore Florida during the war should have heeded such warnings. Dr. Frederick Leitner, a German-born Charleston-based naturalist who, with the help of an Indian guide, had conducted extensive research in Florida in the early 1830s, was killed in battle in 1838. He was considered one of the world's leading experts on Floridian nature, and, as one surgeon wrote, Leitner had been so "anxious to pursue his researches" that, "denied the privilege of doing so alone, by the hostility of the natives," he had joined an army expedition to southern Florida "in the capacity of acting surgeon ... and guide through this 'terra incognita.'" One newspaper correspondent thought Leitner's role as a physician made him a target for Seminole riflemen, as "it could hardly have been a stray bullet that struck him, for we learn 'that his hospital steward was shot down beside him.'" A rumor held that Seminoles had taken him alive "in order to avail themselves of his professional services," but "a young Indian, who had lost a brother" exacted his revenge by shooting Leitner dead.[16]

15. [Smith], *Sketch of the Seminole War,* 273–274 ("pleasure," 273, "I was stooping," 273–274, "I saw," 274); John Bachman to [John J. Audubon], Sept. 14, 1836, reproduced in C. L. Bachman, *John Bachman: The Pastor of St. John's Lutheran Church, Charleston* (Charleston, 1888), 138 ("with regard").

16. Jacob Rhett Motte, *Journey into Wilderness: An Army Surgeon's Account of Life in Camp and Field during the Creek and Seminole Wars, 1836–1838,* ed. James F. Sunderman (Gainesville, Fla., 1953), 184 ("anxious"); O., "From the *New York American;* The Death of Doctor Leitner, a German Naturalist Killed in Florida," *Army and Navy Chronicle,* VI (1838), 181 ("it could hardly"); "Florida War, Charleston, March 5," *Army and Navy Chronicle,* VI (1838), 173 ("in order"); Medicus, "The Climate of South Florida, Washington City, D.C., Feb. 4, 1838," *Army and Navy Chronicle,* VI (1838), 108; George E. Gifford, Jr., "Edward Frederick Leitner (1812–1838) Physician-Botanist," *Bulletin of the History of Medicine,* XLVI (1972), 568–590. Dr. Benjamin Strobel, another Charleston-

After Leitner's death, naturalists and officials "fear[ed] that the fruit of his researches will be lost," thus depriving both the scientific community and the military of valuable natural and geographic knowledge. One reporter observed that Leitner had been preparing for press a work on the botany and "topography of that hitherto *terra incognita,* when it has pleased a mysterious Providence to deprive mankind and the cause of science of the further benefit of his labors." "He had, we believe, also collected a large and choice *herbarium.*" Leitner had carried the three-hundred-page manuscript draft of his "account of the natural productions of Florida" with him on the campaign and planned to publish it on his return to Charleston. This work was considered important enough that the secretary of war asked "the officers of the army ... as well as ... the citizens of the Territory" to do everything in their power to recover it. Like Leitner's body, the manuscript was never found. The death of Dr. Leitner and the loss of his hard-won knowledge caused some whites to wonder whether Florida could ever be known to science or subdued by the army: "It would seem as if Providence had willed, that not only Mars, but science itself, should be frustrated in all attempts to penetrate ... that wayward peninsula." And wars would continue to erase Leitner's scientific work long after his death. Some of the few surviving collections from his Florida expeditions found their way to Berlin where they were destroyed in a World War II bombing raid.[17]

Advanced technology and scientific institutions might have been specific objects of Indian violence during the war. Seminoles often destroyed machinery on sugar plantations, attacks that followed nativist traditions of targeting the material symbols of Euro-American society. The razing of sugar technology was incorporated into Seminole rituals of warfare, such as when one plantation's "sugar works ... were destroyed by Indians by fire, who danced about and around like demons." The so-called Spanish Indians also danced as they torched Dr. Henry Perrine's tropical plant research center in 1840. An agronomist and botanist, Perrine had collected plants from

---

based naturalist who had explored Florida before the war, also volunteered to serve as a regimental physician and guide in order to enjoy the protection of a military escort for his scientific investigations. See Hammond, ed., "Dr. Strobel Reports on Southeast Florida, 1836," *Tequesta,* XXI (1961), 66.

17. Medicus, "Climate of South Florida," *Army and Navy Chronicle,* VI (1838), 108 ("fear[ed]"); O., "From the *New York American;* The Death of Doctor Leitner," *Army and Navy Chronicle,* VI (1838), 181 ("topography," "He had," "It would seem"), emphasis in original; J. R. Poinsett, "Papers of the Late Dr. Leitner," *Army and Navy Chronicle,* X (1840), 390 ("account," "officers"); Gifford, "Edward Frederick Leitner," *Bulletin of the History of Medicine,* XLVI (1972), 586.

throughout the circum-Caribbean and, while serving as the U.S. consul in Campeche, developed a plan to introduce Cuban and Mexican plants into the southern United States. The federal government granted Perrine land on Indian Key in the hope that his experiments would develop new staple crops for southern planters. The Spanish Indians—who, like Perrine, maintained connections with the Spanish Caribbean—might simply have been in search of food and supplies, but it is nonetheless true that Perrine's work would have facilitated the establishment of plantations in a region where the Spanish Indians had remained secure for more than a century. As with Leitner, Perrine's death and the burning of his institution led U.S. officials and men of science to rue the destruction of his research. "Science is robbed of a bright ornament," read one obituary; "His voluminous and valued notes have all been lost; and with the martyrdom of PERRINE, have also perished his labors." Despite the extensive knowledge that army naturalists generated about Florida's nature, the resistance of Florida's Indians significantly impeded white Americans' efforts to know and rule the peninsula.[18]

FIGHTING OVER THE DEAD

Misery and death were the stuff of daily life for native and white fighters during the war. For Florida Indians this was most obvious in the high population loss and the innumerable broken bonds of kinship and community. As for Euro-American soldiers, about 14 percent died from disease, and only 328 of the United States' 1,466 military dead were killed in action. The soldiers and officers were depressed: instead of realizing dreams of martial glory, they were "sent to die, like dogs, in the swamps of Florida, in the ignoble pursuit of vagrant Indians and runaway negroes." They ate horses, drowned, went insane, and committed suicide. "Summer will be at hand," one officer wrote, "and it will require all healthy soldiers to attend to the

18. John Bemrose, *Reminiscences of the Second Seminole War*, ed. John K. Mahon (Gainesville, Fla., 1966), 97 ("sugar works"); "The Late Dr. Perrine," *Army and Navy Chronicle*, XI (1840), 155 ("Science"); Williams, *Territory of Florida*, 138; "Florida War, from the *St. Augustine News*, Aug. 21, Attack on, and Destruction of Indian Key," *Army and Navy Chronicle*, XI (1840), 154; Hester Perrine Walker, "Massacre at Indian Key, August 7, 1840, and the Death of Doctor Henry Perrine," *FHSQ*, V (1926), 27; Nelson Klose, "Dr. Henry Perrine, Tropical Plant Enthusiast," *FHQ*, XXVII (1948), 189–201. The "Spanish Indians" seem to have been descendants of Creeks and Cuban fishermen who began forming a shared community in the early eighteenth century and continued to travel between Cuba and Florida. See John E. Worth, "Creolization in Southwest Florida: Cuban Fishermen and 'Spanish Indians,' ca. 1766—1841," *Historical Archaeology*, XLVI, no. 1 (2012), 142–160.

sick and bury the dead." "And this is warfare! glorious, noble, chivalrous warfare!" For the men who fought it, the war was as much about burying friends as shooting foes.[19]

Indians, blacks, and whites all directed violence against enemy bodies during the Second Seminole War, atrocities that perpetuated an ongoing cycle of revenge that one poet termed "a whirlpool of wrath." This cycle had begun even before the war officially commenced: in August 1835, General Duncan L. Clinch reported that six natives scalped a mail carrier to "revenge the death of a relative of theirs." In December 1835, during the war's first major conflict, native and black Seminoles annihilated Major Francis Dade's company of 110 men and mutilated their remains. Throughout the seven years' war that followed, Seminoles decapitated white officers, dug out the eyes of mail carriers, and scalped dead soldiers in full view of fortified U.S. troops. Many of the most graphic reports of Seminole atrocities involved white women, outrages that Euro-American men believed demanded vengeance.[20]

Whites committed equally brutal attacks against dead Indians, and some

19. "From the *Philadelphia Public Ledger;* The Florida War," *Army and Navy Chronicle,* VIII (1839), 222 ("sent to die"); A Subaltern, "Florida War—No. 3, Tampa Bay, E. F., Jan. 24, 1839," *Army and Navy Chronicle,* VIII (1839), 124 ("Summer"); Mahon, *History of the Second Seminole War,* 325; James Barr, *A Correct and Authentic Narrative of the Indian War in Florida . . .* (New York, 1836), 17; "Statement, Exhibiting, by Regiments, the Names of the Officers, Non-Commissioned Officers, Musicians, Artificers, and Privates of the United States Army, Who Were Killed in Action, or Died of Wounds Received, or Diseases Contracted, during the Late Hostilities with the Florida Indians," in Sprague, *Origin,* 526–548.

20. Thomas Richard Whitney, *The Ambuscade: An Historical Poem* (New York, 1845), 9 ("whirlpool"); Duncan L. Clinch to Generals in Chief, Sept. 1, 1835, "Duncan Clinch Letterbook, 1834–1836," in Kenneth M. Stampp, Randolph Boehm, and Martin Schipper, eds., *Records of Ante-bellum Southern Plantations from the Revolution through the Civil War,* Ser. C: Selections from Holdings of the Library of Congress, pt. 2: Alabama, Georgia, and South Carolina, microfilm, 5 reels (Frederick, Md., 1986), reel 1 ("revenge"); Missall and Missall, *Seminole Wars,* 90–92. Some sources suggest that only black warriors mutilated dead whites after defeating Dade's command. See [Smith], *Sketch of the Seminole War,* 37; Williams, *Territory of Florida,* 218; and Sprague, *Origin,* 91. On Seminoles' mutilation of whites, see "St. Augustine, Nov. 7," *Army and Navy Chronicle,* XI (1840), 332; "Florida War, St. Augustine, April 14," *Army and Navy Chronicle,* VI (1838), 300; and "The Seminole Indians—War in Florida," *Army and Navy Chronicle,* II (1836), 43. Concerning attacks on white women and whites' desire for revenge, see "Florida War," *Army and Navy Chronicle,* XIII (1842), 154; Ransom Clark, *Narrative of Ransom Clark, the Only Survivor of Major Dade's Command in Florida . . .* (Binghamton, N.Y., 1839), 11; and Sprague, *Origin,* 106.

soldiers scalped fallen Seminoles as part of the military tradition of trophy taking. Individuals in the U.S. military had mutilated both European and Indian enemies in previous wars and have continued to take human body parts as trophies in the Civil War, World War II, and at least as recently as the Vietnam War. Such mutilations were so common during the Second Seminole War that one colonel was shocked to find that he could watch his men remove the scalps and ears of eight natives "without a mingled feeling." White fighters often needed no excuse to abuse the Indian dead, yet they also mutilated Seminoles to revenge postmortem violence against whites. When U.S. troops discovered the "baked" head of an express rider in 1841, "the devilish attempt to burn the head ... so incensed the Captain that he vowed [the Indians'] skins should pay the penalty of the deed." After tracking down and killing Waxehadjo, one of the Indians who had supposedly killed the express rider, the soldiers took his "ornaments" (which, the author noted, "would have set up a respectable Jew-shop") and removed "the ornament most precious to an Indian's heart, and upon the preservation of which depends his eternal life in the 'happy hunting-grounds'—his top-knot and the soil that produced it." The officer who penned this account after the war recognized the significance of scalps within Seminole cosmology and scalped Waxehadjo as an act of vindictiveness. As a final insult, he noted that, "if it would be a consolation to any of the surviving members of his illustrious family who may now inhabit the Seminole Reservation on the sunset side of the Mississippi to gaze upon the relic, they can do so by making a pilgrimage to Washington, the city of the Great Father, and applying for the indulgence at the Smithsonian Institute." The officer's revenge impulse made this addition to the Smithsonian's ethnographic collection possible.[21]

Mutilation and trophy taking were effective attacks because all combat-

21. William Foster to Betty Foster, Jan. 2, 1838, in John Missall and Mary Lou Missall, eds., *This Miserable Pride of a Soldier: The Letters and Journals of Col. William S. Foster in the Second Seminole War* (Tampa, Fla., 2005), 128 ("without"); A. T. Lee, "Personal Recollections—A Scout with Ben Beall," in Theophilus F[rancis] Rodenbough, ed., *From Everglade to Canyon with the Second Dragoons* (New York, 1875), 49 ("baked"), 52 ("would have set up"); Richard J. Chacon and David H. Dye, "Introduction to Human Trophy Taking: An Ancient and Widespread Practice," in Chacon and Dye, eds., *The Taking and Displaying of Human Body Parts as Trophies by Amerindians* (New York, 2007), 19–21. The Smithsonian's records do not include any mention of accessioning Waxehadjo's scalp (personal correspondence with Dorothy Lippert [Smithsonian Institution, National Museum of Natural History] and Domonique deBeaubien [Seminole Tribe of Florida, Tribal Historic Preservation Office], June 19, 2014, and July 15, 2014).

ants in the war attached cosmological, social, and personal significances to human remains. As anthropologist Katherine Verdery has argued, burials and reburials translated these attachments into political claims to particular lands. Kinship networks and the larger communities based on them, including nations, consisted of the living, the dead, and the not yet born, and "proper burial"—which, for many communities, meant that *"our 'sons' must be buried on 'our' soil"*—was necessary to establish and maintain connections across these generations. Buried bodies had been part of political contests to claim Florida since the colonial era: Hernando de Soto's men had pilfered and examined Indian tombs during their sixteenth-century entrada, and Spanish colonists dug up the bones of former governors and reburied them in Cuba after they relinquished Florida in 1763. For Indians and whites alike, protecting and repossessing the remains of their dead enabled proper burials and legitimized each group's assertions that Florida was their land, a place in which social bonds between the deceased and their community could be perpetuated.[22]

It was, therefore, imperative for fighters on both sides to recover the corpses and scalps of their fallen comrades. According to one white observer, Seminoles pulled their casualties from the field during heavy fighting "evidently rather to save the scalp than the life of the wounded." A militia officer wrote that "the practice of the Indians, not to abandon their slain, is founded solely in superstition." "For they believe that the scalped cannot enter the hereafter hunting grounds, which constitute their notion of a Heaven." Despite portraying the Seminoles' efforts to recover cadavers as irrational, whites also took great risks to pull their dead from combat and properly bury mutilated remains. One lieutenant was so "highly exasperated by seeing the scalps of his fellow officers gracing their lodges" that he

22. Katherine Verdery, *The Political Lives of Dead Bodies: Reburial and Postsocialist Change* (New York, 1999), 47 (quotes), emphasis in original; Cara Krmpotich, Joost Fontein, and John Harries, "The Substance of Bones: The Emotive Materiality and Affective Presence of Human Remains," *Journal of Material Culture,* XV (2010), 371–384. For a reflection on how the enduring presence of the dead legitimized and limited how the living understood and claimed particular places, see Robert Pogue Harrison, *The Dominion of the Dead* (Chicago, 2003), esp. 17–30. On de Soto and Indian graves, see Luys Hernández de Biedma, "Relation of the Island of Florida," 1539, in Lawrence A. Clayton, Vernon J. Knight, Jr., and Edward C. Moore, eds., *The De Soto Chronicles: The Expedition of Hernando de Soto to North America in 1539–1543,* 2 vols. (Tuscaloosa, Ala., 1993), I, 230–231. On the Spanish governors, see Robert Olwell, "Seeds of Empire: Florida, Kew, and the British Imperial Meridian in the 1760s," in Elizabeth Mancke and Carole Shammas, eds., *The Creation of the British Atlantic World* (Baltimore, 2005), 264.

took off alone after a group of recently departed natives. Failing to catch them, he settled for gathering "all the scalps he could find, which he carefully committed to the earth."[23]

The buried dead were a focal point of white soldiers' duties, native fighters' tactics, and the emotional experience of combatants. U.S. soldiers reburied disinterred comrades and developed new burial techniques to protect against exhumation. An army detachment in the everglades, for instance, staked their dead to the bottom of deep pools "to keep the bodies down so in case the Indians did come back they could not find them and mutilate the bodies." Florida natives seem to have targeted white graves, and, on at least one occasion, Seminoles and U.S. troops literally fought over the burial of white American bodies. A volunteer from New Orleans wrote that, "while burying some of [the army's] dead who had been disinterred and abused by the savages, the left wing was attacked." The emotional stress of concealing white burial sites was also evident in poems like "The Burial of an Infant in Florida":

> And they sprinkled the dust from the oak's old rind,
> And scattered the palm leaf's fan—
> Or friend or foe, that *none* might find
> That grave of scarce a span
>
> . . . .
>
> That thou art buried 'neath this spot
> They say I must not tell,
> Not even to yon little bird
> Which sings so wild and well.
> Nor to the rustling leaf, nor stream
> Which murmurs by thy head—
> Lest they should *prate*, and forms of hate
> Defile thy hallowed bed
>
> . . . .

23. Matthew T. Pearcy, ed., "Andrew Atkinson Humphreys' Seminole War Field Journal," *FHQ*, LXXXV (2006), 226 ("evidently"); Cohen, *Notices of Florida*, 219–220 ("practice," 220); Rowles, "Incidents and Observations," *Southron* (1841), 116 ("highly"), 117 ("all the scalps"). Other examples of Seminoles recovering wounded or dead comrades include E., "Recollections of a Campaign in Florida," *Yale Literary Magazine*, XI (1845), 79; and Motte, *Journey into Wilderness*, ed. Sunderman, 195. Other efforts by white fighters to recover and bury their dead include "Florida War, Jacksonville, Sept. 14," *Army and Navy Chronicle*, IX (1839), 203; and "Florida, Fort Lauderdale, Oct. 5, 1839," *Army and Navy Chronicle*, IX (1839), 310.

They've left no little sign for me
To tell where thou art cast,
On Earth there's not a trace of thee,
My dearest and my last;
The tawny foe may trail the doe
But not thy covert wild—
I lay thee here—my sinless one—
I *leave* thee here, my child.

The poet's emphasis that the living must *"leave"* this infant and never speak of its burial suggests how deeply the threat of disinterment could disrupt the interpersonal bonds between living and dead relatives.[24]

Even after the war, white Americans' sense of separation from countrymen buried in the Florida interior led to some extraordinary efforts to retrieve soldiers' bones and transport them to safer resting places. The recovery of Euro-Americans' remains from imperial frontiers, like the collection of native skulls, reinforced imperial power and national cohesion. The most large-scale example of this was the exhumation of Major Dade's troops and their reburial at a shrine in Saint Augustine, a measure funded through the donations of officers and soldiers in other regiments. The case of General Wiley Thompson was more intimate. Osceola had scalped Thompson at the start of the war, and, after its conclusion, Thompson's wife hired a contractor to find, exhume, and return his bones to her home in Savannah. Mrs. Thompson measured each of the bones to be sure they were her husband's and then kept them under her bed for a year before she could finally bring herself to bury them. Such reburials suggest that, despite seven years of war, white Americans did not yet feel like the Florida interior was a place where they could properly bury, and maintain national and personal ties with, dead kin. Native violence against the white dead was an enduring means of resisting Anglo-American expansion.[25]

24. James D. Elderkin, *Biographical Sketches and Anecdotes of a Soldier of Three Wars* ... (Detroit, Mich., 1899), 22 ("to keep"); Barr, *Correct and Authentic Narrative,* 15 ("while burying"); Lieut. Patten, "From the *Savannah Georgian;* The Burial of an Infant in Florida," *Army and Navy Chronicle,* XI (1840), 216 ("And they sprinkled"), emphasis in original. The editor who published this poem explained that "the Indian custom of violating new made graves, has compelled the troops in Florida to have recourse to every expedient to conceal the spot where a comrade has been buried" (ibid.). See also Woodburne Potter, *The War in Florida* ... (Baltimore, 1836), 143–144; and John W. Phelps, "Letters of Lieutenant John W. Phelps, U.S.A., 1837–1838," *FHSQ,* VI (1927), 75–76.

25. Rodenbough, ed., *From Everglade to Canyon,* 75; Frank Laumer, *Dade's Last*

Like Euro-Americans, Florida's Indians had deep connections with their dead. The propinquity of the living to the deceased and the intactness of human remains, particularly the head, were especially important to perpetuating these bonds. According to historian Susan A. Miller, nineteenth-century Seminoles believed that the spirits of the dead resided at their burial places, and, as long as the community of living Seminoles remained nearby, the dead could continue to interact with them and preside over religious ceremonies. Since the bodies of the dead were tied to the landscape, the living had to dwell near their relations' graves to sustain the community in its entirety. Moreover, anthropologist Patricia R. Wickman has argued that the Seminoles thought the spirit was located in the head, and the spirits of headless corpses were fated to wander away from the community or become malevolent ghosts that could harm the living. In 1839, an army officer who talked with a young Florida native about ghosts found that "he believed in them, and had seen one; he had been hunting, and saw, during the night, his uncle, who had been dead eight days before, with a head as large as his body; he was frightened nearly to death, was sick for two days." On returning to the grave, the youth "found a white crane standing there; most conclusive proof, he thought, of the existence of the ghost." "He was silenced at last by one of the old men who said he was talking foolishness." Although the young man's understanding of ghosts might not have been shared by all Florida Indians, the continued presence of this man's apparition at his grave and interaction with his nephew reflected connections between the land, the dead, and the living while its huge head might have indicated the relationship between the spirit and the cranium.[26]

Command (Gainesville, Fla., 1995), 229–230. On how the collection of soldiers' remains during the Civil War caused similar traumas to social and personal bonds between the living and the dead, see Simon Harrison, "Bones in the Rebel Lady's Boudoir: Ethnology, Race, and Trophy-Hunting in the American Civil War," *Journal of Material Culture,* XV (2010), 385–401. For overviews on death and burial in the early United States, see Gary Laderman, *The Sacred Remains: American Attitudes toward Death, 1799–1883* (New Haven, Conn., 1996); and Nancy Isenberg and Andrew Burstein, eds., *Mortal Remains: Death in Early America* (Philadelphia, 2003). On how Europeans collected the remains of their own dead in colonial territories, see Roque, *Headhunting and Colonialism,* 183–212. On proper burial and maintaining bonds with the dead, see Verdery, *Political Lives of Dead Bodies,* 95–127.

26. "Correspondence of the *Savannah Georgian;* Florida; Copy of a Letter from a Young Officer," *Army and Navy Chronicle,* IX (1839), 93 (quotes); Susan A. Miller, *Coacoochee's Bones: A Seminole Saga* (Lawrence, Kans., 2003), 1–2, 31; Patricia R. Wickman, *Osceola's Legacy* (Tuscaloosa, Ala., 2006), xvi.

Removal from Florida and violence against the Indian dead, particularly scalping and decapitation, posed existential challenges to Florida's natives, and they fought the Second Seminole War largely to remain near and protect their deceased kin. One army surgeon who blamed Anglo-American expansionism for causing the war thought that, "when the territory was ceded to the United States, a flood of emigrants poured in upon them, which threatened soon to consign to the ruthless ploughshare their green and happy hunting grounds, and even the sacred resting places of their dead." "Alarmed at this," he concluded, "they took up arms." The officer and chronicler John T. Sprague argued that "the most zealous and intelligent, believing in the sacredness of the soil ... and reverencing with an idolatrous fanaticism the graves of their men, women, and children, whose spirits they believed hovered around them in their festivals, pertinaciously refused all intercourse with the whites." Another observer drew a direct link between the Seminoles' desire to protect the remains of their dead and their acts of scalping: "The graves and bones of their forefathers have a stronger hold upon their affections than even life itself, and while the means of propagating their race are secured to them, they will wield the bloody tomahawk and scalping knife in ... defence *[sic]* of those relics of by-gone days." Abandoning Florida would mean losing contact and conversation with the dead, and this was a deep regret for those Indians who died abroad or submitted to removal. Before perishing in a South Carolina military prison in 1838, Osceola asked Dr. Frederick Weedon to take his bones to Florida and bury them where "they would not be disturbed." Weedon, however, decapitated Osceola, pickled his head (which would eventually be displayed in a New York museum), and buried his body in South Carolina. After surrendering and agreeing to removal, the Seminole leader Coacoochee was quoted as saying, "In going from Florida, I leave behind me the spirits of the Seminoles, with which I have had many interviews." "Their spirits have taken care of me all my life." The war did not just disrupt and injure the community of living Seminoles; it threatened to destroy the bonds uniting Florida natives with their dead and the land that made their communities complete.[27]

27. E., "Recollections of a Campaign in Florida," *Yale Literary Magazine,* XI (1845), 72–73 ("when the territory," 73); Sprague, *Origin,* 251 ("most zealous"), 328 ("In going," "Their spirits"); "Florida War; From the *Jacksonville Advocate,* March 24; Col. Twiggs' Expedition," *Army and Navy Chronicle,* X (1840), 269 ("graves"); Frederick Weedon, diary entry, Jan. 29, 1838, Frederick Weedon Papers, Alabama Department of History and Archives, Montgomery, quoted in Wickman, *Osceola's Legacy,* 146 ("they would not be"); May McNeer Ward, "The Disappearance of the Head of Osceola," *FHQ,* XXXIII (1955), 193–201; Miller, *Coacoochee's Bones,* 31.

SKULL COLLECTING

Although violence motivated, influenced, and limited the various scientific pursuits of white Americans in wartime Florida, ethnography often depended directly on harming the dead. This was most apparent in the collection and analysis of native remains. Like before the war, one way that collectors obtained native bones was by digging into the burial mounds of Florida's pre-Seminole Indians, and some Euro-Americans thought these mounds provided evidence that legitimated the Seminoles' removal. Some ethnographers in the army viewed mounds as proof that Florida's ancient Indians cared more about their dead than the Seminoles did for theirs, and it was therefore no great crime to rob contemporary Seminole graves or force them to move away from burial grounds. Mound diggers in Florida searched specifically for the skulls and artifacts of groups like the Apalachees and Timucuas who, many whites believed, had been more civilized than the Seminoles and had no biological ties to them. James Gadsden, the officer appointed to negotiate the Seminoles' removal, argued that "the sympathies manifested by many for the red men of Florida as being inhumanly exiled from a country whitened by the bones of their ancestors through successive generations, is altogether misplaced." "The aborigines of Florida have long since disappeared." Even ancient bones had political value, and native military and spiritual leaders recognized this as well: they built new villages near mounds and used at least some mounds to bury Indians who died during the war. Though white Americans stressed that Seminoles had no connection with the remains of long-dead natives in mounds and that their claims to the peninsula were groundless, living Florida Indians intensified their associations with these interred bodies and, thus, their attachment to the land.[28]

Charlestonian militia officers were the most active skull collectors in the first year of the war, and their phrenological analyses emphasized the specific ethnicities of Florida natives instead of lumping them all together as Seminoles. M. M. Cohen described the war leader Mad Wolf as "the Chief of the Ocklawahaw tribe" who "bears in the darkness of his complexion

28. James Gadsden, "To the Hon. Mr. Wise, of the House of Representatives, Wascissa, (Florida,) Feb. 8, 1838," *Army and Navy Chronicle,* VI (1838), 284 (quotes); Weisman, *Like Beads on a String,* 147; D. D. Laxson, "An Historic Seminole Indian Burial in a Hialeah Midden," *Florida Anthropologist,* VII (1954), 111–118; Dr. Samuel Forry to J. W. Phelps, Mar. 4, 1838, in "Letters of Samuel Forry, Surgeon U. S. Army, 1837–1838, Part III," *FHSQ,* VII (1928), 98–99. The classic study of how Anglos used mounds to legitimize Indian removal is Robert Silverberg, *Mound Builders of Ancient America: The Archaeology of a Myth* (Greenwich, Conn., 1968).

the evidence of his people's descent from the Yemasses." Another militia officer from Charleston, William W. Smith, examined the skull of a man he supposed to be Yuchi Billy, whose "head was preserved in the Dr's cabinet, and afforded a fine subject for the speculations of the phrenologists, of whom there were not a few in Camp." These phrenologists supposed that the Yuchis were "the most savage and ill disposed of our Southern Indians" and, unsurprisingly, argued that Billy's cranium exhibited "one leavened mass of destructiveness, and ... every other bad quality appertaining thereto." All of the company's phrenologists agreed with this interpretation because it coincided with what they thought they knew about Billy as a barbarous war leader and the Yuchis as a violence-prone ethnicity. As Smith remarked, "If in other instances the followers of the science agreed as happily in their results as they did in this, Phrenology would not have to contend against so many opponents." The phrenological conclusions that Smith and his colleagues reached were, of course, problematic in several respects, not least of which was that this cranium was not actually Yuchi Billy's, nor is it clear that it belonged to any Yuchi. The real Yuchi Billy would be killed about a year later, and it seems that his head was also cut off, boiled, and added to a phrenological cabinet.[29]

The peninsula-wide context of desecrating the dead put its stamp on how white men fighting in Florida collected and evaluated native skulls. Cohen claimed that he had first observed Mad Wolf's corpse lying dead among "forty or fifty scalps." "How complex the emotions wherewith I gaze[d] upon my handful of small pine sticks, each having at its termination a natural tassel, and a portion of the human scalp cut into flag forms." Evidence of defiling the white dead surrounded Mad Wolf's body, and Cohen's complex emotions—probably sadness, disgust, and the urge for revenge—no doubt influenced his phrenological analysis that destructiveness, combativeness, and acquisitiveness were Mad Wolf's most pronounced faculties. Smith was more explicit about why he felt that turning the body of the man supposed to be Yuchi Billy into an object of study was justifiable: "When it is considered that ... the feelings of our men were excited at the recent loss

29. Cohen, *Notices of Florida,* 169 ("Chief," "bears"); [Smith], *Sketch of the Seminole War,* 249–250 ("head," "If in other," 249, "most savage," "one leavened," 250); Mahon, *History of the Second Seminole War,* 157; N[athan] S. Jarvis, "An Army Surgeon's Notes of Frontier Service, 1833–48," *Journal of the Military Service Institution of the United States,* XXXIX (1906), 277; L. N. Fowler to Dr. Weedon, Sept. 9, 1839, in Wickman, *Osceola's Legacy,* 188. On Yuchis in Florida, see Brent R. Weisman, "A Band of Outsiders: Yuchi Identity among the Nineteenth-Century Florida Seminoles," in Jason Baird Jackson, ed., *Yuchi Indian Histories Before the Removal Era* (Lincoln, Neb., 2012), 215–232.

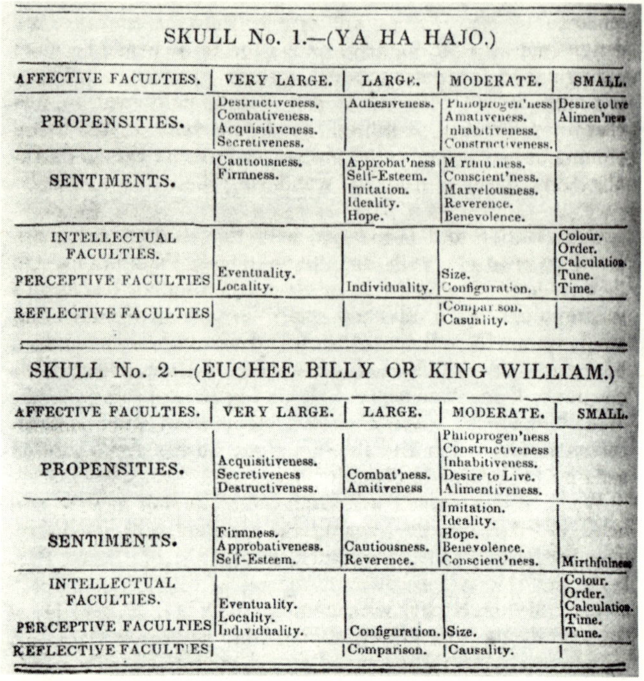

SKULL No. 1.—(YA HA HAJO.)

AFFECTIVE FACULTIES.	VERY LARGE.	LARGE.	MODERATE.	SMALL.
PROPENSITIES.	Destructiveness. Combativeness. Acquisitiveness. Secretiveness.	Adhesiveness.	Philoprogen'ness Amativeness. Inhabitiveness. Constructiveness.	Desire to live Alimen'ness
SENTIMENTS.	Cautiousness. Firmness.	Approbat'ness Self-Esteem. Imitation. Ideality. Hope.	Mirthfulness. Conscient'ness. Marvelousness. Reverence. Benevolence.	
INTELLECTUAL FACULTIES. PERCEPTIVE FACULTIES	Eventuality. Locality.	Individuality.	Size. Configuration.	Colour. Order. Calculation. Tune. Time.
REFLECTIVE FACULTIES			Comparison. Casuality.	

SKULL No. 2—(EUCHEE BILLY OR KING WILLIAM.)

AFFECTIVE FACULTIES.	VERY LARGE.	LARGE.	MODERATE.	SMALL.
PROPENSITIES.	Acquisitiveness. Secretiveness Destructiveness.	Combat'ness. Amitiveness	Philoprogen'ness Constructiveness Inhabitiveness. Desire to Live. Alimentiveness.	
SENTIMENTS.	Firmness. Approbativeness. Self-Esteem.	Cautiousness. Reverence.	Imitation. Ideality. Hope. Benevolence. Conscient'ness.	Mirthfulness
INTELLECTUAL FACULTIES. PERCEPTIVE FACULTIES	Eventuality. Locality. Individuality.	Configuration.	Size.	Colour. Order. Calculation. Time. Tune.
REFLECTIVE FACULTIES		Comparison.	Causality.	

FIGURE 11. Phrenology Chart. From M. M. Cohen, *Notices of Florida and the Campaigns* (Charleston, 1836), 170. P. K. Yonge Library, University of Florida

and injury we had sustained [and for which] we had no revenge, the triumph which some indulged in over this slain Indian ... may be excused perhaps. He was scalped, his body stretched naked upon a pole, and brought into camp for the curious to look at." Revenge, particularly against mutilations of the white dead, could shape and motivate the examination of native remains.[30]

Since the earliest days of the war, army officers and surgeons had exported Indian skulls to nationally recognized phrenologists and craniologists. This was, in part, a response to requests from prominent collectors, particularly Samuel George Morton, the Philadelphia physician whose detailed craniological measurements added supposedly empirical evidence to theories of polygenesis and racial inequality. Morton measured skulls from each of the so-called five races—Caucasian, Ethiopian, Mongolian, Malay,

30. Cohen, *Notices of Florida,* 168 ("forty"); [Smith], *Sketch of the Seminole War,* 247–248 ("When it is," 247). Beyond decapitating Mad Wolf (Ya-Ha Hadjo) and studying his head, Cohen and his comrades enacted revenge against the rest of Mad Wolf's corpse, for the soldiers "abandon[ed] him to the awful fate he has merited, (unredeemed even by his valor) to be hawked by the kites, his flesh gnawed by wolves, and his bones cranched *[sic]* by the bears." See Cohen, *Notices of Florida,* 168–170 ("abandon[ed]," 169).

and American—and averaged out the results by race. For Morton and his followers, a group known as the American School of ethnology, the most significant of these measurements was internal capacity because it suggested that bigger brains equaled smarter people. Morton's data seemed to show that whites had the biggest brains (eighty-seven cubic inches), blacks the smallest (seventy-eight cubic inches), and American Indian brains were in the middle (eighty-two cubic inches). As corresponding secretary of the Academy of Natural Sciences of Philadelphia, Morton drew on a network of skull collectors, particularly army surgeons stationed near Indian groups, to collect hundreds of Indian crania during the 1820s and 1830s. By 1840, collectors in Florida had presented Morton with thirteen "Seminole" crania and at least one skull excavated from an ancient mound. Nevertheless, the cycle of violence against the dead probably did at least as much to motivate the collection of Indian crania as Morton's appeals: only a few of the officers and surgeons who acquired Indian skulls in Florida opted to send their hard-won trophies to Morton.[31]

Morton was thus particularly indebted to one man, army surgeon Dr. E. H. Abadie, for his collection of Florida Indian skulls. In 1838, Abadie sent Morton eight crania, and it is possible that his own experiences with Seminoles abusing his dead friends—including the "mournful duty" of reburying "our comrades whose remains we found scattered and bleaching in the sun"—might have contributed to his zeal for collecting native skulls. These included three children's skulls and "3 specimens in fine order of female heads, one of which evidently of distinction from the elaborate workmanship of her tomb and the many trinkets buried with her." This woman

31. Samuel George Morton, *Catalogue of Skulls of Man, and the Inferior Animals, in the Collection of Samuel George Morton* (Philadelphia, 1840), skull nos. 456, 604, 698, 707, 708, 726–730, 732, 733, 754; Dr. Samuel Forry to Samuel George Morton, June 27, 1839, Morton Papers, Ser. IV, microfilm, APS; Stanton, *Leopard's Spots*, 28. By 1871, institutions in Philadelphia, Boston, Washington, and New York had more than three hundred skulls from Florida natives. Many of these have been repatriated as a result of the Native American Graves Protection and Repatriation Act (NAGPRA) of 1990. See Aleš Hrdlička, *The Anthropology of Florida* (Deland, Fla., 1922), 72; and Ann Fabian, "The Curious Cabinet of Dr. Morton," in Leah Dilworth, ed., *Acts of Possession: Collecting in America* (New Brunswick, N.J., 2003), 132. An early example of exporting Indian crania during the war is Barr, *Correct and Authentic Narrative*, 6. For averaged cranial capacities, see Samuel George Morton, *Crania Americana* (Philadelphia, 1839), 249–261. On polygenesis and the American school of ethnology, see Bruce R. Dain, *A Hideous Monster of the Mind: American Race Theory in the Early Republic* (Cambridge, Mass., 2002), 197–206; and Fabian, *Skull Collectors*, 83.

would be recorded in Morton's catalogue as "Seminole woman of rank," but no stories remain to explain the entry for another of Abadie's specimens, a "Seminole warrior of Florida: woman." Morton's 1849 catalogue simply noted that this warrior was thirty years old, had a cranial capacity of ninety-one cubic inches, and a facial angle of seventy-three degrees.[32]

From Morton's perspective, the most useful crania that Abadie sent him were accompanied with identity-creating stories about the decapitated individuals. Yet Abadie's stories could include misleading information, and Morton sometimes obscured Abadie's specifics and imposed his own system for identifying particular Indians. Abadie told Morton that one child's skull belonged "to the tribe of the Black Dirt (Hola-TepEmathla)," another was from a young girl of "the Fuke-luste-Hadjo-tribe," and that he had collected the cranium of a boy from "the party of Seminoles headed by John Cavollo *[sic]* or Cow-a-gee." John Cavallo was the leader of a group of black Seminoles; but Abadie never mentioned this in his description, and Morton catalogued this cranium as "Seminole boy." This boy was most likely of African descent, but, since his skull was (and probably still is) classified as Indian, Morton could incorporate its measurements into craniological data about the Seminoles as an ethnic group and Indians as a race. Even when Abadie's descriptions were more precise, Morton used "Seminole" as the default ethnic classification for all of the remains that Abadie collected in Florida (though he did sometimes note their tribal differences). There was nothing new in using Seminole as a catchall term for Florida Indians, but Morton added scientific legitimacy to this classification by assigning this ethnically disparate alliance of Indians a single cranial characterization, turning Seminole into a biologically determinative category.[33]

32. E. H. Abadie to Samuel George Morton, Feb. 3, 1838, Morton Papers, Ser. 1, APS ("mournful," "our comrades," "3 specimens"); Samuel George Morton, *Catalogue of Skulls of Man and the Inferior Animals in the Collection of Samuel George Morton*, 3d ed. (Philadelphia, 1849), skull nos. 726 ("Seminole woman") and 708 ("Seminole warrior"). This woman might have been listed as a warrior because Abadie collected her skull after a battle, perhaps one of the "two fine sculls *[sic]*" he obtained after the Battle of Okeechobee. See Abadie to Morton, Feb. 3, 1838, Morton Papers.

33. Abadie to Morton, Feb. 3, 1838, Morton Papers ("to the tribe," "the party"); Morton, *Catalogue* (1840), skull nos. 729 ("Fuke-luste-Hadjo") and 728 ("Seminole boy"). On how Morton benefited from detailed descriptions of the individuals whose skulls he collected, see Robert E. Bieder, *Science Encounters the Indian, 1820–1880: The Early Years of American Ethnology* (Norman, Okla., 1986), 67; and Fabian, *Skull Collectors*, 36–43. Black Seminoles, including the slaves of native Seminoles, usually lived in separate communities with their own leaders. On black Seminoles and the remarkable career of John

The stories that skull collectors in Florida sent to Morton along with various crania reflected the perspectives of the whites that recorded them and associated brutality, particularly an irrational urge to defile the white dead, with the Seminoles as an ethnicity. In *Crania Americana* (1839), Morton published the account of why one of the Seminole skulls illustrated in the text had a bullet hole. The story emphasized the heroism of Captain Justin Dimmick who, after having his horse shot out from under him, succeeded in dispatching two Indian fighters. The author was careful to note that the Seminoles "rushed towards [Dimmick] to scalp him" but the captain was able to recover just in time to shoot them down before his own mutilation. This was the only episode from a Seminole's life that Morton described in any detail, and both the text and image reinforced the message that the Seminoles were a ferocious people whose lust for scalps overrode concerns for their own safety. Dimmick's gift to Morton was one of at least two Seminole skulls with bullet holes in Anglo-American collections, and the visible evidence of marksmanship written on these crania suggest that they were initially collected as personal trophies before being appropriated as scientific relics.[34]

Phrenologists who examined the three detailed lithographs of Seminole skulls in *Crania Americana* perceived evidence that the Seminoles were inherently violent. Based on an analysis of these images, one anonymous phrenologist argued in 1841 that Seminole crania evinced an "animality" that was equal only to the Caribs, who were "beyond all other tribes, wild and indomitable, ferocious and sanguinary. . . ." "Rather than submit to conquest and slavery, or to any form of civil restraint," the writer concluded, "they covet extermination." The Caribs, a Caribbean native group that Anglo-Americans thought had been completely wiped out, set the biological and historical precedent for the Seminoles: "With an organisation and development of brain, and a condition of the mind not dissimilar [to the Caribs], the Seminoles are pursuing at present a course of warfare, which, if not abandoned, must lead in the end to a like result." In contrast, "possessed of brains, as appears from their skulls, more liberally supplied with moral, religious, and intellectual organs, the Creeks, Choctaws, and Cherokees, though brave, warlike, and proverbially artful, have shown themselves

---

Cavallo, see Kevin Mulroy, *Freedom on the Border: The Seminole Maroons in Florida, the Indian Territory, Coahuila, and Texas* (Lubbock, Tex., 1993), esp. 30.

34. Morton, *Crania Americana,* 166–167 (quote, 166). On the other such cranium, see J. S. Wright, "Some Measurements of Skulls and Heads," *Johnston's Dental Miscellany,* VI, no. 65 (May 1879), 182–183.

less inexorably cruel, and less brutally devoted to havoc and blood" than the Seminoles. The skulls of Creeks somehow portrayed their capacity for civilization and survival while those of Seminoles—who, like the Creeks, were an amalgam of multiple southeastern peoples—revealed an ethnicity made to disappear.[35]

By motivating, enabling, and influencing phrenological and craniometric investigations, violent encounters in the borderlands shaped how Indians got to be read. White officials and phrenologists looked to the remains of Florida Indians for proof that the Seminoles were a unified ethnic group, one that lacked meaningful attachments to Florida and whose brains were so geared toward destruction that their extermination seemed inevitable. Yet the Seminoles did endure, largely because they, too, developed a new sense of their own ethnic coherence during the war. The collection of white scalps advanced and reinforced this process.

SEMINOLE ETHNOGENESIS AND THE "SCIENCE OF SCALPING"

The path toward becoming Seminole began when the so-called proto-Seminoles, who were primarily Hitchiti-speaking minority groups within the Creek Confederacy, migrated to Florida in the mid-eighteenth century. They established independent, ethnically diverse towns that became increasingly unified politically in response to the First Seminole War and Anglo-Americans' efforts to enforce removal to a reservation in central Florida and, eventually, west of the Mississippi in the 1820s and 1830s. The cosmologies and social organization of these natives had some important continuities with older Creek traditions. During the Second Seminole War, however, violence—both against and by Florida Indians—and the influence of nativism encouraged Seminoles to see themselves as a coherent ethnicity with a unique culture and shared history.[36]

35. "Remarks on the Cerebral Organisation of the American Indians and Ancient Peruvians," *American Phrenological Journal and Miscellany,* III (1841), 209–210 (quotes, 210).

36. The first waves of proto-Seminole migrants might have arrived in Florida as early as the 1720s. Their descendants maintained core aspects of older Creek beliefs but, by the 1840s, had developed new rituals and sacred objects around which they based their religious practices. See Sturtevant, "Creek into Seminole," in Leacock and Lurie, eds., *North American Indians in Historical Perspective,* 101–117. Based on an admittedly limited study of native bodily remains, one bio-archaeologist has suggested that the "proto-Seminole were in many cases the very same ethnic groups that had lived in Florida during the 17th century [like Apalachees and Timucuas], and as such, Seminole

The collection, exchange, and ritual uses of scalps contributed to the Florida Seminoles' ethnogenesis during the Second Seminole War. As gory trophies of victory, acquiring and displaying scalps symbolized the Seminoles' pride in resistance and their nativist rejection of white people and culture. The circulation of scalps also brought Florida's Indians together in ceremonies that made the fusion of hitherto disparate bands possible. Most importantly, the presentation of scalps to ritual leaders appeased the dead and, therefore, enabled the perpetuation of bonds between the ethnically diverse dead and the germinal Seminole nation. Burial rites—and the connections between the dead, the living, and the land that they engendered—have contributed to the cohesion of many groups, including new ethnicities consisting of individuals whose primary communities had been shattered by death and displacement. Such mortuary rites typically involved the remains of a group's own dead; the knowledge and ritual practices that promoted Seminole ethnogenesis, however, also depended on the remains of the enemy dead.[37]

The social and spiritual significances of scalps made them the most highly prized of all war trophies. This was nothing new: scalps and severed heads had been important aspects of southeastern warfare and religious

---

ethnogenesis should be viewed as a conscious return [from Creek Country] to ancestral lands by peoples whose identities were replaced in the colonial record but preserved in the physical remains of their bodies." See Christopher M. Stojanowski, *Bioarchaeology of Ethnogenesis in the Colonial Southeast* (Gainesville, Fla., 2010), 10 ("proto-Seminole"), 131. The flexible sense of ethnicity that made the Seminoles' ethnogenesis possible was itself part of a deeper pattern by which southeastern natives realigned their identities to reflect novel political arrangements. See Sattler, "Remnants, Renegades, and Runaways," in Hill, ed., *History, Power, and Identity*, 36–69. Based on historical and archaeological evidence, anthropologist Brent R. Weisman has argued convincingly that Seminole ethnogenesis was a "'creative adaptation' to violent change" during the Second Seminole War. See Weisman, "Nativism, Resistance, and Ethnogenesis," *Historical Archaeology*, XLI, no. 4 (2007), 199 (quote), 202, 205–207; and Brent Richards Weisman, *Unconquered People: Florida's Seminole and Miccosukee Indians* (Gainesville, Fla., 1999), 43–65.

37. On the role of death and burial in constructing group identities, see Brown, *Reaper's Garden;* Verdery, *Political Lives of Dead Bodies;* Patricia Kay Galloway, *Choctaw Genesis, 1500–1700* (Lincoln, Neb., 1995), 290–303; and Weisman, "Nativism, Resistance, and Ethnogenesis," *Historical Archaeology*, XLI, no. 4 (2007), 207. The circulation of enemy remains between the leaders of different native and Euro-American groups sometimes reinforced strategic and political alliances between them. See Andrew Lipman, "'A Meanes to Knitt Them Togeather': The Exchange of Body Parts in the Pequot War," *WMQ*, LXV (2008), 13–14.

ceremonies for centuries, and Florida Indians had been scalping European intruders since at least 1540. Like craniology and phrenology, "the science of scalping" also followed established methods to achieve specific goals, and the ways that Seminoles processed and used scalps hearkened back to older Creek practices and ends. The physical removal and preservation of scalps required skill, and Indian trader James Adair described eighteenth-century Creeks as "barbarous artists" who could remove scalps in two minutes. They then stretched "their speaking trophies of blood in a small hoop, to preserve it from putrefaction, and paint[ed] the interior part of the scalp, and the hoop, all around with red, their flourishing emblematical colour of blood." It is not clear why Adair referred to scalps as "speaking," but natural objects could be active agents in the worldviews of southeastern Indians, and it seems likely that scalps, which were associated with human spirits, would continue to have agency after death and, perhaps, even speak to religious experts.[38]

One of the ritual purposes of scalping—placating the souls of the dead—was particularly significant during the Second Seminole War. As white troops and surgeons mutilated the bodies and took the scalps and skulls of recently killed and long-buried Indians and, by enforcing removal, drove many Seminoles to abandon the graves of their kin, the need for Florida's Indians to appease the spirits of their dead was greater than ever. Euro-American expansion posed such a threat to the natives' spiritual and social worlds that one army officer was all too correct in claiming that the Seminoles were fighting against "vampyre-like pioneers of civilization." Many southeastern natives considered the physical intactness of the body to be

38. "For the N. H. Gazette; Seminole War—First Campaign; Extracts from the Journal of a Private," *New-Hampshire Gazette,* May 16, 1837 ("science of scalping"); James Adair, *The History of the American Indians,* ed. Kathryn E. Holland Braund (Tuscaloosa, Ala., 2004), 382–383 ("barbarous," 382, "their speaking," 383); James A. Brown and David H. Dye, "Severed Heads and Sacred Scalplocks: Mississippian Iconographic Trophies," in Chacon and Dye, eds., *Taking and Displaying of Human Body Parts,* 278–298, Keith P. Jacobi, "Disabling the Dead: Human Trophy Taking in the Prehistoric Southeast," 299–338; James Axtell and William C. Sturtevant, "The Unkindest Cut, or Who Invented Scalping," *WMQ,* XXXVII (1980), 451–472. On the agency of natural objects, see Clara Sue Kidwell, "Native American Systems of Knowledge," in Philip J. Deloria and Neal Salisbury, eds., *A Companion to American Indian History* (Oxford, 2002), 87–89. My interpretation of "speaking" scalps is conjectural, but it fits with scholarly views of human remains as both subjects and objects that impact the world of the living. See Krmpotich, Fontein, and Harries, "Substance of Bones," *Journal of Material Culture,* XV (2010), 371–384.

crucial to the spirit's persistence in the afterlife, and scalps taken in war might have transferred the spiritual powers of scalped victims to the triumphant warrior and his relations. It is possible that nativist spiritual and military leaders in Florida saw scalping as a way of garnering the attachment of their own dead while also causing psychological, social, and spiritual harm to living and dead whites. Considering how deeply many whites experienced the disinterment and mutilation of their dead comrades and family members, they were probably right. Scalping was a multivalent act of resistance to the destruction of bonds between the living, the dead, and Floridian space caused by the threat of removal and Euro-American acts of skull collecting.[39]

Taking, preserving, and displaying scalps seems to have been part of Creek and Seminole rituals for appeasing their dead. One late eighteenth-century traveler described the Creeks' belief in "the Ghosts of their departed Heroes who have either unfortunately lost their Scalps, or remain unburied." These ghosts were sentenced to haunting the wilderness "until the Indignity shall be retaliated on the Enemy, by some of his surviving Friends." Similarly, Adair detailed how the Creeks "cut the scalps into several pieces, fix them on different twigs of the green leaved pine, and place them on the . . . houses of their deceased relations—whose deaths (if by the hands of an enemy) they esteem not revenged till then, and thus their ghosts are enabled to go to their intermediate, but unknown place of rest, till, after a certain time, they return again to live for ever in that tract of land which pleased them best, when in their former state." After properly preserving and displaying the scalps, the Creeks would "dance for three days and nights, rejoicing before the divine presence, for their victory; and the happiness of sending the spirits of their killed relations from the eaves of their houses which they haunted." Revenge motivated these acts of scalping, yet Adair made it clear that the purpose of this revenge was to allow the spirits of the dead to reside near the "land which pleased them best," presumably the same locales where their communities and relatives continued to live. Revenge through scalping made it possible to maintain relationships with the dead that were doomed to dissolve if southeastern Indians

39. Phelps, "Letters of Lieutenant John W. Phelps," *FHSQ,* VI (1927), 68 (quote); John Reed Swanton, *Social Organization and Social Usages of the Indians of the Creek Confederacy* (Washington, D.C., 1928), 405, 424, 438; Weisman, *Like Beads on a String,* 110; Robert P. Mensforth, "Human Trophy Taking in Eastern North America during the Archaic Period: The Relationship to Warfare and Social Complexity," in Chacon and Dye, eds., *Taking and Displaying of Human Body Parts,* 224, Jacobi, "Disabling the Dead," 312.

left their lands or let the killings and postmortem mutilations of people in their communities go unavenged.[40]

During the Second Seminole War, Florida's Indians participated in scalp rituals that were probably meant to fulfill these same spiritual and social functions of revenging and appeasing the dead. Although no accounts of scalp rituals during the war were as explicit as Adair's, glimpses of these ceremonies suggest striking similarities. After discovering the stretched and preserved scalps in Mad Wolf's village, Cohen mentioned that "these tasselled sticks from which the scalps depend, are triumphantly flourished by the Indians in their dances, and *at feasts!*" In another vague sketch, surgeon W. P. Rowles witnessed in a deserted Indian town "a large quantity of the herb from which they decoct their *black drink:* a number of recent scalps, and other appendages of a grand dance." A young officer provided a more detailed description of these rituals by interviewing Toney, a black translator who had lived among Florida's Indians. The officer wrote that his troops "received an invitation from Sam Jones to attend the green corn dance to-morrow, although Toney (our interpreter) says we had better not go." "From what I can learn," the officer observed, the ceremony proceeded "nearly as follows: On the first day they fast, separating themselves from one another, and, as Toney expresses it, only '*study.*' Every man thinks over the affairs of the nation,... and counts up his scalps and all his exploits of daring in war." White phrenologists were not the only people in Florida to study human remains in the context of interethnic violence.[41]

Warriors also circulated the scalps they collected to prominent medicine men in centers of ritual and knowledge. A few days after the start of the war, a group of Florida Indians held a ceremony to celebrate their victory over Major Francis Dade's company and Osceola's killing and mutilation of General Wiley Thompson. According to the officer and historian John T. Sprague:

40. John Pope, *A Tour through the Southern and Western Territories of the United States of North-America; the Spanish Dominions on the River Mississippi, and the Floridas; The Countries of the Creek Nations; and Many Uninhabited Parts* (New York, 1888 [1792]), 63 ("Ghosts"), 64 ("until"); Adair, *History of the American Indians,* ed. Braund, 390 ("cut the scalps").

41. Cohen, *Notices of Florida,* 169 ("tasselled"), emphasis in original; Rowles, "Incidents and Observations," *Southron* (1841), 116 ("large quantity"), emphasis in original; "Correspondence from the *Savannah Georgian;* Florida; Copy of a Letter from a Young Officer of the Army," *Army and Navy Chronicle,* IX (1839), 93 ("received"), emphasis in original.

> Osceola returned on the night of the 28th of December [1835] to the Wahoo Swamp. His party were loaded with all kinds of goods, and their bodies decorated with some trophy, to make known their atrocious acts. Scalps were suspended from their girdles, the warm blood still dripping; others hung them on their heads and necks, besmearing their persons with blood .... The night was spent in a boisterous and joyful manner. The scalps were given up to the great medicine-chief, Illis-higher-Hadjo, who arranged them upon a pole ten feet in height, around which they exultingly danced till daylight, accompanying their frantic mirth by songs, ridiculing and defying the white men .... Speeches were addressed by the most humorous of the company to the scalp of General Thompson, imitating his gestures and manner of talking to them in council.

Although Sprague might have been right to consider the "speeches" made to and through Thompson's scalp as parody, it is possible that this performance incorporated the "speaking" scalps mentioned by Adair. More importantly, scalps were central to Osceola's killings and the ritual that followed it, and he and his men gave all of their scalps to "the great medicine-chief" for processing and display. The parallels with the exchanges and performances underlying white skull collecting are clear: fighters collected remains and presented them to leading medicine men who displayed them in centers of ceremony and calculation and gave them meaning. Skull collectors and medicine men did, to be sure, have different intellectual goals. Nevertheless, a shared context of violence and similar sorts of exchanges (from fighters to physician-philosophers) and public displays conditioned how they each acquired and interpreted enemy body parts.[42]

The intellectual, spiritual, and military leaders who interpreted and displayed scalps during the war drew on both Creek traditions and Redstick nativism. This was most evident in the person of "Illis-higher-Hadjo" (now written as Hillis Haya Hadjo), the ritual leader to whom Osceola and his men presented their scalps in 1835. "Hillis haya" was a Muskogee term for a "medicine maker" who healed with herbs and sacred objects, led rituals and dances, and, though not formally a prophet, interceded between the natural and extranatural worlds. Hadjo—alternately translated as "visionary" or "mad or furious in battle"—was one of the titles that young men who earned a war name could choose for themselves. Florida Indians had delivered scalps to Redstick hillis hayas during the First Seminole War, and it

42. Sprague, *Origin*, 91.

is likely that the material circulation and interpreted significance of scalps would have been similar during the Second Seminole War.[43]

The same men who incorporated white scalps into rituals were also the spiritual and military leaders of the small but diverse communities of Indians that formed within Florida's swamps and other redoubts during the Second Seminole War. The most well known of these leaders, the hillis haya Sam Jones and "the Prophet" Otulka Thakko, both had Redstick roots. Sprague related how "this Prophet recited to his listeners his exploits in the Creek war of 1836, . . . inducing them to believe that the Great Spirit came to him, in the form of an Indian." "He was told to flee to Florida, and enjoined to revenge the wrongs of those who had suffered in his own land." Otulka Thakko incited and united Florida's Indians "by midnight fires, by dances, songs, and the use of roots as medicine," and, though Sprague was not explicit about it, these rituals often involved scalps. Sam Jones' band of about two hundred warriors would remain in Florida after the Second and Third Seminole Wars and became the main population from which today's Florida Seminoles are descended.[44]

Scalp collecting encouraged the fusion process that made both living and dead Florida Indians into Seminoles. This was most noticeable among the diverse "Seminoles, Creeks, and Mickasukies" that coalesced around Sam Jones, the war leader Billy Bowlegs, and Otulka Thakko in the Big Cypress Swamp. Based on the testimony of an Indian spy, Sprague wrote that "here were concentrated the desperate characters of all the tribes once inhabiting Georgia and the neighboring states, with whom the government had, from time to time, been at war for thirty years past." Their desire for vengeance and to conciliate the spirits of their dead had brought these peoples

43. Ibid., 91 ("Illis-higher-Hadjo"); Lee Irwin, *Coming Down from Above: Prophecy, Resistance, and Renewal in Native American Religions* (Norman, Okla., 2008), 195–198 ("medicine maker," 196, "visionary," 198); Wright, *Creeks and Seminoles,* 30 ("mad"), 200, 263; Martin, *Sacred Revolt,* 123–124. On how Redstick nativism motivated the Seminoles' resistance and shaped their ethnic identity, see Weisman, "Nativism, Resistance, and Ethnogenesis," *Historical Archaeology,* XLI, no. 4 (2007), 205. The scalp rituals and war dances of Florida Indians during the Second Seminole War contained echoes of the Shawnee and Redstick rituals performed during the Creek and First Seminole Wars. See Martin, *Sacred Revolt,* 146–148; and Claudio Saunt, *A New Order of Things: Property, Power, and the Transformation of the Creek Indians, 1733–1816* (Cambridge, 1999), 243–244.

44. Sprague, *Origin,* 295–296 ("the Prophet," 295, "this Prophet," 296), 270 ("by midnight"); Weisman, *Unconquered People,* 59. On Sam Jones as a hillis haya, see Wright, *Creeks and Seminoles,* 254, 263.

together: "Sympathy with each other in the wrongs they conceived had been inflicted upon them and their ancestors, caused a concert of action, excited by revenge, and emboldened by the most wild and fanatical superstitions." These were the same goals to which Creek medicine men had long devoted their scalp rituals, and their "fanatical superstitions" probably incorporated scalps. Ritual centers like the Big Cypress Swamp camp brought disparate Indians together because they offered individuals the opportunity to appease their dead and, thus, perpetuate personal and social bonds with them. In the process, living Indians forged new connections with each other and assimilated their dead relations as integral members of the small communities that endured in the peninsula. The presence of their collective dead grounded the Seminoles' social and spiritual attachments with Florida and political claim to it.[45]

Anthropologists have argued that the war's multiple stresses inspired the Seminoles' ethnogenesis as the approximately five thousand Seminoles in Florida before the war fused into a group of three hundred by the war's end. Yet, from the perspective of Florida Indians, the full Seminole community most likely consisted not only of the three hundred survivors but also their dead relations. Collecting, circulating, and analyzing scalps seems to have encouraged living Florida Seminoles to understand themselves and the peninsula's diverse Indian dead as part of a coherent ethnic group with transgenerational ties to Florida. Whites might not have been the only people in the Gulf South who redefined the boundaries between us and them through violence against the dead.[46]

• • •

Seminoles and whites both developed new knowledge about the Seminoles as a discrete ethnic group as they harmed and studied human remains. On the one hand, Euro-Americans looked to native bones, particularly skulls,

45. Sprague, *Origin,* 295–296 (quotes, 296). On the role of burial practices in reconfiguring the spatial and temporal perspectives that delimited national groups, see Verdery, *Political Lives of Dead Bodies,* 95–127. For Florida natives during the Second Seminole War, killing and scalping whites, and the rituals associated with these practices, might have filled some of the social and emotional functions that captive taking had satisfied in earlier centuries. Whereas living captives could sometimes broker peace between their birth nations and their adoptive ones, human remains, however, drove a cycle of vengeance and a deepening sense of difference. On captives, see Christina Snyder, *Slavery in Indian Country: The Changing Face of Captivity in Early America* (Cambridge, Mass., 2010), 101–126, 213–243.

46. Weisman, "Nativism, Resistance, and Ethnogenesis," *Historical Archaeology,* XLI, no. 4 (2007), 198.

to add scientific legitimacy to assertions that the Seminoles were a clearly defined ethnicity whose supposed predisposition for violence and lack of ancestral bonds to Florida's pre-Seminole Indians justified their removal. On the other hand, the collection and circulation of white scalps strengthened the Seminoles' understanding of themselves as a distinct people and allowed them to rebuild complete communities—ones that integrated the living, the multiethnic dead, and Floridian land—despite the extreme trauma of the war. This new identity helped Florida Seminoles continue to resist subjugation through the nineteenth century and beyond.

By the early 1840s, the United States had effectively completed its conquest of the Gulf South. But, as in any imperial power, domination was never absolute. Enslaved African Americans continued to find ways both large and small to resist forced labor, commodification, and charges of inherent inferiority. French and Spanish creoles adapted to conquest by developing new identities that embraced southern Americanness but also refused to discard unique cultural and intellectual traditions. Native communities persisted throughout the Gulf South, including the hundreds of Seminoles who learned to survive in the harsh environment of southern Florida and fought U.S. expansion again during the Third Seminole War of 1855–1858. Even the conquerors themselves, southern Anglo-American males, increasingly emphasized their distinct sectional identity and, in the 1860s, challenged federal rule when it seemed to threaten the very ambition that had made them lords of the Gulf South: the aggressive transformation of America's borderlands into Anglo-dominated slave country.

U.S. expansion into western borderlands during the 1840s and 1850s heightened the sectional tensions over the spread of African American slavery that would lead to the Civil War. It also ensured that imperialism and, therefore, multinational encounters remained as fundamental to knowledge production in North America as they had been since the 1500s.

{ EPILOGUE }

HOW THE WEST WAS KNOWN

• • •

No sooner had U.S. officials and Anglo-American planters consolidated their rule over the southeast borderlands than they turned their attention to northern Mexico. Aggressive expansionism in the southwest borderlands initiated the U.S.-Mexican War (1846–1848) and the Treaty of Guadalupe Hidalgo (1848), which added more than five hundred thousand square miles of Mexican and native territory to the United States.

Historians of the United States have seen 1846 as a momentous year. The U.S.-Mexican War led to the acquisition of gold-rich California, the incorporation of diverse Mexicans as a racialized underclass, the beginning of the end of western Indians' independence, the shift from a West in which Euro-American males exercised influence through racially mixed families to one where Anglo men enforced dominance through state-sanctioned violence, and the dawn of a series of crises over the expansion of slavery that culminated in the Civil War. Historians of American science have likewise insisted that 1846 was pivotal. That year witnessed the birth (after a ten-year gestation) of the Smithsonian Institution and an increased commitment by the federal government to gathering knowledge from distant territories. Soldiers, army engineers, and civilians supported by the military performed the majority of this work in the Southwest, often with the assistance of the Smithsonian.[1]

1. D. W. Meinig, *The Shaping of America: A Geographical Perspective on 500 Years of History*, II, *Continental America, 1800-1867* (New Haven, Conn., 1993), 128–218; Thomas R. Hietala, *Manifest Design: American Exceptionalism and Empire*, rev. ed. (Ithaca, N.Y., 2003), 132–272; Brian DeLay, *War of a Thousand Deserts: Indian Raids and the U.S.-Mexican War* (New Haven, Conn., 2008); Anne F. Hyde, *Empires, Nations, and Families: A History of the North American West, 1800-1860* (Lincoln, Neb., 2011); Amy S. Greenberg, *Manifest Manhood and the Antebellum American Empire* (Cambridge, 2005); James M. McPherson, "What Caused the Civil War?" *North and South*, IV, no. 1 (January 2000), 12–22; A. Hunter Dupree, *Science in the Federal Government: A History of Policies and Activities* (Baltimore, 1986), 66–114; William H. Goetzmann, *Army Exploration in the American West, 1803-1863* (New Haven, Conn., 1959); Robert V. Bruce, *The Launching of Modern American Science, 1846-1876* (Ithaca, N.Y., 1988); Michael F. Conlin, "Manifest Destiny, the Far West, and the Smithsonian Insti-

The developments in U.S. imperialism and American science that began in 1846 were no doubt significant, yet the more things changed, the more they stayed the same. The ways territorial expansion and the pursuit of knowledge reinforced each other after 1846 were consistent with, and deeply marked by, fifty years of U.S. imperialism in the southeast borderlands and, more broadly, three hundred years of European encroachments in North America. The rise of the Smithsonian Institution and the extension of U.S. governance were interrelated processes: territorial expansion influenced the Smithsonian's foundational mandate and early activities, while the Smithsonian—which historian Robert V. Bruce called the "fulcrum of scientific power in antebellum America"—organized, facilitated, and patronized an array of expansion-promoting scientific projects in collaboration with federal officials. The relationship between the conquest of the Southwest and the emergence of the Smithsonian reflects that violence, competition, exchange, and encounters with the environment and history were still inextricable from knowledge production at both the local and imperial levels.[2]

• • •

If, as Major General Thomas Jesup declared, the Second Seminole War was "the first instance ... of a nation employing an army to explore a country," U.S. officials wasted little time before employing this strategy again in Mexico, where Anglo-Americans relied on military violence to study terrain, plants, animals, and peoples during the U.S.-Mexican War. This was especially evident in the work of the Army Corps of Topographical Engineers, professional soldier-scientists who explored enormous swaths of the Southwest as the United States invaded Mexico from 1846 to 1848. Although many of their observations focused on Mexico's northern territories, officers of the topographical corps followed U.S. troops deep into Mexico's interior; their geological observations near Mexico City even fueled congressional debates over the potential colonization of the rest of Mexico. Leading civilian naturalists also took advantage of U.S. troops to collect specimens in North America's war zones. Botanists Asa Gray, of Harvard University, and George Engelmann, a physician based in Saint Louis, desired "to render the occupation of New Mexico by the United States troops subservient to the advancement of science," so, with the assistance of the

tution: The Boundless Empire for American Science, 1848–56," in Daniel McDonough and Kenneth W. Noe, eds., *Politics and Culture of the Civil War Era: Essays in Honor of Robert W. Johannsen* (Selinsgrove, Pa., 2006), 72–95.

2. Bruce, *Launching of Modern American Science,* 187.

secretary of war, they funded a naturalist, supplied him with instructions, and, in 1846, sent him off to Santa Fe with a military escort.[3]

Of course, wartime exploration was not simply a story of Anglo meets desert; encounters with earlier imperial powers and diverse inhabitants bolstered Anglo-Americans' scientific and expansionist missions. Lieutenant William H. Emory, a Maryland-born slaveowner in the topographical corps, drew maps of California for the federal government that incorporated older British and Spanish observations. He also recorded that Pueblo Indians had "a tradition ... that the white man would come from the far east and release them from the bonds and shackles which the Spaniards had imposed, not in the name, but in a worse form than slavery." In one terse ethnographic remark, Emory managed to use Indian prophecy to legitimate U.S. conquest, frame Anglo-Americans as liberators of peoples oppressed by Spain, contrast Spaniards with whites, and argue that African American slavery was comparatively benevolent.[4]

U.S. expansion shaped the pursuit of knowledge far beyond the borderlands. This was particularly evident in the debates over what to do with English chemist James Smithson's bequest of five hundred thousand dollars for founding an "establishment for the increase and diffusion of knowledge" in Washington, D.C. Congress voted to accept the gift in 1836, but the vagaries of Smithson's wording led to disagreements over the the Smithsonian's structure and goals, and these disagreements often revolved around how the institution would best serve imperialism. One Philadelphian naturalist told Congress that the main function of the Smithsonian should be accumulating details about America's vast interior territories so as to ensure the

3. Th[omas] S. Jesup to J. R. Poinsett, Feb. 11, 1838, in J. T. Sprague, *The Origin, Progress, and Conclusion of the Florida War* ... (New York, 1848), 200 ("first"); Asa Gray, "Plantae Fendlerianae Novi-Mexicanae ...," *Memoirs of the American Academy of Arts and Sciences,* New Ser., IV (1849), 1 ("to render"); W. H. Emory, *Notes of a Military Reconnaissance from Fort Leavenworth, in Missouri, to San Diego, in California* ... (Washington, D.C., 1848), 7, 8, 20–22, 127–158, 180–385; *Congressional Globe,* New Ser., 31st Cong., 1st Sess., XXI, part 1 (Washington, D.C., 1850), 390–391; Goetzmann, *Army Exploration in the American West,* 127–146. On other expeditions in which naturalists enjoyed military protection and federal support, see *Fourth Annual Report of the Board of Regents of the Smithsonian Institution* ... (Washington, D.C., 1850), 16–25.

4. Emory, *Notes of a Military Reconnaissance,* 33 (quote), 114–115. A similar account of pro-Anglo Indian myths is in John T. Hughes, *Doniphan's Expedition Containing an Account of the Conquest of New Mexico* (New York, 1973), 63–64. For topographical engineers using native guides and acquiring native botanical knowledge, see "Notes of Lieutenant J. W. Abert," in Emory, *Notes of a Military Reconnaissance,* 390–399.

United States avoided the fate of postindependence Latin America, where the "courts of Madrid and Lisbon spread for three centuries a midnight of ignorance" that left inhabitants unable to exploit their environments. This naturalist had, it seems, forgotten that Anglo-Americans had aspired to emulate Spanish scientific institutions only a few decades earlier. Politician Richard Rush, for his part, argued that the Smithsonian should focus on organizing the scientific undertakings of military officers, U.S. consuls, and foreign ministers, a plan that would further the United States' growing might. "By their physical resources and power," Rush crowed, "the United States are well known." "Their resources of intellectual and moral strength have been more in the back ground; but may not an auspicious development of them be aided by an institution like this, rising up in their capital simultaneously with this new condition of things[?]" he queried. Even those who promoted using Smithson's bequest to found a national university looked to imperialism to support their cause. One southerner held that the Smithsonian should become "a Central School of NATURAL SCIENCE" because Anglo-Americans had "the duty of developing on a scale hitherto unknown in this world the resources of the giant empire . . . ." "The men are *now* born who will hear the loud snort of the locomotive in the deserts beyond the Rocky Mountains," he continued; "no system of education that has ever yet been tried will meet their wants." Expansion required raising a new kind of American man, one fit to order a continent.[5]

Congress finally passed a bill for establishing the Smithsonian in April 1846; by the time President James K. Polk signed the bill into law in August, the United States had declared war on Mexico. The war was never far from the minds of the men launching the Smithsonian. For one, it impinged on Smithsonian secretary Joseph Henry's attempts to initiate the institution's scientific agenda; he worried Congress would use the war as an excuse to

5. "Will of James Smithson," in William J. Rhees, ed., *The Smithsonian Institution: Documents Relative to Its Origin and History* (Washington, D.C., 1879), 2 ("establishment"), "Memorial of Prof. Walter R. Johnson," May 21, 1838, 172–186 ("courts," 173), "Letter from Richard Rush," Nov. 6, 1838, 849–856 ("physical," 855), [Delta] to Th. W. White, *Southern Literary Messenger*, VI (1840), 25, quoted ibid., 872–890 ("Central," 883, "duty," 885), emphasis in original. Rush even placed the rise of the Smithsonian in a context of empire reaching back to the early Spanish era: "The continent that Columbus found was a desert, or overspread with barbarous people and institutions. The continent that steam has found teems with civilization, fresh, advancing, and unavoidably innovating upon the old world . . . . It is at such a point in the destinies of America that the Smithsonian Institution comes into being." See "Letter from Richard Rush," Nov. 6, 1838, ibid., 855.

cut off his funding. But Henry also entertained South Carolina College professor Francis Leiber's advice about how the Smithsonian might "shed the lustre of Knowledge upon our gallant blades and set a stamp of civilisation, or an additional one, upon the rude work of war." Leiber told Henry his wish that "government could be persuaded to make some important scientific enquiries in Mexico—enquiries which require power and means we may now possess," and added that Henry ought to "devise something"—perhaps a "problem connected with light or sound"—that "could be solved with the help of our soldiers." At least some learned Anglos believed that the Smithsonian should be involved with the business of imperialism even if Henry had to invent some means of doing so.[6]

And Henry did just that. He successfully challenged policymakers and regents who aimed to make the Smithsonian a university or library and, instead, stressed original research on large-scale projects—such as meteorology, natural history, and ethnology—that benefited from the collection and synthesis of data on a continental or global scale. To accomplish these grand projects, Henry looked to men tasked with projecting U.S. power, including military personnel, Indian agents, and consuls serving overseas. The Smithsonian's leaders requested specific materials from these informants, including "skulls, skeletons, [and] antiquities" necessary to delineating "the physical history of our species." Henry believed that the Smithsonian's essential mandates were to acquire knowledge from distant territories and "spread the benefits to be derived from the institution" throughout the wider world. The Smithsonian would thus institutionalize the simultaneously national and multinational dimensions that had long characterized American science while also establishing the United States as a laboratory on a hill that would diffuse enlightenment to the world.[7]

6. Joseph Henry to Harriet Henry, [Jan. 16–17], 1847, in Nathan Reingold et al., eds., *The Papers of Joseph Henry,* 12 vols. (Washington, D.C., 1972–), VII, 10, 13, Francis Leiber to Joseph Henry, Oct. 12, 1847, 199 (quotes). Polk attended the elaborate cornerstone ceremony for the institution's new building in full Masonic regalia, and other Freemasons at the event drew clear associations between righteous expansionist violence and the increase and diffusion of knowledge. One head Mason thanked God for giving Americans the strength to vanquish the enemies who "fled before them" and praised the founding of an institution that might "be as a central sun of science, about which systems may revolve, and from which light and knowledge may be reflected through every clime and kingdom of the globe." See *Report of the Board of Regents of the Smithsonian Institution* (Washington, D.C., 1848), 133, 134 ("fled"), 136 ("central sun"), 139.

7. Robert Dale Owen, Regents' Report, [n. d.], Smithsonian Institution Board of Regents, Minutes, RU 1, box 1, 75, SI ("skulls"), Joseph Henry, "Report of the Secretary

By the early 1850s, Henry and Assistant Secretary Spencer F. Baird boasted that the Smithsonian was flourishing hand in hand with U.S. expansion. Baird reported that "the number of important scientific explorations embraced in this period, mark it conspicuously in the history of American discovery," adding proudly that "every expedition of any magnitude has received more or less aid from the Smithsonian Institution." This aid included all of the functions that metropolitan scientific institutions in European empires had performed since the 1500s, including supplying distant observers with instruments and instructions, identifying and training experts to participate in territorial surveys, and receiving and processing data and specimens. As Baird put it in 1853, the Smithsonian had equipped several explorers who were "collecting grist for my mill" in New Mexico, California, the Great Basin, Mexico, Amazonia, and Chile.[8]

The Smithsonian supported many experts who were simultaneously engaged in studying nature and expanding the United States. For one, the institution became a magnet for Anglo-American men of science anxious to build reputations in the territories seized from Mexico: Henry complained in 1849 that "naturalists of late have become quite plenty." "I have now on my table six applications," he noted, "for assistance in obtaining situations

---

of the Smithsonian to the Board of Regents," Dec. 8, 1847, 177–178 ("spread"), Report of Regents, May 20, 1854, 250–251, 481, Joseph Henry, "Second Annual Report of the Secretary of the Smithsonian Institution," Dec. 13, 1849, 203, Minutes, Dec. 4, 1846, 43–44; Conlin, "Manifest Destiny," in McDonough and Noe, eds., *Politics and Culture of the Civil War Era*, 77–79; Joseph Henry to Asa Gray, Nov. 6[–11], 1852, in Reingold et al., eds., *Papers of Joseph Henry*, VIII, 402. On the Humboldtian methods of these large-scale scientific projects, see Laura Dassow Walls, *The Passage to Cosmos: Alexander von Humboldt and the Shaping of America* (Chicago, 2009), 123–129.

8. S. F. Baird, "Report on American Explorations in the Years 1853 and 1854," in *Ninth Annual Report of the Board of Regents of the Smithsonian Institution* (Washington, D.C., 1855), 79 ("number"); Baird to My Dear Father [Sylvester Churchill?], Mar. 8, 1853, Baird Papers, box 2, folder 4, SI ("collecting"). For how Henry and Baird supplied experts in the field with instructions and instruments to standardize the collection and circulation of knowledge, see Baird to Joseph Henry, May 26, 1850, Baird Papers, box 2, folder 2; Baird, "Directions for Collecting, Preserving, and Transporting Specimens of Natural History," *Annual Report of the Board of Regents of the Smithsonian Institution* (Washington, D.C., 1857), 235–253; Henry to Alexander Hugh Holmes Stuart, Aug. 4, 1852, in Reingold et al., eds., *Papers of Joseph Henry*, VIII, 377; Henry, "Second Annual Report of the Secretary of the Smithsonian Institution," Dec. 13, 1849, Smithsonian Institution Board of Regents, 254; J. M. Gilliss, "An Account of the Total Eclipse of the Sun on September 7, 1858, as Observed Near Olmos, Peru," *Smithsonian Contributions to Knowledge*, XI (1859), 1, 13.

as explore[r]s of California." Funding for transportation also encouraged army officers to export specimens to the Smithsonian, such as a captain in Texas who sent Baird a keg of "all sorts of creeping, jumping, running, and swimming things" packaged "in accordance with your request." This captain voiced his hope that "if the Institution can pay for [the shipment of] these things, well and good, for corporations have no souls." Since many congressmen remained reluctant to fund scientific work, especially when performed by nonmilitary personnel, Henry and Baird kept their eyes open for opportunities to turn officers into field collectors, to make positions for civilian naturalists in federal expeditions, and to get the military and steamship companies to cover the costs of transporting specimens. Although the polycentric networks of earlier decades did not disappear, the Smithsonian and the federal government were becoming increasingly influential nodes around which information and experts gravitated. One lecturer at the institution pointed to the interrelated successes of imperialism and knowledge production to argue that the United States had (almost) become the heir to the Spanish Empire in knowing and ruling the Americas.[9]

The Smithsonian had, in fact, commenced organizing research on South America and the Pacific almost immediately after its founding. For example, the institution helped support navy lieutenant James Gilliss's astronomical expeditions to Chile (1849–1852) and Peru (1858) and, despite Henry's ini-

9. Joseph Henry to Asa Gray, Jan. 18, 1849, in Reingold et al., eds., *Papers of Joseph Henry,* VII, 455 ("naturalists"), John Torrey to Henry, Dec. 1, 1849, 642–643; Stewart Van Vliet to Spencer Baird, Apr. 28, 1853, Baird Papers, box 35, folder 4 ("all sorts"), John P. Kennedy to Henry, February 1853, box 61, folder 1; Minutes, Mar. 8, 1856, Smithsonian Institution Board of Regents, Minutes, RU 1, box 1, 564, Minutes, Feb. 15, 1859, 657. This lecturer was German geographer Johann Georg Kohl. His 1857 talk described how imperial competitions, such as those among "Spanish, English, and Russian navigators," and encounters with natives had long generated data about American geography. He argued that Spain had once dominated the Americas and geographical information about them but that the torches of empire and knowledge alike had passed to the United States. The United States thus needed scientific institutions modeled after those of early modern Seville where geographers could gather historical maps and process the information arriving daily from "exploring expeditions performed by government officers." "Since the destruction and dispersion of the American chartographical collection of King Ferdinand at Seville," Kohl claimed, "the concentrating of all American maps ... into one focus is now, for the first time, made possible again ...." "The whole continent of America," he concluded, "finds in the United States a central power nearly in as high a degree as formerly in Spain." See Kohl, "Substance of a Lecture Delivered at the Smithsonian Institution on a Collection of the Charts and Maps of America," in *Annual Report of the Board of Regents of the Smithsonian Institution,* 105 ("Spanish"), 107, 145 ("exploring").

tial reluctance, accessioned the massive collection that Lieutenant Charles Wilkes's U.S. Exploring Expedition (1838–1842) had gathered in South America, Oregon, and the South Pacific. The Smithsonian also obtained materials from navy lieutenant William Lewis Herndon's 1851 expedition to the Amazon. This expedition was the brainchild of Matthew Fontaine Maury, oceanographer and superintendent of the United States' Naval Observatory in Washington, D.C., who dreamed of extending the slaveocracy of the lower Mississippi valley into South America. Herndon reported on Amazonia's hydrology and fitness for plantation agriculture. He also, like Maury, emphasized that the political and economic worlds of the Mississippi Valley and Amazonia were inextricably intertwined because currents carried the waters of the Amazon and Mississippi Rivers alike into the Gulf of Mexico. When Herndon crossed into the Amazon basin from the Peruvian Andes, he "musingly dropped a bit of green moss [into the Amazon River]." "As it floated along," he "followed it, in imagination, . . . to the mouth of the great river; thence across the Carribbean *[sic]* sea . . . into the Gulf of Mexico." There, he "fancied it might meet with the silent little messengers cast by the hands of sympathizing friends and countrymen high upon the head-waters of the Mississippi, or away in the 'Far West,' upon the distant fountains of the Missouri." In the imaginations of at least some officials, the "Far West" and South America alike were tied by nature to the plantation world of the lower Mississippi Valley and open to Anglo-American exploitation.[10]

Still, collaborations between the Smithsonian and federal officials were especially common in western North America, including during the boundary surveys that ran the two-thousand-mile line—which had to be moved further south thanks to the 1854 Gadsden Purchase—between the United States and Mexico from 1849 to 1855. Secretary Henry secured federal funding to publish tables that would facilitate "the computations for the occultations of . . . all the stars which will be used in our Mexican boundary surveys"

10. [Lewis] Herndon, *Exploration of the Valley of the Amazon . . .*, 3 vols. (Washington, D.C., 1853–1854), I, 62 (quotes); Whitfield J. Bell, Jr., "The Relation of Herndon and Gibbon's Exploration of the Amazon to North American Slavery, 1850–1855," *HAHR*, XIX (1939), 494–503; John P. Harrison, "Science and Politics: Origins and Objectives of Mid-Nineteenth Century Government Expeditions to Latin America," *HAHR*, XXXV(1955), 175–202; Matthew Karp, *This Vast Southern Empire: Slaveholders at the Helm of American Foreign Policy* (Cambridge, Mass., 2016), 143–146. Other U.S. Navy explorers compared South American rivers with the Mississippi and South American resources with those of the West. See Thomas J. Page, *La Plata, the Argentine Confederation, and Paraguay . . .* (New York, 1859), 164, 225.

into "accurate determination[s] of the longitude of important places on the continent." Henry and Baird also took advantage of the presence of a corps of scientifically trained men with an armed escort to gather meteorological and magnetic data and, more dramatically, specimens of cacti, fishes, reptiles, insects, fossils, meteorites, birds, bears, and prairie dogs. Baird argued that "the Mexican Boundary Survey has imperishably identified itself with the history of the progress of science in the collecting of perhaps a larger number of new species of North American animals and plants than any one party ever gathered before, or will again," and he was proud to have aided this success by attaching naturalists to the expedition and providing them with instructions, books, instruments, and preservative alcohol (which was occasionally lost to the thirstier members of the commission).[11]

In the Southwest, as in the Gulf South, astronomy was one of the first sciences that brought men of science from the United States and its southern Hispanic neighbor together. The main duty of the boundary commissioners for the United States and Mexico was to survey and map the borderline, but their mandates also extended to studying the nature and peoples of a region that was as distant and little known to federal officials in Mexico City as it was to those in Washington, D.C. Although the Mexican and U.S. commissions shared many of the same challenges, they also competed with each other to demonstrate the scientific prowess of their respective nations. Mexican astronomer José Salazar Ylarregui, for example, rejoiced that each commission would conduct their operations independently of the other since Anglo supervision would be an affront to his "national pride and self-respect." Mexico had just lost a devastating war and its northern prov-

11. Joseph Henry to Alexander Dallas Bache, Sept. 9, 1848, in Reingold, et al., eds., *Papers of Joseph Henry,* VII, 394 ("computations"), John Torrey to Henry, June 20, 1850, VIII, 62, Henry to Stuart, Aug. 4, 1852, 377, Samuel Steman Haldeman to Henry, Mar. 13, 1849, VII, 490; Henry, "Second Annual Report of the Secretary," Dec. 13, 1849, Smithsonian Institution Board of Regents, 248 ("accurate"); Spencer F. Baird, "Appendix to the Report of the Secretary," in *Tenth Annual Report of the Board of Regents of the Smithsonian Institution* … (Washington, D.C., 1856), 43 ("Mexican"); John Russell Bartlett, *Personal Narrative of Explorations and Incidents in Texas, New Mexico, California, Sonora, and Chihuahua, Connected with the United States and Mexican Boundary Commission* … , 2 vols. (New York, 1854), I, vii–viii, 292, II, 297, 409, 458, 548–564; William H. Emory, *Report on the United States and Mexican Boundary Survey* …, 2 vols. (Washington, D.C., 1857–1859), II; Asa Gray, "Plantae Wrightianae Texano-Neo-Mexicanae: Part II," *Smithsonian Contributions to Knowledge,* V (1852), 1–119; "Quarterly Return of Articles Received and Used during the Quarter Ending Dec. 31, 1854, for Collecting, Preserving, and Transporting Specimens of Natural History," Baird Papers, box 66, folder 15.

inces to the United States, and these blows drove Mexico's commissioners to prove that military might did not equal intellectual capability. Observers on both sides agreed that the Mexican commission's leader, General Pedro García Conde, was a craftier negotiator than U.S. commissioner John Bartlett and, more importantly, that Mexico's surveyors outclassed those of the United States. García Conde was pleased to declare that Bartlett was "a fine fellow, but without an idea of the work that we have to do; he brought 120 engineers, of whom not one knows his duty in determining the line ... excepting two or three fairly good ones." Surprisingly enough, Bartlett agreed. He confessed that Mexico's ten engineers possessed "more science than the whole American commission." "This I would not say openly," he admitted, "but *it is certainly true*." U.S. commissioners were perhaps so confident in their nation's superiority after the victory in 1848 that they were willing to concede that Mexican experts sometimes outdid Anglo ones in competitions for prestige.[12]

In terms of government support, however, the United States' commission far exceeded Mexico's. This was a dramatic change from the days of the Florida boundary survey of 1798–1800, when Anglo-American astronomers looked on Spain's patronage of the sciences with envy. The U.S. government provided its commissioners on the Mexican border with state-of-the-art instruments, and Anglo astronomers were proud that these offered a major advantage over their otherwise accomplished counterparts. Mexican officials, for their part, claimed to have dedicated ten thousand pesos for purchasing European instruments, but, apparently, their contacts in

12. José Salazar Ylarregui, *Datos de los trabajos astronómicos, y topográficos ...* (Mexico, 1850), 13 ("national"); Alberto María Carreño, *México y los Estados Unidos de América* (Mexico City, 1922), 278, quoted in Paula Rebert, *La Gran Línea: Mapping the United States-Mexico Boundary, 1849–1857* (Austin, Tex., 2001), 24 ("fine fellow"); John Russell Bartlett to Thomas Ewing, Dec. 23, 1850, LIV, Ewing Papers, quoted in Joseph Richard Werne, *The Imaginary Line: A History of the United States and Mexican Boundary Survey, 1848–1857* (Fort Worth, Tex., 2007), 57 ("more science"); Rachel St. John, *Line in the Sand: A History of the Western U.S.-Mexico Border* (Princeton, 2011), esp. 22–24; Rebert, *La Gran Línea,* 1–12, 17–20; Goetzmann, *Army Exploration in the American West,* 158, 160, 176. García Conde's claim that Bartlett's team included 120 engineers was an exaggeration; this number probably reflects both the United States' engineers (39) and soldiers (105). Anglos attached to the survey also voiced opinions about the inadequacy of the United States' surveyors: one naturalist told Baird that "there are more peculiarities about these Topogs than any other set of men I ever had to deal with. It is frequently remarked here that they are all more or less crazy." See J. H. Clark to Spencer Baird, Jan. 11, 1852, Baird Papers, box 17, folder 21.

Paris shipped defective apparatus instead of the "magnificent" ones they had ordered. Salazar Ylarregui embraced the shortcomings of Mexico's instruments as proof of the Mexican astronomers' talents, stressing that "the quality of [their] operations" was equal to those of the United States despite the disparities between their equipment. Still, the United States funded almost twice as many engineers and military escorts, and the Mexican commissioners often lacked tents, food, shoes, or anything beyond "the desire to defend national honor" to sustain them.[13]

Contests with Mexico for prestige and territory were central to the boundary survey, but international competitions with European powers also remained a driving force behind U.S. scientific expeditions, including such seemingly innocuous activities as gathering and naming the flora and fauna of the Southwest. From his office at the Smithsonian, Baird endeavored "to inoculate" as many men on the boundary survey as possible "with the Nat. Hist. virus" and secure positions for his pupils on the expedition because, as he declared, "I want *everything*." And, as he told Emory in 1853, he wanted it before "English, French, and German" naturalists could form collections that would "nibble by degrees" at the novelty of the specimens flowing in from the boundary survey. "The honor of first describing," he insisted, should be reserved for the United States.[14]

The competition over southwestern nature was partially a race to obtain local knowledge before other nations could do so, including the ethnographic, topographic, meteorological, astronomical, botanical, and zoological collections of the "Swiss Savant" Dr. Louis Berlandier. Berlandier had arrived in Mexico in 1826, was employed by the Mexican government to explore Texas in 1827, and eventually settled in Matamoros where he made maps of the Rio Grande for the Mexican army and continued to amass and describe specimens until his death in 1851. Stewart Van Vliet, an army captain stationed in southern Texas, aimed to buy this collection from Berlandier's widow and forward it to the Smithsonian, but he made it clear that other nations were vying to purchase it as well. Van Vliet admitted this

13. Salazar Ylarregui, *Datos de los trabajos,* 9 ("magnificent"), 13, 15, 16 ("quality"), 37; Francisco Jiménez, "Jimenez Memoria," quoted in Werne, *Imaginary Line,* 100–102 ("desire," 101); Emory, *Report on the United States and Mexican Boundary Survey,* I, 5; St. John, *Line in the Sand,* 24–25; Goetzmann, *Army Exploration in the American West,* 157–158.

14. Spencer Baird to Geo[rge] B. McClellan, Nov. 6, 1852, Baird Papers, box 2, folder 4 ("to inoculate"), emphasis in original, Baird to William H. Emory, Jan. 10, 1853, box 2, folder 3B ("English").

competition would drive up the price but assured Baird that "to set about forming such a collection would cost much more." By early 1853, Van Vliet began to think that "there will be no chance of obtaining the collection of the late Dr. Berlandier unless I head a party of filibusters and sack Matamoros." As it turned out, money and not bullets won Berlandier's life's work for the United States in 1853 when Lieutenant Darius Nash Couch bought it for $500 on Baird's promise to reimburse him and cover transportation costs. Couch made it clear that competition from Mexico made this price a steal since "people on the other side are realizing the loss they have suffered in allowing it to go out of the country—one asked if $1500 would purchase it." The acquisition of the collection—"a labor of 24 years [that] ranges from the Sabine to California" and included manuscripts in Spanish and French by a Swiss naturalist—reveals that knowledge of the borderlands still emerged through exchanges that were simultaneously local and transnational.[15]

Beyond observing stars and collecting plants and animals, the surveyors and naturalists in the U.S. boundary commission also described the peoples they encountered. Commissioner Bartlett, who studied natives on both sides of the new border, represented the persistent notion that environmental and social conditions had made Indians uncivilized and, therefore, they could be uplifted by Anglo influence. Bartlett's solution to the question of "what is to be done with the large tribes of Indians on the Mexican frontier" was to pacify them by demonstrating "the overwhelming superiority of force on our side" before sending in philanthropists, mechanics, and agriculturalists to civilize them. Bartlett's romantic appreciation of Indians rarely extended to Mexicans, however, and he argued that entrepreneurial Anglos were pushing Mexicans out of towns like San Antonio because "Mexican indolence cannot stand by the side of the energy and industry of the Americans." But this did not mean that Mexicans were incapable of emulating

15. D. N. Couch to Spencer Baird, Jan. 30, 1855, Baird Papers, box 18, folder 8 ("Swiss Savant"), Stewart Van Vliet to Baird, Sept. 1, 1852, box 35, folder 4 ("to set about"), Van Vliet to Baird, Jan. 13, 1853, box 35, folder 4 ("there will be no chance"), Couch to Baird, Mar. 14, 1853, box 18, folder 8 ("people"). On Berlandier, see Russell M. Lawson, *Frontier Naturalist: Jean Louis Berlandier and the Exploration of Northern Mexico and Texas* (Albuquerque, N.Mex., 2012). Baird failed to secure congressional funding to translate Berlandier's manuscripts, so, while Smithsonian naturalists did make some efforts to examine Berlandier's work on birds and mammals, his texts on indigenous mummies, medicinal plants, and astronomy, among many other topics, mostly got buried beneath the overwhelming mass of data flowing in from ongoing southwestern explorations. See Jean Louis Berlandier Papers, RU 7052, esp. box 1, folder 13; box 8, folder 2; box 8, folder 5; box 14, folder 4; box 14, folder 16; and boxes 15 and 16, SI.

Anglos' experimental energies, for "some of the Mexicans have the good sense to fall in with the spirit of progress." Other Anglos on the boundary survey found evidence that seemed to support the increasingly popular theory that Indians and Mexicans were inherently barbarous. Emory, who became the new boundary commissioner in 1851, railed against racial mixing with southwestern peoples because he believed it could "never work any beneficial change" and, instead, promoted "exterminating or crushing out the inferior races, or placing them in slavery." Emory claimed that his opinions about natives were based on firsthand observation and not prejudice, since, "in the early stages of my experience with these Indians, I was inclined . . . to consider their present reckless condition the result of the encroachments of the white people upon their rights; but . . . experience proved to me that no amount of forbearance or kindness could eradicate or essentially modify the predominant savage element of character." Whereas Bartlett had endorsed uplift, Emory looked forward to the day "when the Indians are exterminated." The ethnic cleansing of California natives during the 1850s suggests that Emory's point of view was all too common.[16]

Anglos in the Spanish American borderlands did co-opt and value the local knowledge of Hispanics and Indians. But Anglo-Americans also characterized these same peoples as ignorant of local nature, geography, and history, and, with surprising frequency, did so by describing how Hispanic and native interlocutors responded to queries with the words "*Quién sabe?*" (who knows?). Commissioner Bartlett claimed this was Mexicans' "universal reply" and one that he had heard "a hundred times" from Indians he questioned about ruins like Casa Grande. An army officer in search of the famed ruins of Gran Quivira noted: "All the way from Albuquerque we have asked the people of the country where the ruins were situated? How they looked? Who built them? etc., etc. To all these questions we could seldom get a more definite reply than *Quien sabe?*" The officer even conflated this response with southwesterners' physiognomy, remarking that "the national expression of *Quien sabe* appeared deeply written on every face." Even as far away as the Argentine Confederation, a U.S. Navy captain expressed exasperation that his queries about the origin of the region's peach trees "were

16. Bartlett, *Personal Narrative*, I, 39, 40 ("Mexican," "some"), 191, 268–269, II, 7, 29, 66, 92, 227–338, 384 ("what"), 388 ("overwhelming"), 389, 551; Emory, *Report on the United States and Mexican Boundary Survey*, I, 65 ("early stages"), 69–70 ("never," "exterminating," "when the Indians," 70); Benjamin Madley, *An American Genocide: The United States and the California Indian Catastrophe, 1846–1873* (New Haven, Conn., 2016).

invariably answered by *Quien sabe?*" More generally, the writings of these Anglos tended to quote (or misquote) the Spanish words of inhabitants if they emphasized locals' ignorance, such as when the navy captain noted that Argentine Indians would "exclaim '*Que animal!*'" when they saw his steamship. Gone were the days when Anglo-Americans looked to the Spanish world for expertise and model scientific institutions. Thanks largely to the collapse of the Spanish Empire, Spanish American intellectual life had fallen so low in the estimation of many Anglo-Americans that they used the Spanish language as the voice of ignorance.[17]

Anglo relations with Hispanics and natives would remain vital to intellectual and political developments in the southwest borderlands and the continent as a whole, yet, between 1846 and 1860, Anglo-Americans were most concerned with determining what place African Americans would have in the lands acquired from Mexico. This was apparent during the six surveys initiated by Congress to determine a potential transcontinental railroad route, the preferred latitude of which corresponded directly with one's pro- or antislavery convictions. But even Anglos involved in less explicitly slavery-related expeditions were quick to record observations that could weigh for or against the expansion of slavery. Such observations included "the capacity of the soil and climate" of the Southwest "to produce any of

17. Bartlett, *Personal Narrative,* I, 277 ("*Quién sabe?,*" "universal"), 279 ("a hundred times"), 420, II, 248; James Henry Carleton, "Diary of an Excursion to the Ruins of Abó, Quarrá, and Gran Quivira, in New Mexico . . . ," in *Ninth Annual Report of the Board of Regents of the Smithsonian Institution,* 298 ("national expression"), 306 ("All the way"); Page, *La Plata,* 67 ("invariably"), 370 ("exclaim"); emphases and spellings of *quién* in originals. See also Gilliss, "Account of the Total Eclipse of the Sun," *Smithsonian Contributions to Knowledge,* XI (1859), 9–10. One naturalist on the boundary survey readily admitted that "nobody [in the U.S. commission] seems to know anything about the country," and the surveyors found themselves relying on the expertise of Mexican army officers, French creole mountain men, villagers who described fossil bones that had belonged to giants, and Indians who told stories about ancient ruins. See C. B. R. Kennerly to Spencer Baird, Jan. 16, 1855, Baird Papers, box 26, folder 17 ("nobody"). For instances in which Anglos learned from natives and Hispanics, see Kennerly, "Notes on the Zoology of the Country Traversed by the U.S. and Mexican Boundary Commission (Fish and Reptiles)," Caleb Burwell Rowan Kennerly Papers, RU 7202, folder 10, 2, 14, SI; Bartlett, *Personal Narrative,* II, 28, 85, 247–248, 277, 283; Goetzmann, *Army Exploration in the American West,* 185; Duff C. Green, "Exploring the Rio Grande: Lt. Duff C. Green's Report of 1852," ed. Ronnie C. Tyler, *Arizona and the West,* X (1968), esp. 46; Emory, *Report on the United States and Mexican Boundary Survey,* I, 70; Stewart Van Vliet to Spencer Baird, Feb. 28, 1853, Baird Papers, box 35, folder 4; and James H. Simpson, *Navaho Expedition . . . ,* ed. Frank McNitt (Norman, Okla., 1964), 47–55.

the great Southern crops," the relative efficiency and happiness of Latin American peons and enslaved blacks, and the economic viability of importing black slaves instead of exploiting local laborers. Based on his experience along the border, even the proslavery astronomer Emory came to believe that planters would not be able to extract enough wealth from southwestern soils to offer what he considered proper care for enslaved blacks; only "slavery, as practised by the Mexicans, under the form of peonage . . . enables their master to get the services of the adult while in the prime of life, without the obligation of rearing him in infancy, supporting him in old age, or maintaining his family." Emory, who brought one of his own enslaved laborers to assist him while running the boundary, argued against extending slavery into the arid Southwest in a way that both reinforced the fantasy of southern slavery as benevolent paternalism and denigrated Mexican society. The sectional conflict over the expansion of slavery was not simply about ideological differences that free labor northerners and proslavery southerners projected onto empty space. Encounters with the lands and peoples of the borderlands informed arguments about what sorts of ethnic hierarchies and labor regimes should hold sway in the West.[18]

Such encounters were rarely safe: death, hardship, and violence shaped and thwarted Anglo-Americans' research in the Southwest well after the conclusion of the U.S.-Mexican War. Captain Van Vliet notified Baird that the specimens he had promised to the Smithsonian would be late because he was overwhelmed by grief after burying his "only child, a bright little boy," along the Rio Grande. Hunger and threats of mutiny slowed and sometimes halted the astronomical work of the U.S. and Mexican boundary commissions. Unsurprisingly, many southwestern Indians resented Anglo-Americans' new presumption of sovereignty and, by stealing hundreds of mules and threatening surveyors' lives, challenged U.S. efforts to order the region. In 1853, Paiutes annihilated "nearly the whole of the scientific corps" exploring a potential railroad route near the 38th parallel, prompting the Smithsonian's assistant secretary to lament that "science has much to deplore in the loss of these gentlemen." Such killings were rare, but the threat of violence did hamper and alter several western expeditions, including the boundary commission. Commissioner Emory excused himself for not using triangulation—the slower but more accurate surveying method favored by the Mexican commission—because hostile "bands of savages" made it im-

18. Bartlett, *Personal Narrative,* I, 28 ("capacity"); Emory, *Notes of a Military Reconnaissance,* 98 ("slavery"). On the railroad surveys and their contributions to geography and natural history, see Goetzmann, *Army Exploration in the American West,* 262–304.

possible to do so unless "every party that went out [was] escorted by ten or fifteen armed men."[19]

Still, the U.S. boundary commission traveled with a formidable military escort, and their capacity to enact violence enabled the expedition to perform their scientific tasks and formalize U.S. territorial claims. When Indians surrounded Emory's party and stole "some of his animals," the "report was treated lightly, and with the addition of a small escort he ... proceeded with his work." According to one naturalist on the line, Emory employed "twelve fighting men, who he calls Surveyors—we [call them], Indian hunters by way of derision." Emory boasted about his party's ability to overcome Indian hostilities that crippled the progress of Mexico's surveyors. He wrote that he "never trusted" Indians and "gave orders to permit none to come into any camp ... and to kill them at sight," a measure that had helped ensure that his commission "passed the entire width of the continent and returned with the loss of only two men and without losing a single animal ... at the same time that our co-operators on the Mexican commission were twice robbed of every hoof by the Apaches." Violence buttressed scientific pursuits, and this included violence against the dead. Anglos in the U.S. army collected Indian and Mexican crania throughout the Southwest that often ended up in the Smithsonian.[20]

The ability of Anglo-Americans to transmute violence into knowledge during the boundary expedition stands in stark contrast to the vulnerability of naturalists who journeyed beyond the United States without U.S. military support. Lieutenant Couch, who was enjoying a leave of absence from

19. Stewart Van Vliet to Spencer Baird, Apr. 19, 1853, Baird Papers, box 35, folder 4 ("only child"); Baird, "Report on American Explorations in the Years 1853 and 1854," in *Ninth Annual Report of the Board of Regents of the Smithsonian Institution,* 83 ("nearly"); Emory, *Report on the United States and Mexican Boundary Survey,* I, 12, 140 ("bands"); Bartlett, *Personal Narrative,* I, 152; St. John, *Line in the Sand,* 13–14, 34; Robert V. Hine, *Bartlett's West: Drawing the Mexican Boundary* (New Haven, Conn., 1968), 21; Goetzmann, *Army Exploration in the American West,* 285.

20. J. H. Clark to Spencer Baird, Oct. 3, 1852, Baird Papers, box 17, folder 21 ("some," "report"), Clark to Baird, Oct. 22, 1854, box 17, folder 21 ("twelve"); Emory, *Report on the United States and Mexican Boundary Survey,* I, 25 ("passed"), 88 ("never," "gave orders"). On skull collecting, see D. N. Couch to Baird, Feb. 15, 1853, Baird Papers, box 18, folder 8, C. B. R. Kennerly to Baird, Nov. 25, 1853, box 26, folder 17; and Carleton, "Diary of an Excursion," in *Ninth Annual Report of the Board of Regents of the Smithsonian Institution,* 314. John James Audubon had sent the skulls of Mexicans killed in the Texas Revolution to Samuel George Morton in 1836. See *Catalogue of Skulls of Man, and the Inferior Animals in the Collection of Samuel George Morton* (Philadelphia, 1840), skull nos. 555–558, 690.

the army, traveled to northern Mexico in 1853 with "instructions and apparatus" from the Smithsonian but without any formal government protection. Couch saw this as a scientific expedition, but, according to Captain Van Vliet, Couch's "departure from the outskirts of civilization (the Mexicans are semi-barbarous)" had more in common with a quixotic quest through a dangerous land. He told Baird that Couch rode off

> on a piebald horse in the direction of the 'Sierra Madre.' He was undecided which road to take . . . and I believe had determined when he got outside Matamoros to let his horse select his own road. His animal looks so much like the celebrated 'Rocinante' and one of his men like 'Sancho Panza' that we likened the entire 'turn out' to one that it was said might have been seen some years ago in Old Spain. I trust that it will become as well known and at the same time prove more useful to society than its predecessor. I think he is running some risk for our amigos have no great love for the barbarians of the North and besides have a disagreeable way of knifing a person on very short notice and for very small consideration. Its true they very considerately place a cross over the spot on which you fall but I doubt it is a fair equivalent for ones life, unless you happen to be a Catholic.

For Van Vliet, the physical space, inhabitants, and historical and literary legacies of the former Spanish Empire contextualized Couch's expedition and made it perilous.[21]

Violence permeated Couch's writing and limited his observations in Mexico. He resented that the "continual alarm from the incursions of Northern Indians . . . greatly impeded my labor," and he eventually had "to give up the journey and return to the United States" after his hired Mexican servants deserted for fear of Indian attack. Couch, however, usually felt more threatened by Mexicans than natives. After several weeks in Mexico, he decided to return his instruments to the Smithsonian because he came to think Mexicans would steal them. Worse yet, he found that many Mexicans believed the goal of his expedition was military reconnaissance for another U.S. invasion. A priest in Monterey "openly said that my intentions were not pacific," Couch complained, and "that my true object [was] to search out their strong points in the mountains" (Couch did, in fact, note Monterey's defenses in his diary). When another priest publicly decried him as a

21. *Ninth Annual Report of the Board of Regents of the Smithsonian Institution*, 15 ("instructions"); Van Vliet to Baird, Feb. 28, 1853, Baird Papers, box 35, folder 4 ("departure").

"heretic—it was deemed advisable to shorten my stay as any one of his flock would have been serving their holy religion by putting an uncomfortable amount of steel in my body." Couch offered many examples of how Mexicans were "not in love with Los Americanos," but he was particularly appalled to discover that "the graves of the American troops" buried near Monterey had "been *open[ed]* either for pillage or desecration." These desecrated graves, like those of U.S. troops in Florida during the Second Seminole War, proved that Mexico remained beyond U.S. control.[22]

Like other Anglos, Couch was interested in ascertaining whether the peoples of the border region were capable of citizenship in the United States, and, as in the Gulf South, intellectual capacity became a measuring stick of inhabitants' fitness for incorporation. He considered the disdain for innovation he supposedly observed among the farmers of northern Mexico evidence that they lacked the mental abilities required to be full participants in U.S. society. However, he suggested that there were other ways to add Mexican land to the United States other than forming territories that would eventually achieve statehood. He argued that the United States should reach out a "friendly dictatorial arm" and annex all of Mexico—"(not as free and independent states—God forbid)"—in order to lift Mexicans out of their "brutal stupid ignorance." It was, he concluded, "fortunate for her that she has so magnanimous a neighbor as the United States," since the other powers competing for influence in Mexico, England and France, would aim to limit Mexico's "advancement" instead of promoting it.[23]

Couch hoped Mexico would be the frontier of a new kind of U.S. imperialism, one founded on dominating distant places as colonies. U.S. officials in the mid-1800s rejected this vision of a U.S. Empire, but expansion into the Southwest was nevertheless imperial because, as in the Gulf South decades earlier, the United States did the same sorts of imperial things that

22. Couch to Baird, Jan. 30, 1855 ("continual," "give up"), Baird Papers, box 18, folder 8, D. N. Couch to Baird, May 14, 1853 ("openly," "true object"), Couch to Baird, Apr. 17, 1853 ("not in love"), emphasis in original; Darius Nash Couch, "Notes of Travel," 50 ("graves"), 51 ("been *open[ed]*"), 70 ("heretic"), emphasis in original, Old Colony History Museum, Taunton, Mass.; Baird, "Report on American Explorations in the Years 1853 and 1854," in *Ninth Annual Report of the Board of Regents of the Smithsonian Institution*, 87. An "old rancher" who guided Couch to the officers' cemetery explained that "bad people came from all parts of the neighboring country, dug open graves scattered the bones subjected the remains to unimaginable outrage." See Couch, "Notes of Travel," 51 ("old rancher"). For Couch's itinerary, see Roger Conant, "Zoological Exploration in Mexico—the Route of Lieut. D. N. Couch in 1853," *American Museum Novitates*, no. 2350 (Oct. 4, 1968), 1–14.

23. Couch, "Notes of Travel," 34 (1) ("friendly"), 34 (2) ("fortunate").

other imperial powers did. U.S. officials applied diplomacy and military force for the purpose of acquiring territory, they depended on the cooperation of local informants and power brokers, and they imposed hierarchies of racial difference in which Anglos shared power at the expense of other peoples. The extension of authority over distant territories and peoples is an imperial act, as is their forcible incorporation into an expanding republic. London and Queen Victoria, Moscow and Czar Nicholas I, Washington and President Polk, with their different titles and definitions of self, all ruled parts of the North American West formerly claimed by Spain. From Alaska to Arizona, inhabitants of the West felt the weight of power imposed from afar.

The United States did not colonize Mexico in the manner Couch had envisioned, but it would begin to implement a similar colonial agenda in the late nineteenth century. Expansion into the Pacific and the Caribbean after the Spanish American War of 1898 was, in many ways, a new phase of developing U.S. imperialism, one in which the United States ruled distant lands as colonies instead of flooding them with Anglo citizens and turning them into politically equal states. Despite this different approach to governance, the occupation of the Phillipines and Puerto Rico was the latest chapter in the interwoven stories of science and expansion in the United States. For example, peaceful and violent interactions in the Philippines affected how Anglo-Americans and Filipinos alike constructed hierarchies of race and place while campaigns to uplift Puerto Rico's inhabitants through agronomic and sexual education influenced the work of both Puerto Rican and Anglo-American scientists. In short, imperialism remained integral to knowledge production in the United States well into the twentieth century.[24]

• • •

Since the sixteenth century, imperial expansion into the Gulf South had fostered violence, geopolitical competitions, exchanges, and engagements with nature and history. These encounters generated knowledge among all the peoples involved. Spanish adelantados set an influential precedent

24. Paul A Kramer, *The Blood of Government: Race, Empire, the United States, and the Philippines* (Chapel Hill, N.C., 2006); Warwick Anderson, *Colonial Pathologies: American Tropical Medicine, Race, and Hygiene in the Philippines* (Durham, N.C., 2006); Stuart McCook, *States of Nature: Science, Agriculture, and Environment in the Spanish Caribbean, 1760–1940* (Austin, Tex., 2002); Laura Briggs, *Reproducing Empire: Race, Sex, Science, and U.S. Imperialism in Puerto Rico* (Berkeley, Calif., 2002); Megan Raby, *American Tropics: The Caribbean Roots of Biodiversity Science* (Chapel Hill, N.C., 2017).

by using violence to observe Florida's environment, experiment with commodities, and enforce colonial hierarchies. England and France also employed violence against natives and blacks to know and rule southeastern colonies, and competitions among European powers and native groups—many of which gravitated around the Indian slave trade—spurred the work of Indian and European cartographers. Spanish, French, British, and native inhabitants, moreover, all valued exotic goods and knowledge, and, by the eighteenth century, this shared interest culminated in a web of exchange that integrated indigenous networks of gifts and information with Atlantic networks of commodities and specimens.

Spain achieved sovereignty over the entire Gulf South in the late 1700s, but Spanish efforts to govern and profit from the region were consistently frustrated by their dependence on a wide variety of local experts and power brokers with their own agendas. Blacks, natives, and even Europeans also found that they, too, were too weak to control the flow of information or benefit from specialized knowledge. The Hitchiti storyteller Yaolaychi, for example, risked the wrath of Spanish officials, Creek leaders, and perhaps even a ferocious beast when he attempted to profit from circulating narratives about valuable minerals. This context of shared weakness persisted for decades after the United States became the latest nation to compete for the Gulf South in the 1790s.

Imperialism kept on conditioning social and political relationships in the Gulf South during and after the transition to U.S. rule, and, as in the colonial era, these relationships affected the production and circulation of natural knowledge. Some of this knowledge assisted U.S. officials' in their geopolitical competitions or bolstered Anglo-Americans' power over other peoples. Eastern officials and men of science developed expansion-promoting techniques in dialogue with local astronomers, and some Anglo astronomers hoped federal scientific institutions would emulate Spanish ones. The borderland inhabitants who benefited most from the growth of the United States were men like William Dunbar, Anglo slaveowners who leveraged violence against blacks and their position between competing powers to win support and recognition from U.S. officials and institutions. Indeed, many Anglo-American ethnographic observers became increasingly convinced that Anglos were the only people in the region capable of doing science or participating in a republican government. Their conclusions helped reify a social and political hierarchy in which whites shared power while excluding, evicting, or enslaving other inhabitants.

Violent encounters continued to undergird the creation and application of natural knowledge in the Gulf South after Spain, France, and Britain ceased competing for the region. Slavery—which was, at heart, grounded in violence—mobilized resources, labor, and networks that supported geological research. The planter-geologist Rush Nutt even drew on new geotheories to legitimize plantation slavery and ease its spread. Warfare sustained science as well. The Second Seminole War enabled the exploration of the Gulf South's last borderland and the collection of Indian remains that, in the eyes of some ethnographers, proved that Seminoles were not only inferior to Anglos but also an ethnicity destined for extinction. Imperialism had, in sum, already transformed the United States' social and intellectual worlds well before the conquest of northern Mexico or the era of overseas colonialism.

Of course, not all knowledge promoted Anglo-American power. U.S. expansion also produced knowledge and resistance among the diverse peoples of the borderlands. Some white men of science, such as Thomas Power, directly contested U.S. rule. Others, like Barthélémy Lafon, opted to share their expertise with Spain and other polities when it became clear that they had few opportunities for patronage within the United States. Black, native, and white writers in the Gulf South, moreover, argued against the notion that non-Anglos were inherently incapable of intellectual excellence. They drew on both Enlightenment-era notions of human improvability and the history of the Spanish circum-Caribbean to argue that environment and education, not biology, determined differences among peoples. Even traumatic wars encouraged ongoing resistance. Florida Seminoles developed a new sense of themselves as a distinct ethnic group in part by collecting, studying, and circulating the remains of white invaders. This identity has helped Florida Seminoles keep portions of the peninsula Indian Country despite the influx of whites. The endurance of the Seminoles and other native groups has meant that they have continued to encounter (and become) scientists, and these relationships still lead to new approaches to doing science. For example, the 1990 Native American Graves Protection and Repatriation Act (NAGPRA)—a law inspired by native activism that obliges federal institutions to return human remains and artifacts to descent communities—has altered how anthropologists research the bodies and cultures of Indian peoples. In other words, many scientists are still reckoning with the legacy of imperialism.[25]

There could be even broader consequences to recovering the role of im-

25. On how the study of native remains has changed in the wake of NAGPRA, see Samuel J. Redman, *Bone Rooms: From Scientific Racism to Human Prehistory in Museums* (Cambridge, Mass., 2016), 277–290.

perialism and borderlands in early American intellectual life. If America's intellectual history continues to be told essentially as a story about Anglos working in isolation from the rest of the continent, then it will remain easy to conclude that other American peoples have never really mattered to knowledge and culture in America. It will also remain easy for huge portions of the population to rally behind a narrow vision of America's cultural past and use this vision to justify ongoing exclusion and inequality. If, however, scholars and teachers strive to elaborate the many ways in which imperialism and multinational encounters have long been central to American intellectual life, then a new set of stories about which peoples and places mattered to American culture can be normalized. More Americans might then start drawing on a vision of history that inspires greater inclusivity and humility.

The history of natural knowledge suggests that imperialism itself is the constant that reveals the big picture of American history in a way that recognizes, but is not limited by, political, cultural, and temporal divisions. Imperialism and resistance to it contextualized how all the continent's peoples understood nature through observation, experimentation, circulation, and inspiration. This is because territorial expansion—of European empires and the United States—brought hitherto separate individuals and groups into intellectually productive contact while also shaping the power dynamics, environments, and social contexts in which these encounters occurred. If there is a unifying thread that runs through the history of natural knowledge in America, it is not the influence of liberty but the persistence of imperialism.

INDEX

FSC
www.fsc.org
MIX
Paper from responsible sources
FSC® C013483